THE BIOLOGY OF MUTUALISM

The Biology of Mutualism

ECOLOGY AND EVOLUTION

Edited by
Douglas H. Boucher

Oxford University Press
New York

© 1985 Douglas H. Boucher

First published in 1985 by Oxford University Press, Inc.,
200 Madison Avenue, New York, NY 10016

First issued as an Oxford University Press, Inc., paperback, 1988

Oxford is a registered trademark of Oxford University Press

Library of Congress Cataloging in Publication Data

Main entry under title:

The Biology of mutualism

1. Mutualism (Biology)   2. Evolution.   I. Boucher, Douglas H.
QH548.3.B56   1985   574.5′2482   85-7264
ISBN 0-19-505392-3 (pbk.)

2 4 6 8 10 9 7 5 3 1

Printed in the United States of America

CONTENTS

LIST OF CONTRIBUTORS

John F. Addicott: Department of Zoology, University of Alberta, Edmonton, Alberta T6G 2E9, Canada (and Rocky Mountain Biological Laboratory, Crested Butte, Colorado 91224, USA).

Douglas H. Boucher: Département des Sciences Biologiques, Université du Québec a Montréal, Case Postale 8888, Succursale 'A', Montréal, Québec H3C 3P8, Canada.

Clayton B. Cook: Bermuda Biological Station for Research, Ferry Reach 1–15, Bermuda.

Antony M. Dean: Department of Genetics, Washington University School of Medicine, Box 8031, 4566 Scott Avenue, St Louis, Missouri 63110, USA.

D.L. DeAngelis: Environmental Sciences Division, Oak Ridge National Laboratory, Oak Ridge, Tennessee 37831, USA.

Lawrence E. Gilbert: Department of Zoology, University of Texas, Austin, Texas 78712, USA.

Brian Hazlett: Division of Biological Sciences, University of Michigan, Ann Arbor, Michigan 48109–1048, USA.

D.H. Janzen: Department of Biology, University of Pennsylvania, Philadelphia, Pennsylvania 19104, USA.

Kathleen H. Keeler: Ecology and Evolutionary Biology, School of Biological Sciences, University of Nebraska, Lincoln, Nebraska 68588–0118, USA.

Patricia A. Lane: Department of Biology, Dalhousie University, Halifax, Nova Scotia B3H 4J1, Canada.

Richard Law: Department of Biology, University of York, Heslington, York Y01 5DD, UK.

D.H. Lewis: Department of Botany, University of Sheffield, Sheffield S10 2TN, UK.

Jorge Soberon Mainero: Departamento de Botanica, Instituto de Biologia, Universidad National Autonoma de Mexico 04510, Mexico, DF, Mexico.

W.M. Post: Environmental Sciences Division, Oak Ridge National Laboratory, Oak Ridge, Tennessee 37831, USA.

Carlos Martinez del Rio: Departamento de Botanica, Instituto de Biologia, Universidad National Autonoma de Mexico 04510, Mexico, D.F, Mexico.

Beverly Rathcke: Division of Biological Sciences, University of Michigan, Ann Arbor, Michigan 48109–1048, USA.

Alan R. Templeton: Department of Biology, Washington University, St Louis, Missouri 63130, USA.

C.C. Travis: Health and Safety Research Division, Oak Ridge National Laboratory, Oak Ridge, Tennessee 37831, USA.

John Vandermeer: Division of Biological Sciences, University of Michigan, Ann Arbor, Michigan 48109–1048, USA.

Carole L. Wolin: Department of Zoology, University of California, Davis, California 95616, USA.

PREFACE

> Between these extremes of the close mutual association of organisms in symbiosis, and the general interdependence of the plant and animal members of a biotic community, lie a host of reciprocal relationships that are usually described under some general term such as 'mutualism' or 'commensalism'. I have not seen any general review of these associations, though I should think they would form an interesting and instructive subject for a book.
>
> Marston Bates,
> *The Nature of Natural History*
> (Scribner's, New York, 1950), p. 134

This is, I hope, the book that Bates foresaw three decades ago. It deals with those interactions between living organisms which are of mutual benefit, and attempts to describe some recent theoretical attempts to cover mutualisms in a general fashion. Our emphasis is on ecological and evolutionary topics and the chapters are arranged deliberately by theory and concept rather than by taxonomic group. In doing this, we hope to be able to discern general patterns relating to mutualistic interactions which apply widely, and are not restricted to particular taxa or habitats.

After introductory chapters on the history of the subject, questions of terminology, and the natural history of mutualisms, we cover theories of the evolution of mutually beneficial interactions. These include cost-benefit and population genetics approaches and concepts relating to sex and environmental stability. The following chapters deal with interactions of mutualists with third species, such as competitors and parasites, followed by models of the population dynamics of mutualistic species. These are expanded from two to many species in the following chapters, which present ideas relating to the concepts of indirect and community mutualisms. The final chapter reviews the importance of mutualism in agriculture, with brief consideration of related fields such as silviculture and fisheries.

The idea for the book had its origin in a symposium entitled 'Mutualism: New Ecological Theories', presented at the Annual

Meetings of the American Association for the Advancement of Science in January, 1982. Financial support for its preparation was provided by l'Université du Québec a Montréal. Acknowledgements are to be found to the ends of chapters, but special thanks are owed to Louise Bouthillier, who did the final typing of all the manuscripts, and to Tim Hardwick for his patient editing.

As the reader will see, there are a variety of ways of approaching mutualistic interactions, and different authors often adopt divergent or even contradictory positions. Rather than try to impose an artificial uniformity on the chapters, I have chosen to let the differences show clearly, in the hope of stimulating discussion and comparison. In this way, we can find which are the most promising paths into this new and expanding field of study.

1 THE IDEA OF MUTUALISM, PAST AND FUTURE

Douglas H. Boucher

Don't know much about history
Don't know much biology
Don't know much about a science book
Don't know much about the French I took
But I do know that I love you
And I know that if you love me too
What a wonderful world this would be

the late great Sam Cooke

Renaissance of an Idea

Different kinds of organisms help each other out. This, in brief, is the idea of mutualism, and it is an idea which has been reborn in the last decade. Never entirely absent from ecological thought, it none the less fell out of favour as modern ecology grew, and only since the early 1970s have we begun to find it important again. No one can be sure that its recent renaissance will not in turn fade away, but at least today it is an idea which is steadily gaining ground. Ecologists once again find it interesting.

As an example of this new interest, consider Table 1.1 and Figure 1.1. They illustrate a rather amazing convergence of biologists' thoughts about mutualism toward a particular way of representing it, called the phase-plane model. The phase-plane depicts the population densities of two species along vertical and horizontal axes, producing a plane which is separated by lines called isoclines into four regions called phases. The four possible phases correspond to what will happen to the two species' population densities in the four regions of the plane: both densities will increase, both densities will decrease, species 1 will increase while species 2 decreases, or species 2 will increase while species 1 decreases. The isoclines which separate the phases are the sets of points for which one or the other species' density will remain constant, and their point of intersection, the combination of densities for which neither species' density will

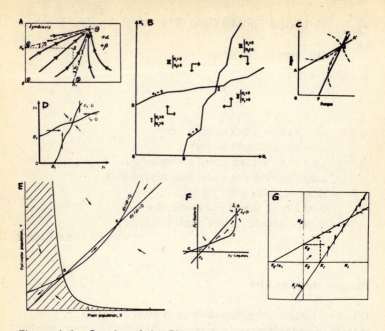

Figure 1.1: Graphs of the Phase-plane Model of Mutualism (a) Gause and Witt 1935; (b) Rescigno and Richardson 1967; (c) Boulding 1972; (d) Christiansen and Fenchel 1976; (e) May 1976a; (f) Vandermeer and Boucher 1978; (g) Hutchinson 1978; (h) Goh 1979; (i) Travis and Post 1979; (j) Viera da Silva 1979; (k) Freedman 1980; (l) Heithaus *et al.* 1980; (m) Soberon and Martinez del Rio 1981; (n) Post, Travis and DeAngelis 1981; (o) Wells 1983.

change, is called an equilibrium point. If the isoclines are straight lines, as for several of the graphs reproduced in Figure 1.1 (a, c, g, h, k, l,), we have a particular kind of phase-plane model called the Lotka-Volterra model, named after Alfred Lotka and Vito Volterra, who first applied it to competition and predation in the 1920s and 1930s (Hutchinson 1978; Scudo and Ziegler 1977; Kingsland 1981).

The phase-plane model, in its Lotka-Volterra form, was applied to mutualism (called 'symbiosis') only a few years after it was developed for the antagonistic interactions of competition and predation (Gause and Witt 1935, Figure 1.1a). Gause and Witt pointed out

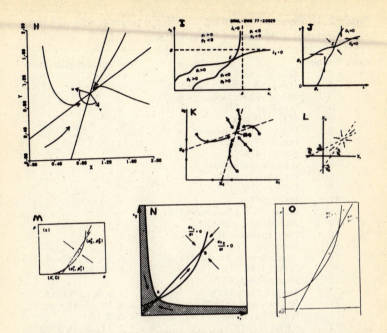

some of the basic properties of the Lotka-Volterra model of mutualism: the isoclines have positive slopes, a stable equilibrium exists if the product of the mutualism coefficients (representing how much an individual of one species helps an individual of the other) is less than one; both species attain higher densities together than when they exist alone; and when the product of the mutualism coefficients exceeds one, both species will grow to infinity. They also suggested a solution to this last, clearly paradoxical prediction: that the mutualism coefficients are not constants, but rather that they decrease (mutual aid diminishes) as the populations grow larger. Thus the basic analysis of the phase-plane mutualism model was done half a century ago, and was published in a major ecological journal in an article that was later reprinted in an important book of readings (Hazen 1964).

Nevertheless the phase-plane mutualism model was to be rediscovered repeatedly, decades later. Nearly 30 different papers have presented what is essentially the same model, often with almost the same graph (Figure 1.1, Table 1.1). And at least as far as one can judge by citations, few of the authors were aware of each other's work, and only two of them of Gause and Witt's original paper.

Table 1.1: Papers Presenting the Phase-plane Model of Mutualism and their Citations of Other Papers

Paper Number	Reference	Numbers of Papers Cited
1	Gause & Witt (1935)	—
2	Haldane & Jayakar (1966)	—
3	Rescigno & Richardson (1967)	—
4	Boulding (1972)	—
5	May (1973a,b)	—
6	Rescigno & Richardson (1973)	3
7	Hirsch & Smale (1974)	3
8	Siljak (1974)	—
9	Albrecht et al. (1974)	3
10	Whittaker (1975)	—
11	Siljak (1975)	—
12	May (1976a)	5, 10
13	Christiansen & Fenchel (1976)	—
14	Albrecht et al. (1976)	—
15	Levin & Udovic (1977)	—
16	Hutchinson (1978)	1
17	Vandermeer & Boucher (1978)	5, 10, 12, 13
18	Goh (1979)	5, 6, 7, 9, 12
19	Viera da Silva (1979)	—
20	Travis & Post (1979)	7, 8, 9, 10, 11, 12
21	Heithaus et al. (1980)	5, 12, 16, 17
22	Freedman (1980)	3, 9, 14
23	Hallam (1980)	5, 11, 12, 18, 20
24	Post, Travis & DeAngelis (1981)	10, 12, 17, 18, 20, 21
25	Soberon & Martinez del Rio (1981)	10, 12, 17, 18
26	Addicott (1981)	1, 5, 10, 12, 13, 17, 18, 20
27	Ewing et al. (1982)	5, 12, 17, 20
28	Gilpin et al. (1982)	5
29	Wells (1983)	20

Were it not for the long period over which such articles have continued to appear, this would seem to be a rather extraordinary case of 'simultaneous rediscovery'.

Now, the application of the phase-plane model to mutualism is 'hardly a revolutionary approach', as Vandermeer and I said (1978), incorporating a sceptical reviewer's comment verbatim. Indeed, it is the logical next step in theoretical ecology from the Lotka-Volterra competition model, which has been a standard part of ecology textbooks for over three decades. (One simply reverses the signs of the interaction coefficients to turn the competition model into one of mutualism.) Indeed, it is precisely *because* it is such an ordinary development, rather than a Nobel-Prize-winning breakthrough, that its 'simultaneous rediscovery' is significant. For it indicates that a simple, logical next step suddenly became worth taking for a lot of ecologists at about the same time.

The first few examples can perhaps be dismissed as marginal to the field of ecology: Haldane and Jayakar's application is in the field of population genetics, and furthermore is highly mathematical, as are the treatments of Rescigno and Richardson (1967, 1973), Siljak (1974, 1975), and Albrecht *et al.* (1974, 1976); Hirsch and Smale's 'presentation' is actually a problem assigned at the end of a chapter in a mathematics text (1974: 273–4). But from 1973 on, the model appears in the very core of the ecological literature: in the standard journal, *Ecology* (May 1973a), in an important monograph series (May 1973b), and then in several textbooks (Whittaker 1975; May 1976a; Christiansen and Fenchel 1976; Hutchinson 1978; Viera da Silva 1979). Furthermore, the model begins to be presented not only as part of a general treatment of all possible kinds of interactions, perhaps simply for completeness, but also as the basis for papers in and of itself, and is often applied to a particular pair of mutualistic species (Vandermeer and Boucher 1978; Goh 1979; Travis and Post 1979; Heithaus *et al.* 1980; Hallam 1980; Post, Travis and DeAngelis 1981; Soberon and Martinez 1981; Addicott 1981; Ewing *et al.* 1982; Wells 1983).

Thus in the early 1970s the phase-plane model of mutualism was taken up by ecology and repeatedly applied by researchers with apparently little knowledge of each other's work (see the citation data in Table 1.1). That this is not merely a question of coincidence, but rather a reflection of a change in ecologists' thinking, is suggested by several lines of evidence. First, consider the statements about mutualism in successive papers by Robert May, the first important ecologist to deal with the model since the 1930s. In 1973, May's analysis of the Lotka-Volterra model led him to the generalization that

competitive and symbiotic mutualistic relations are less compatible with overall community stability than are commensal, amensal and predator-prey relations. It is tempting to speculate that such community stability considerations may play a role in explaining why mutualism 'is a fascinating biological topic, but its importance in populations in general is small (Williamson 1972: 95)'. (May 1973a; comparable statement in May 1973b: 73)

Note that exactly the same logic would lead to the conclusion that competition is unimportant in nature also, but this conclusion is not drawn.

However in only a few years (contribution to Whittaker 1975; May 1976a, 1976b), May's appreciation of mutualism changed considerably; it is now seen as 'a conspicuous and ecologically important factor in most tropical communities' (1976a). Indeed, in recent years May has become one of the leaders in encouraging work on mutualism, which he sees as 'likely to be one of the growth industries of the 1980s' (May 1982).

Nor is May unique in coming to see mutualism as worthy of study in the early and middle 1970s. The year 1975, for example, saw the publication of two major articles arguing that the importance of mutualism had been seriously underestimated (Gilbert 1975; Colwell and Fuentes 1975), soon followed by the development of two fitness models which help to disprove Williams's (1966) argument against mutualism on evolutionary grounds (Roughgarden 1976; Wilson 1975; 1976). Several of the last-cited researchers were working in the San Francisco Bay area at the time and hence it might be suggested that mutualism was a local interest of this 'group', but the same interest was shown by other workers in far-distant places. For example, three 1976 papers showed how interactions which appear to be predatory or competitive might actually be mutualisms (Levine 1976; Owen and Wiegert 1976; Porter 1976), and the *Bulletin* of the Ecological Society of America published a commentary in the same year suggesting that ecologists had neglected to look for mutualism and thus had failed to note its importance (Risch and Boucher 1976). In recent years, in fact, mutualisms have seemed to turn up in all sorts of unexpected places: in mid-ocean diatom mats (Martinez *et al.* 1983); in 'poisonous' sulphur-rich waters in the deep sea (Cavanaugh *et al.* 1981), inshore (Blakeslee 1984) and in rice paddies (Joshi and Hollis 1977); between dogs and langurs (Sharma 1977), mangroves and root borers (Simberloff *et al.* 1978), spiders and parasitic wasps (Valerio 1975), trees and epiphytes (Nadkarnia 1981), invertebrates and their epibionts (Bloom 1975; Vance 1978), corals and the fish which excrete on them (Meyer *et al.* 1983), and even in 'mixed flocks' of sea urchins (Duggins 1981). Add to these cases the numerous possible 'indirect mutualisms' suggested recently (Levine 1976; Vandermeer 1980; Boucher *et al.* 1982; Chapters 13 and 14, this volume), and it is clear that ecologists have started to look for mutualisms, and to find them.

We thus have a phenomenon which, to a historian of science,

requires an explanation — a rapid shift in ecological thinking, so that what was previously thought unworthy of study, discussion and theorizing is suddenly a hot topic. As with other shifts in opinion, the historian must ask why this development occurred at this time and this place, whence it sprang, and whether it is likely to endure.

I will try to answer these questions in this chapter, beginning with a review of the history of the idea of mutualism on both science and society, and then focusing on the immediate background of the renaissance — what I call the 'Newtonian Ecology' of the 1950s and 1960s. My approach is unashamedly Whiggish — that is, I search the past for precursors of an idea that is important today. In doing so, I clearly risk committing the error of seeing past thought in terms of the present, rather than in its real historical context. Furthermore I inevitably concentrate on Western thought, to the exclusion of ideas of mutualism in other societies and traditions. My justifications for these biases are three: that the concrete problem at hand is to explain the modern renaissance of the idea of mutualism; the limitations of space; and a commitment to offer a more well-rounded treatment in the future.

The Origins of Mutualism

The Balance of Nature

The balance of nature, as Egerton has said, is the oldest ecological theory (1976: 323). It is the idea that there are tendencies in nature which prevent any species from either becoming too abundant or going extinct. Many expressions of the concept can be found in different cultures, and in the West it can be traced back to some of the earliest ancient sources (Egerton 1973). While not the same as the idea of mutualism, it is clearly related, and various ancient writers used mutualisms as examples of nature's balance. Herodotus' *History* (Book 3, Chapters 108–9) describes a plover taking leeches out of the mouth of a crocodile, noting that 'The crocodile enjoys this, and never, in consequence, hurts the bird'. (Egerton 1968: 180) Aristotle was sufficiently impressed by this tale to repeat it in three different works, and he added a story about a mutualism between a bivalve and a crustacean (Egerton 1973: 328). Cicero and Aelian gave similar descriptions, drawing the moral that humans should learn friendship from nature (Egerton 1968: 186). Pliny tells how friendships occur between between peacocks and pigeons, turtle-doves and parrots, blackbirds and turtle-doves, the crow and the little heron in a joint emnity against the fox kind, and the goshawk and kite against the buzzard. Why, are there not signs of affection even in snakes, the most hostile kind of animals? We have mentioned the story that Arcady tells about the snake that saved his master's life and recognized him by his voice. (Book 10, Chapter 96, *Natural History*; Rackham 1956)

In all these cases, mutualism is part of the adaptation of species to each other, but it is generally not contrasted with other sorts of interactions — predators and prey, or competitors, maintain the balance of nature as well. As Egerton notes, 'What was needed was some relative assessment of the importance of mutualism versus competition in nature. However, before any answer could be given, the question itself would have to be discovered' (1973: 329).

Aristotle's and Pliny's influences remained predominant in natural history through the Middle Ages, with the balance of nature being seen in hierarchical terms, analogous to the structure of medieval society. The idea of crop plants and animals being made for humanity fits in well, both with the balance of nature concept and with Christian dogma, and agriculture was often cited as a mutually beneficial relation. Animals and plants were thought to provide food for each other, on the one hand by being eaten and on the other by dying or excreting nutrients and so fertilizing the soil. The harmony of nature was seen as a mirror of the harmony of society, in which each person, from peasant to king, had a preordained role assigned by the Creator (Worster 1977; Glacken 1967).

The scientific revolution of the seventeenth century was accompanied by some profound changes in views of the natural world (Merchant 1980), but as concerns mutualism its effect seems to have been to strengthen the prevailing concept. Mutualisms became favourite examples of Divine Providence in the natural theology of the seventeenth and eighteenth centuries, which found its maximum expression in the works of Linnaeus on the polity and economy of nature (Limoges 1972; Stauffer 1960). He notes that

The general received opinion has been that vegetables were created for the food and uses of animals; but attending to the order of nature, we discover that animals were created upon account of plants . . . Animals serve in the first place to preserve a due proportion among vegetables: secondly to adorn the theatre of nature and consume every thing superfluous and useless: thirdly to remove all impurities arising from animal and vegetable putridity; and lastly, to multiply and disseminate plants and serve them in many other respects . . . There are some animals likewise in the Police of Nature who are appointed as watchmen to warn other animals of their danger, as the *Charadrius spinosus, Lannius, Grus, Meleagris,* and others, which give notice to the birds of a hawk being out in search of prey: nor would the larger conchs ever escape the *Sepia loligo* unless they had the *Cancer pinnotheres* as a guard. Thus we see Nature resemble a well regulated state in which every individual has his proper employment and subsistence, and a proper gradation of offices and officers is appointed to correct and restrain every detrimental excess. (Linnaeus 1977: 137–8 and 163–4)

Natural theology clearly sees the natural and human worlds as analogous, and views both as being in balance. This balance is maintained by the adaptation of each individual to a place in a hierarchy, an adaptation which was seen as a reflection of the Wisdom of the Creator. This made it difficult to conceive of any species going extinct, and even massive die-offs of individual species were seen as preventing them from overpopulating the globe and thus preserving conditions for other species. Natural theology, in other words, saw a Panglossian universe (Gould and Lewontin 1979), in which mutualisms were simply examples of the perfect adaptation created by God.

The Industrial Revolution

This universe was shattered by the Industrial Revolution of the nineteenth century. The hierarchy of society was overthrown politically and economically, and at the same time the well-ordered universe of the natural theologians began to be challenged by new ideas. Thomas Malthus is an important transitional figure, on the one hand demonstrating the mathematical necessity of a struggle for existence in which most of the population would inevitably suffer, and on the other hand arguing that this struggle tended to create a balanced society. Malthus simply took one element of natural theology — the idea that deaths created harmony — and expressed it in terms of the industrial society developing around him. As with Adam Smith's invisible hand, the result of competition is the greatest possible welfare of all (Silvertown 1984).

At the same time, competition began to be recognized as an important factor in nature, most importantly among plants by de Candolle (Egerton 1977). De Candolle may have got this idea from talking with Malthus, and he in turn influenced Charles Lyell, from whose *Principles of Geology* it was picked up by Darwin. Competition was seen as a source of what would today be called selection pressure, as well as a factor explaining geographical distribution. It also represented a potential cause of extinctions, now being admitted as a possibility by some (Ruse 1979).

A contemporary of Darwin's, and in some ways an even more influential figure through most of the nineteenth century, was the philosopher Herbert Spencer (Wiltshire 1978). Spencer's ideas of struggle and competition as the basic elements of the universe were elaborated in both the natural and social realms, and indeed he saw no fundamental difference between them. But while Spencer is justly recognized as a champion of *laissez-faire* and individualism, he felt that the final result of competition is an increasing integration of all members of a society into a differentiated, more efficient social organism. This differentiation, like the development of an individual organism, led inevitably to a higher state of society, in which violent competition (for example, war) was replaced by the peaceful competition of the free market (Silk 1983). Thus competition was ultimately the source of human progress.

The idea that progress has its roots in struggle found its biological expression in the *Origin of Species*. While Darwin devoted a good deal of time to studying such mutualisms as insect pollination of flowers and the decomposition of leaf litter by earthworms, and mentioned the dispersal of mistletoe seeds by birds in the beginning of the *Origin*, there is no doubt that his fundamental contribution was the demonstration that adaptation and speciation can be explained by competition. He shied away from political affairs himself, but his followers continued the tradition of Malthus and Linnaeus and drew the parallels between the natural and social worlds. This is the source of what has come to be called Social Darwinism (Ruse 1979; Silvertown 1984). Its basic precept is basically the same as Spencer's — the inevitability, and indeed the desirability, of competition. Certainly competition leads to much death and suffering, for the world would always be too small for a population with an ineluctable tendency to grow. But this suffering, by producing fitter individuals through natural selection, would ultimately produce a better race. Thus progress was inextricably linked to pain.

The Socialists and Mutualism

While competition and progress-through-struggle were the dominant themes of both natural and social science in the nineteenth century, an undercurrent of political and intellectual opposition was developing in concert. In Britain it was expressed in the growth of working-class organizations such as trade unions, Chartism, and the 'Friendly Societies' which workers organized to pool their resources. These were associations formed to allow workers to deal with catastrophes such as illness or funerals; they functioned as small co-operative insurance companies. They were also a useful cover for trade unions and a way of amassing strike funds, and in the 1830s blossomed into great co-operative federations inspired by the philanthropist Robert Owen (Thompson 1963).

The analogous organizations in France, where workers' organizations were outlawed by the Loi Le Chapelier in 1791 (Caceres 1967), were the mutual aid associations. Beginning in the period of the Revolution, the different *Mutualité* societies were in theory totally apolitical. But in fact, they were a hotbed of socialist ideas and became a major force with the revolt of the silk-weavers of Lyon (*les canuts*) in 1831 (Rude 1969). With their slogan 'To live working or die fighting', these *mutuellistes* provided the inspiration for the development of revolutionary socialism.

French *mutuellisme*'s most famous exponent was Pierre-Joseph Proudhon, a young working-class student who became famous with his book *What is Property?*, a question which he answered simply, 'property is theft'. Proudhon is regarded as a founder of both the socialist and anarchist movements, which did not become clearly distinct until the 1870s (Woodcock 1962). His mutualism, or federalism, is an antiauthoritarian ideology, based on the abolition of governments and the reconstruction of society as a great federation of workers' co-operatives. For Proudhon, political revolution was unnecessary and even dangerous to liberty; rather a system of mutual credit, through which workers could borrow the funds to amass capital and create co-operatives which would eventually replace capitalism, was the path to socialism. Nevertheless Proudhon was regarded by the French government as an exceedingly dangerous man, and spent many years in exile in Belgium, dying in 1865 just as his ideas came to be predominant in the French working-class movement and in the International (Caire 1971).

The Proudhonians' idea of socialism was basically one of just and equal exchange, made possible by eliminating the unfair advantage of the capitalists which came from their inheritance of property. Mutual credit would allow workers to produce on equal terms with capitalists, since the advantage of accumulated wealth would disappear; credit would be freely available to all. Thus Proudhonian mutualism built on both the experience of industrial workers' organizations and a somewhat romantic vision, rooted in peasant life, of exchange among individual producers (Woodcock 1962).

The ruling class's worst fears of mutualism were realized in the spring of 1871 when the Commune took control of Paris. In the short time it had to begin to reorganize the economy before being bloodily repressed, the Commune manifested a clear Proudhonian inspiration. Thus the Metal-Workers and Mechanics Unions of Paris expressed their objectives as 'The abolition of the exploitation of man by man, last vestige of slavery; The organization of labour in mutual associations with collective and inalienable capital.' (Schulkind 1972: 164). The Association of Women, led by the revolutionary socialist Elizabeth Dimitrieff, went further and proposed an organizational plan to provide work for women, which began

> Given this fundamental point, namely credit, the problem of work for women is now only a question of organization. Now, the essential thing is to facilitate the setting up of genuine homogeneous groupings; to look after their development and, while leaving them free and autonomous, to instil into them a strong consciousness of mutualism. (Schulkind 1972: 177)

With the massacre of the Communards and the subsequent repression of French workers' organizations, mutualism in its Proudhonian form would decline in importance, to be progressively replaced by Marxism. But in the 1870s it was still widely influential; the First Republic in Spain in 1873 was led by Francisco Pi y Margall, Spanish translator of Proudhon's *Du Principe Federatif*. Like the Commune, the Republic fell after only a few months (Carr 1966).

Mutualism thus developed as a political idea in counterpoint to the predominant emphasis on struggle. Rooted in the working class, it was a revolutionary threat to the system defended by Spencer and the Social Darwinists. But it also shared certain ideas with the dominant ideology: progress, the liberty of the individual, and the virtues of free and equal exchange of products. The great difference was that for Spencer, these goals were being realized under capitalism, while for Proudhon they required a revolutionary transformation of society.

A third vision was that of Marxian socialism. While now viewed as considerably more revolutionary than the Proudhonian version, in some ways Marxism tried to incorporate both competition and mutualism. Thus Engels, in the *Dialectics of Nature*: 'The struggle for life'. Until Darwin, what was stressed by his present adherents was precisely the harmonious co-operative working of organic nature, how the plant kingdom supplies animals with nourishment and oxygen, and animals supply plants with manure, ammonia and carbonic acid. Hardly was Darwin recognized before these same people saw everywhere nothing but *struggle*. Both views are justified within narrow limits, but both are equally one-sided and prejudiced. The interaction of bodies in non-living nature includes both harmony and collisions, that of living bodies conscious and unconscious co-operation as well as conscious and unconscious struggle. Hence, even in regard to nature, it is not permissible one-sidedly to inscribe only struggle on one's banners' (Parsons 1977: 139). As a theory of class struggle, Marxism was sympathetic with Darwinism, but as a critique of capitalism, it was necessarily opposed to the Social Darwinists. As a materialist ideology, it insisted on the fundamental unity of society and nature, but at the same time its dialectical basis led it to distinguish the qualitatively new aspects of human society produced by evolution.

Crystalization of the Biological Concept

The Definition

It was in this context that the term *mutualism* was brought into biology by the renowned Belgian zoologist Pierre Van Beneden, in a communication to the Royal Academy of Belgium on 16 December 1873. Van Beneden's lecture, titled 'A word on the social life of lower animals' (Van Beneden 1873), was later incorporated with others of his lectures (Van Beneden 1869) into a book entitled *Animal Parasites and Mess-Mates* (Van Beneden 1875), a popular work published simultaneously in French, English and German. His style is quite reminiscent of natural theology; drawing the parallel between the lower animals and human society, he says that:

> Every kind of industry exists in nature, and if some are honest, there are others who deserve a different name. In the Old World as in the New, more than one animal plays the knight of industry, living the life of a *grand seigneur*, and it is not uncommon to find, beside the modest pickpocket, the daring brigand of the highway, who lives only on blood and carnage. (Van Beneden 1873: 783-4)

In the English version of this passage in the 1875 book, the term 'knight of industry' (*chevalier d'industrie*) is translated 'sharper', and the ironic implication that the industrialist is a thief is present in the French version also.

Van Beneden goes on, however, to distinguish other species which, far from being parasites, actually assist other species by keeping them clean; he cites Aristotle's description of the plover and the crocodile, noting that it 'has been verified since' (Van Beneden 1873: 795). Thus 'There is mutual aid in many species, with services being repaid with good behaviour or in kind, and *mutualism* can well take its place beside *commensalism*'. (Van Beneden 1873: 785)

The use of analogies from human society to describe nature in this passage is not at all unusual in contemporary zoology, if somewhat more colourful in Van Beneden. But it is interesting to note the possible resonances of the particular parallels he chooses. Use of the term 'mutualism' may well have evoked thoughts of Proudhon and the mutualists of the Commune (repressed only two years previously) in the minds of some of Van Beneden's listeners, and the ironic comparison of parasites and 'knights of industry' is tantalizingly close to 'Property is theft'. Given Van Beneden's deep Catholic faith it seems unlikely that he himself was a socialist, although his biographer quotes him as reminiscing in later years about his participation in the Belgian national revolution of 1830, in which 'more than once I found myself going into combat on the walls of Antwerp with a fossil shell in one hand and a cartridge in the other' (Kemna 1897). It is simply hard to believe that he would have used such words without some knowledge of Proudhonian thought.

Van Beneden's definition was quickly taken up: a doctoral thesis presented at the University of Paris by Alfred Espinas in 1877, called *On Animal Society* (Espinas 1935), presents tick-birds and rhinoceroses, ants and aphids, mixed flocks of birds, and especially domesticated animals, as examples of mutualism. Espinas' work shocked one university authority with its long introductory chapter discussing the philosophy of Auguste Comte, 'whom no one at that time dared to mention in the University, and, fearing intervention by the ecclesiastical authorities, he demanded the suppression of the Introduction' (Espinas 1935). In the end a compromise was reached: the first published edition of the thesis, without the Introduction, would be printed in at least 500 copies, and any later editions could restore the original text.

Ecology Becomes a Science

While the meanings of both 'mutualism' and 'symbiosis', defined by De Bary in 1879 to apply to any association of dissimilar organisms living closely together, whether parasitic, commensal, or mutualistic, were quite clear and distinct, confusion between them soon set in. The young American botanist Roscoe Pound, in a presentation before the Botanical Seminar of the University of Nebraska on 17 December 1892, noted that already by then 'Symbiosis in the strict sense and mutualism are often confounded' (Pound 1893). His review, published in the following year in the *American Naturalist*, mentions humans and wheat, *Yucca* and yucca-moths, legumes and *Rhizobium*, and the recent discovery of mycorrhizae, but most of it is concerned with lichens. (The title of the paper is simply 'Symbiosis and Mutualism' on its first page, but 'Symbiosis and Mutualism in Lichens' in the volume's table of contents.) Pound argues that

> It is not necessary, as Frank seems to think, in order to establish mutualism to show that the organisms do no injury to each other. Mutualism of the kind we meet with in the vegetable kingdom involves sacrifices on the part of the host. The parasite is not there gratuitously. It is there to steal from its host the living it is hereditarily and constitutionally indisposed to make for itself. If the host gains any advantage from the relation, it can only do so by sacrificing — by giving the parasite the benefit of its labor that it may subsist. If the plant or the plant colony benefits as a whole, it purchases the benefit by the sacrifice of certain parts or individuals . . . Ethically, there is nothing in the phenomena of symbiosis to justify the sentimentalism they have excited in certain writers. Practically, in some instances,   symbiosis seems to result in mutual advantage. In all cases it results advantageously

to one of the parties, and we can never be sure that the other would not have been nearly as well off, if left to itself. (Pound 1893: 519–20)

Pound's article shows a considerably more hard-headed approach than does Van Beneden or Espinas, and he and his co-workers (including Charles Bessey and Frederick Clements) were engaged in a serious effort to make botany a real science rather than an extension of natural history (Tobey 1981). Emphasizing experimental methods and quantification, they had a very considerable influence on the then-being-born science of ecology. Pound himself, after collaborating with Clements on the *Phytogeography of Nebraska*, gave up science and went into law, eventually becoming Dean of the Harvard Law School and a founder of the sociological school of jurisprudence (Glueck 1965). A Quaker, he opposed the Welfare State but welcomed the co-operation of churches and fraternal organizations with the courts, saying 'This cooperation of organized religion and organized morality with the law is the more gratifying, because if individual, self-reliant, free enterprise has been an American characteristic, *cooperation* has not'. (Glueck 1965: 5)

Mutualism is given consideration by Warming in his 1895 book *Oecology of Plants*, described by Goodland as the first ecology textbook (Goodland 1975). But it has surprisingly little place in the works of Pound's colleague Frederick Clements, who became the dominant figure in American ecology in the first decades of the twentieth century (Tobey 1981). This seems contradictory, since Clements' name is associated with his description of the plant community as a superorganism, but in fact the superorganism concept derives not from a philosophy of mutualism, but rather from the ideas of Herbert Spencer and the American Social Darwinist Lester Frank Ward (Tobey 1981). It is basically a theory of development (succession in plant ecology), in which the competition between parts produces a more mature differentiated adult state. Thus Clements' texts devote very little space to pollination, mycorrhizae, or other mutualisms, nor to the concept itself; the same is true of the early works of Arthur G. Tansley, an ally of Clements and first president of the British Ecological Society. Tansley had collaborated with Spencer on the revision of the latter's *Principles of Biology*, and he was influenced to take up ecology by reading Warming's book; while he broke with the superorganism concept in the 1930s, in the early years of the century he helped to make it the basis of plant ecology in Britain, just as Clements did in America (Tobey 1981).

Mutual Aid

The tradition of mutualism had in fact become important in biology, but in quite another form. It was made the basis of a best-selling book, *Mutual Aid*, by the Russian anarchist Peter Kropotkin (1902). Published originally as a series of articles in the magazine *The Nineteenth Century* beginning in 1890, as a response to an article by Thomas Huxley on natural selection and morality, the book dealt with mutual aid in both animals and humans. In fact most of the discussion is not about mutualism at all, strictly speaking, since the majority of the examples concern co-operation within species rather than between species. But Kropotkin did not make a clear distinction, being more interested in showing that co-operation as well as competitiveness can help organisms to survive and reproduce, and thus be favoured in evolution.

Kropotkin was influenced in this by a lecture given in St. Petersburg in 1880 by a Professor Kessler, in which it was argued that 'the struggle for existence would be insufficient to explain the progress in organic life, if another law, that of sociability and of mutual help did not powerfully work for the improvement of the organisms and for strengthening the species' (Anon. 1880). He was also encouraged to publish his reply to Huxley by Henry Bates, a fellow geographer whose work on mimicry had been a major argument in favour of Darwin's ideas in the 1860s (Burkhardt 1982). Bates' friend and Darwin's co-discoverer Alfred Wallace had also come to feel the need for something besides natural selection to explain human evolution, although he had turned more towards spiritualism and the single-tax theories of the American Socialist Henry George (Marchant 1916). Thus Kropotkin's ideas, though unorthodox, were scientifically respectable, and indeed the contention that mutual aid can be a means of increasing fitness has become a standard part of modern sociobiology.

More problematic, however, was Kropotkin's fame as an anarchist, which at the time was associated in the public mind with bomb-throwing and assassination (Woodcock 1962). While this characterization is not at all appropriate for the majority of anarchists, and certainly not for Kropotkin, it was doubtlessly a serious handicap in winning acceptance. Furthermore Kropotkin's style of argument, with its sweeping analogy between human and animal society, was a drawback with those scientists who wished to make ecology an experimental, unphilosophical discipline (Tobey 1981). So *Mutual Aid*, while popular with the public in many countries, never became an accepted part of the scientific literature on mutualism.

In fact, mutualism in the first few decades of the twentieth century had a somewhat paradoxical status. Major discoveries of new mutualisms (for example, mycorrhizae, nitrogen-fixing nodules) had been made in the 1880s and 1890s, and the literature on such interactions had rapidly grown to substantial size (Schneider 1897). Philosophical works incorporating ideas about mutual aid, such as Jan Smuts' holism and W.C. Allee's theories of co-operation (Farley 1977; Caron 1978) were regularly popular with the general public. And substantial confusion was being introduced into the literature through discussions of the terminological problems with words such as 'symbiosis' and 'mutualism'. But yet mutualism did not become a major element of either Clementsian plant ecology or emerging population ecology (Kingsland 1981, 1982; Scudo and Ziegler 1977).

Newtonian Ecology

A New Synthesis Emerges

The beginnings of the school which was to grow to prominence in ecology in the second half of the century were inauspicious. Although mathematical, population-dynamics modelling of species had its origins in practical problems such as malaria control and fisheries management, its development by researchers such as Lotka, Volterra and Gause in the 1920s and 1930s had little immediate impact (Kingsland 1981). Most ecologists found the mathematics difficult and not very useful, and the Clementsian tradition in plant ecology encouraged looking at communities, not populations. The logistic equation became popular, partly because of the constant pushing of Raymond Pearl (Kingsland 1982), but the phase-plane models of competition and predation remained the purview of a small number of specialists (Kingsland 1981). One of those who did take an interest, however, was the zoologist Warder C. Allee, a Quaker with a lifelong interest in applying his ecological ideas to the problem of war and peace (Caron 1978). Allee wrote several books on the question of co-operation in animal and human society, and he came to want to put his philosophical feelings on a firmer scientific basis. He found that the physiological well-being of a population often would increase with density to an optimum, and then decrease (Odum and Allee 1954). Thus mathematical models could help to predict the population density which was best for the population.

Allee and various colleagues pushed the mathematical, population-dynamic approach to ecology throughout the 1930s and 1940s, but their success came only after the war. Their text *Principles of Animal Ecology* (Allee *et al*. 1949) marked both the end of an era and the beginning of a new one. On the one hand the superorganism view was predominant, and indeed co-operation and mutualism also were given substantial treatment. On the other hand a strong pitch was made for mathematical models as a useful tool in analysing population dynamics, and for the ideas of the Neo-Darwinian synthetic theory of evolution, developed during the same years as mathematical ecology.

These same years saw growing criticism of the superorganism metaphor, and a resurgence in plant ecology of the idea that species distributions are determined independently, rather than as parts of integrated communities (Egler 1952; Tobey 1981). The trophic-dynamic approach to ecosystems, using energy as the basic unit of analysis, had also been developed (Lindemann 1942). So a range of new ways of looking at ecology were available to American ecology as it began a phase of rapid growth after the war.

Economic and social factors were also favourable for a change. The development of computers and of atomic energy during the war provided new tools which could be used to analyse complex ecosystems. Funding was rapidly increasing, particularly after the creation of the Atomic Energy Commission and the National Science Foundation in the early 1950s. And more and more ecologists were being trained, as reflected in a rapid growth in the membership of the Ecological Society of America (which had been static since the 1920s) after about 1950 (Burgess 1977).

The transformation of ecology was clear with the publication of Eugene Odum's text *Fundamentals of Ecology* in 1953. Energy flow was a basic theme, as was the use of mathematics. The Lotka-Volterra models were presented in detail. The superorganism analogy was downplayed, and a mechanistic view of the ecosystem was presented as a sort of substitute. Other works of the same period carried the transformation into related disciplines, giving rise to the concepts of evolutionary ecology and population biology. Plant ecology was transformed from a study of the development of plant communities to a study of the distribution, and later the population dynamics, of plant species. And finally, the unification of plant and animal ecology into a single field with a common set of concepts and methods became a reality, exemplified by the title of Odum's text.

The World-view of Newtonian Ecology

Newtonian ecology has done for living nature what Newtonian physics did for the inanimate universe — explain it, using mathematics, according to a few basic theories about the interactions of individual entities. The theories include natural selection, Lotka-Volterra and similar models of population dynamics, difference equations, and general systems theory, and all use the numbers of individuals in a species as their basic measure. The species is seen as the sum of its individual members, and the community as the sum of its individual species. While a variety of interactions are possible, predation and competition are by far the most important.

The world-view of Newtonian ecology is thus one of a living world in which individual fitness-maximizing organisms compete for resources which regulate their growth according to simple laws, leading to a balance of numbers determined by intra- and interspecific competition and a flow of energy determined by the laws of thermodynamics. The 'community structure' resulting from these interactions will ultimately be predictable from the laws of natural selection acting on individuals (Orians 1973), supplemented by the laws of physics and chemistry.

This summary is of course a personal appreciation, necessarily overgeneralized and perhaps caricatural. But I believe it captures the main elements of ecology's development since about 1950, with allowance made for the diversity of schools within the field. While the intellectual battles within ecology are often fierce, on a number of points post-war ecologists have been considerably more unified than their predecessors. They regard ecology as a unified field without a basic plant-animal split, and indeed find plant-animal interactions one of the most interesting things to study. They recognize the value of mathematics, in some form, to their work. Their basic outlook is mechanistic, rather than organicist (Merchant 1980; Tobey 1981). They see whatever balance may exist in nature as being maintained by competition for resources, such as energy or food.

They believe that ultimately the 'emergent properties' of ecological systems can be explained at the individual level.

Or rather, they *used* to more or less agree. For I will argue that, since the early 1970s, Newtonian ecology has entered a period of crisis, during which some of the most deeply believed theories are being challenged. The rebirth of interest in mutualism about the same time is simply one manifestation of this, and a less controversial one at that, but its ultimate fate will depend on how the crisis is resolved.

The Crisis

The challenge to the world view of Newtonian ecology takes several forms. Some are clearly recognized as challenges to received belief, for example in the importance of competition in determining community structure, or the power of the adaptationist programme to explain all aspects of organisms' phenotypes. Others take the form of a gradual decline of interest in an approach, for example in the mathematically sophisticated multivariate techniques used in ordination, which are ultimately based on the individualistic concept of the plant community. Similarly, Lotka-Volterra modelling is not so much being challenged as wrong, but rather as uninteresting or useless, and the criticisms of systems modelling are not of the meaningfulness of the concept of system as such, but rather of its value in predicting anything.

These challenges have allowed a variety of heretical or bypassed ideas to come to the fore — group selection, shifting balance theory, lack of equilibrium, density-independence, and as part of the same constellation, mutualism. The crisis continues because the alternative ideas can neither gain enough adherence to replace Newtonian ecology, nor even be unified into a single alternative world-view. Thus the future of mutualism as an ecological concept will depend not only on the evidence amassed showing its value, but also the fate of Newtonian ecology and of the ideas which are challenging it.

The Ecological Crisis and the Crisis of Newtonian Ecology

How can we explain this crisis? One possible theory is that, like all scientific ideas, Newtonian ecology has simple reached the end of its usefulness — it has explained all that it is able to. This can even been formulated mathematically, in terms of a logistic curve of productivity (Tobey 1981), or metaphorically in Spencerian-Clementsian terms as the growth, maturity, and now decline of the Newtonian intellectual superorganism. But such explanations leave unanswered the question of why is the crisis occurring here and now? Is there simply a fixed lifetime for ecological ideas, or does the crisis have something to do with other crises in society?

I propose that the crisis of Newtonian ecology is closely linked to the emergence of the environmental movement around 1970 (Worster 1977). For while many leaders and participants in the environmental movement have also been academic ecologists, the ideas necessary in the political sphere have been very different from those provided by Newtonian ecology. Received ecological theory provided a mechanistic view of nature to those fighting to restore organic wholeness, and measures of fitness, calories, or births and deaths to a movement seeking to express the *value* of the environment. While arguing that nature is an integrated whole and that everything is connected to everything else, we continued researching with theories that said that communities are no more than sets of individual organisms. The problem, in other words, is one of cognitive dissonance — the difficulty of working with two sets of ecological ideas, based on different fundamental assumptions and ultimately in conflict.

Thus the rapid increase in the number of PhDs in ecology in the late 1960s and early 1970s (Farley 1977) is a generational expression of change in ideas — the coming into ecology of a large number of researchers, well trained in Newtonian ecology, who were to find it wanting for the needs of the *political* 'ecology movement'. This has led to more and more challenging of the conventional wisdom until virtually no corner of Newtonian ecology remains unchallenged.

Nature, Green in Root and Flower?

Is mutualism destined to be part of a *new* new synthesis, in which Newtonian ecology is replaced by a more organicist, integrated, value-laden view of the natural world? Several aspects of our present theories of mutualism argue against it, at least in the near future. As demonstrated in the following chapters, our present theories of mutualism are still basically mechanistic, mathematical, fitness-maximizing, and individualistic. It is all well and good to point this out, but up to now we simply do not see any alternative. Like all science, ecology must work with what it has inherited; truly original ideas are rare.

But it is possible to extrapolate what one would have to demonstrate to make the case that, say, mutualism is the major organizing principle in nature. Following up on recent advances, one could try to show:

(1) Survival and reproduction are generally mutualist-limited, for example, seed set is pollinator-limited.

(2) Mutualisms whose value remains unclear have definite fitness values, for example, seed dispersal.

(3) In cases such as the above, mutualism is generally a 'better' (more efficient, more dependable, less costly) solution than alternatives such as wind.

(4) Resource uptake is often dependent on mutualists, such as mycorrhizae, *Rhizobium*, actinorhizae, or zooxanthellae.

(5) The relative abundance of different members of a community is radically altered if mutualists are removed or introduced.

(6) Mutualism can explain apparent contradictions of the Competitive Exclusion Principle (Hutchinson 1978).

(7) Competition in nature often produces indirect mutualisms.

(8) Carnivore-plant and similar alternate-trophic-level relationships are often indirect mutualism.

(9) Mutualistic environments tend to expand (Law, this volume).

(10) Antagonistic interactions often cause mutualisms to evolve (Thompson 1982).

(11) Conversely, sharing of mutualists more often produces 'friends' friends' indirect mutualisms rather than competition for mutualism (Boucher *et al.* 1982).

(12) Biomass production of natural or artifical communities is increased by the addition of mutualisms (for example, inoculation with mycorrhizae) and decreased by their elimination.

(13) Mutualisms can evolve, despite the 'group selection problem', even with large trait groups (Wilson 1980).

(14) There are many more mutualisms in nature than we have recognized.

This is effectively a programme to replace Newtonian ecology's 'competition is the basic organizing principle of nature' with 'mutualism is the basic organizing principle of Nature'. Instead of being red in tooth and claw, nature is seen as green in root and flower. I am not convinced that such an assertion would be any more valid, or even more meaningful, than the one it replaces. Nevertheless, there is a value in making the effort to demonstrate it even if the end result is, as I suspect, that neither competition nor mutualism is 'the basic organizing principle'. For at least we will know that, if we do not find mutualism to be always and everywhere important, our conclusion will be based on a much broader exploration than has been done in the period of Newtonian ecology.

The contributors to this book, while disagreeing with much, perhaps most, of what I and each other say, are nevertheless implicitly engaged in a programme something like that outlined above. We share, at the least, the feeling that mutualism is playing a role in nature that is worth looking at, and that, using ideas, concepts and theories such as those we present here, ecologists can better understand what difference it makes that different organisms help each other out.

Acknowledgements

My thanks to John Vandermeer, Don Strong, Dick Lewontin, and my fellow members of the New World Agriculture group for their comments and criticisms of my ideas, and to Louise Bouthillier and Roger Bernard for their help in preparing the manuscript. Also my apologies to the many persons whose ideas I may have misrepresented, disregarded, or forgotten.

References

Addicott, J.F. (1981) 'Stability Properties of Two-Species Models of Mutualism: Simulation Studies', *Oecologia*, *49*, 42–9
Albrecht, F., H. Gatzke, A. Haddad and N. Wax (1974) 'The Dynamics of Two Interacting Populations', *Journal of Mathematical Analysis and Applications*, *46*, 658–70
Allee, W.C., A.E. Emerson, O. Park, T. Park and K.P. Schmidt (1949) *Principles of Animal Ecology*, W.B. Saunders, Philadelphia
Anonymous (1880) 'Notes', *Nature*, *21*, 284–6
Blakeslee, S. (1984) 'Strange Animals Thrive on Poison In Shallow Water', *New York Times*, 28 February, p. C–1
Bloom, S.A. (1975) 'The Motile Escape Response of a Sessile Prey: Sponge-Scallop Mutualism', *Journal of Experimental Marine Biology*, *17*, 311–28
Boucher, D.H., S. James and K.H. Keeler (1982) 'The Ecology of Mutualism', *Annual Review of Ecology and Systematics*, *13*, 315–47
Boulding, K.E. (1972) 'Economics as a Not Very Biological Science' in J.A. Behnke

(ed.), *Challenging Biological Problems*, Oxford University Press, Oxford, pp. 357–75

Burgess, R.L. (1977) 'The Ecological Society of America. Historical Data and Some Preliminary Analysis: in F.N. Egerton (ed.), *History of American Ecology*, Arno Press, New York, unpaginated

Burkhardt, F. (1982) 'Darwin and the Biological Establishment', *Biological Journal of the Linnean Society*, *17*, 39–44

Caceres, B. (1967) *Le Mouvement Ouvrier*, Seuil, Paris

Caire, G. (1971) *Les Syndicats Ouvriers*, Presses Universitaires de France, Paris

Caron, J. (1977) *La Théorie de la Cooperation dans l'Ecologie de W.C. Allee*, MSc thesis, Université de Montréal

Carr, R. (1966) *Spain 1808–1939*, Oxford University Press, Oxford

Cavanaugh, C.M., S.L. Gardener, M.L. Jones, H.W. Jannasch, and J.B. Waterhouse (1981) 'Prokaryotic Cells in the Hydrothermal Vent Tube Worm *Riftia pachyptila* Jones: Possible Chemoautotrophic Symbionts', *Science*, *213*, 340–2

Christiansen, F.B. and T. Fenchel (1976) *Theories of Biological Communities*, Springer-Verlag, New York

Colwell, R.K. and E.R. Fuentes (1975) 'Experimental Studies of the Niche', *Annual Review of Ecology and Systematics*, *6*, 281–310

Duggins, D.O. (1981) 'Interspecific Facilitation in a Guild of Benthic Marine Herbivores', *Oecologia*, *48*, 157–63

Egerton, F.N. (1968) 'Ancient Sources for Animal Demography', *Isis*, *59*, 175–89

Egerton, F.N. (1973) 'Changing Concepts of the Balance of Nature', *Quarterly Review of Biology*, *48*, 322–50

Egerton, F.N. (1976) 'Ecological Studies and Observations Before 1900' in B.J. Taylor and T.J. White. (eds), *Issues and Ideas in America*, University of Oklahoma Press, Norman, OK, pp. 311–51

Egerton, F.N. (1977) 'A Bibliographical Guide to the History of General Ecology and Population Ecology', *History of Science*, *15*, 189–215

Egler, F.E. (1954) 'A Commentary on American Plant Ecology, Based on the Textbooks of 1947–1949', *Ecology*, *32*, 673–95

Espinas, A. (1878) *Des Societés Animales*, Bailliere, Paris

Ewing, M.S., S.A. Ewing, M.S. Keener and R.J. Mulholland (1982) 'Mutualism Among Parasitic Nematodes: A Population Model', *Ecological Modelling*, *15*, 353–66

Farley, M. (1977) *Formation et Transformations de la Synthese Ecologique aux Etats-Unis (1949–1971)*, MSc thesis, Université de Montréal

Freedman, H.I. (1980) *Deterministic Mathematical Models in Population Ecology*, Marcel Dekker Inc., New York

Gause, G.F. and A.A. Witt (1935) 'Behavior of Mixed Populations and the Problem of Natural Selection', *American Naturalist*, *69*, 596–609

Gilbert, L.E. (1975) 'Ecological Consequences of a Coevolved Mutualism between Butterflies and Plants' in L.E. Gilbert and P.H. Raven (eds), *Coevolution of Animals and Plants*, University of Texas Press, Austin, pp. 210–40

Gilpin, M.E., T.J. Case and E.A. Bender (1982) 'Counterintuitive Oscillations in Systems of Competition and Mutualism', *American Naturalist*, *119*, 584–8

Glacken, C.J. (1967) *Traces on the Rhodian Shore*. University of California Press, Berkeley

Glueck, S. (ed.) (1965) *Roscoe Pound and Criminal Justice*, Oceana Publications, Dobbs Ferry, NY

Goh, B.S. (1979) 'Stability in Models of Mutualism', *American Naturalist*, *113*, 261–75

Goodland, R.J. (1976) 'The Tropical Origin of Ecology: Eugen Warming's Jubilee', *Oikos*, *26*, 240–5

Gould, S.J. and R.C. Lewontin (1979) 'The Spandrels of San Marcos and the

Panglossian Paradigm', *Proceedings of the Royal Society of London, B205*, 489–511

Hallam, T.G. (1980) 'Effects of Co-operation on Competitive Systems', *Journal of Theoretical Biology, 82*, 415–23

Hazen, W.E. (1964) *Readings in Population and Community Ecology*, W.B. Saunders, Philadelphia

Heithaus, E.R., D.C. Culver, and A.J. Beattie (1980) 'Models of Some Ant-Plant Mutualisms', *American Naturalist, 116*, 347–61

Hirsch, M.W. and S. Smale (1974) *Differential Equations, Dynamical Systems, and Linear Algebra*, Academic Press, New York

Hutchinson, G.E. (1978) *An Introduction to Population Ecology*, Yale University Press, New Haven, CT

Joshi, M.M. and J.P. Hollis (1977) 'Interaction of Beggiatoa and Rice Plant: Detoxification of Hydrogen Sulfide in the Rice Rhizosphere', *Science, 195*, 179–80

Kemna, A. (1897) *P.J. Van Beneden*, J.E. Buschmann, Anvers

Kingsland, S. (1981) *Modelling Nature: Theoretical and Experimental Approaches to Population Ecology, 1920–1950*, PhD thesis, University of Toronto

Kingsland, S. (1982) 'The Refractory Model: The Logistic Curve and the History of Population Ecology', *Quarterly Review of Biology, 57*, 29–52

Levin, S.A. and J.D. Udovic (1977) 'A Mathematical Model of Coevolving Populations,' *American Naturalist, 111*, 657–75

Levine, S.H. (1976) 'Competitive Interactions in Ecosystems', *American Naturalist, 110*, 903–10

Limoges, C. (1972) *L'Equilibre de la Nature*, Vrin, Paris

Lindemann, R.L. (1942) 'The Trophic-Dynamic Aspect of Ecology', *Ecology, 23*, 399–418

Linnaeus, C. (1977) *Select Dissertations from the Amoenitates Academicae*, Arno Press, New York

Marchant, J. (1916) *Alfred Russel Wallace: Letters and Reminiscences,* Harper and Brothers, New York

Martinez, L., M.W. Silver, J.M. King and A.L. Alldredge (1983) 'Nitrogen Fixation by Floating Diatom Mats: A Source of New Nitrogen to Oligotropic Ocean Waters', *Science, 221*, 152–4

May, R.M. (1973a) 'Qualitative Stability in Model Ecosystems', *Ecology, 54*, 638–41

May, R.M. (1973b) *Stability and Complexity in Model Ecosystems*, Princeton University Press, Princeton, NJ

May, R.M. (1976a) 'Models for Two Interacting Populations' in R.M. May (ed.), *Theoretical Ecology: Principles and Applications*, W.B. Saunders, Philadelphia, pp. 47–71 (2nd edn, 1981)

May, R.M. (1976b) 'Mathematical Aspects of the Dynamics of Animal Populations' in S.A. Levin (ed.), *Studies in Mathematical Biology IV*, American Mathematical Society, Providence, RI

May, R.M. (1982) 'Mutualistic Interactions Among Species', *Nature, 290*, 401–2

Merchant, C. (1980) *The Death of Nature: Women, Ecology and the Scientific Revolution*, Sierra Club Books, San Francisco

Meyer, J.L., E.T. Schultz, and G.S. Helfman (1983) 'Fish Schools: An Asset to Corals', *Science, 220*, 1045–7

Nadkarnia, N.M. (1981) 'Canopy Roots: Convergent Evolution in Rainforest Nutrient Cycles', *Science, 214*, 1023–4

Odum, E.P. (1953) *Fundamentals of Ecology*, W.B. Saunders, Philadelphia

Odum, H.T. and W.C. Allee (1954) 'A Note on the Stable Point of Populations Showing Both Intraspecific Cooperation and Disoperation,' *Ecology, 35*, 95–7

Orians, G.H. (1973) 'A Diversity of Textbooks: Ecology Comes of Age,' *Science, 181*, 1238–9

Owen, D.F. and R.G. Wiegert (1976) 'Do Consumers Maximize Plant Fitness?', *Oikos, 27*, 488–92

Parsons, H. (1977) *Marx and Engels on Ecology*, Greenwood Press, Westport, CT

Porter, K.G. (1976) 'Enhancement of Algal Growth and Productivity by Grazing Zoo plankton', *Science, 192*, 1332–4

Post, W.M., C.C. Travis and D.L. DeAngelis (1981) 'Evolution of Mutualism Between Species' in K.L. Cooke and S. Busenberg (eds), *Differential Equations and Applications in Ecology, Epidemics, and Population Problems*, Academic Press, New York, pp. 183–201

Pound, R. (1893) 'Symbiosis and Mutualism', *American Naturalist, 27*, 509–20

Rackham, H. (1956) *Pliny's Natural History*, Harvard University Press, Cambridge, MA.

Rescigno, A. and I.W. Richardson (1967) 'The Struggle for Life: I, Two Species', *Bulletin of Mathematical Biophysics, 29*, 377–88

Rescigno, A. and I.W. Richardson (1973) 'The Dynamics of Two Interacting Populations' in R.R. Rosen (ed.), *Foundations of Mathematical Biology*, Academic Press, New York, pp. 463–82

Risch, S. and D. Boucher (1976) 'What Ecologists Look For', *Bulletin of the Ecological Society of America, 57 (3)*, 8–9

Roughgarden, J. (1975) 'Evolution of Marine Symbiosis: A Simple Cost-Benefit Model', *Ecology, 56*, 1201–8

Rude, F. (1969) *L'Insurrection Lyonnaise de Novembre 1831*, Editions Anthropos, Paris

Ruse, M. (1979) *The Darwinian Revolution: Science Red in Tooth and Claw*, University of Chicago Press, Chicago

Schneider, A. (1897) 'The Phenomena of Symbiosis', *Minnesota Botanical Studies*, 1, 923–48

Schulkind, E. (ed.) (1972) *The Paris Commune of 1871: The View from the Left*, Jonathan Cape, London

Scudo, F.M. and J.R. Ziegler (1977) *The Golden Age of Theoretical Ecology: 1923-1940*, Springer-Verlag, New York

Sharma, I. (1977) 'Development of Commensalism Between Prey Predator', *American Naturalist, 111*, 1009–10

Siljak, D.D. (1974) 'Connective Stability of Complex Ecosystems', *Nature, 249*, 280

Siljak, D.D. (1975) 'When is a Complex Ecosystem Stable?, *Mathematical Biosciences, 25*, 25–50

Silk, L. (1983) 'The Spencer Prophecies', *New York Times*, 29 July, p. 26

Silvertown, J. (1984) 'Ecology, Interspecific Competition, and the Struggle for Existence' in L. Birke and J. Silvertown (eds.), *More than the Parts*, Pluto Press, London, pp. 177–96

Simberloff, D., B.J. Brown and S. Lowrie (1978) 'Isopod and Insect Root Borers May Benefit Florida Mangroves', *Science, 201*, 630–2

Soberon, J.M. and C. Martinez del Rio (1981) 'The Dynamics of a Plant-Pollinator Interaction', *Journal of Theoretical Biology, 91*, 363–78

Stauffer, R.C. (1960) 'Ecology in the Long Manuscript Version of Darwin's *Origin of Species* and Linnaeus' *Oeconomy of Nature*', *Proceedings of the American Philosophical Society, 104*, 235–41

Thompson, J.N. (1982) *Interaction and Coevolution*, Wiley-Interscience, New York

Tobey, R.C. (1981) *Saving the Prairies*, University of California Press, Berkeley, CA

Travis, C.C. and W.M. Post III (1979) 'Dynamics and Comparative Statics of Mutualistic Communities', *Journal of Theoretical Biology, 78*, 553–71

Valerio, C. (1975) 'A Unique Case of Mutualism', *American Naturalist, 109*, 235–8

Van Beneden, P.J. (1869) 'Le Commensalisme dans le Regne Animal', *Bulletin de l'Academie Royale de Belgique, serie 2, 28*, 621–48

Van Beneden, P.J. (1873) 'Un Mot sur la Vie Sociale des Animaux Inferieurs,' *Bulletin de l'Academie Royale de Belgique*, serie 2, *36*, 779–96

Van Beneden, P.J. (1875) *Les Commensaux et les Parasites*, Bibliotheque

Scientifique Internationale, Paris

Vance, R.R. (1978) 'A Mutualistic Interaction Between a Sessile Marine Clam and its Epibionts', *Ecology*, 59, 679–85

Vandermeer, J.H. (1980) 'Indirect Mutualism: Variations on a Theme by Stephen Levine', *American Naturalist*, 116, 441–8

Vandermeer, J.H. and D.H. Boucher (1978) 'Varieties of Mutualistic Interactions in Population Models', *Journal of Theoretical Biology*, 74, 549–58

Viera da Silva, J. (1979) *Introduction à la Theorie Ecologique*, Masson, Paris

Wells, H. (1983) 'Population Equilibria and Stability in Plant-Animal Pollination Systems', *Journal of Theoretical Biology*, 100, 685–99

Whittaker, R.H. (1975) *Communities and Ecosystems*, 2nd edn, Macmillan, New York

Williams, G.C. (1966) *Adaptation and Natural Selection*, Princeton University Press, Princeton, NJ

Wilson, D.S. (1975) 'A Theory of Group Selection', *Proceedings of the National Academy of Science of the U.S.A.*, 72, 143–6

Wilson, D.S. (1976) 'Evolution on the Level of Communities', *Science*, 192, 1358–60

Wilson, D.S. (1980) *The Natural Selection of Populations and Communities*, Benjamin/Cummings, Menlo Park, CA

Wiltshire, D. (1978) *The Social and Political Thought of Herbert Spencer*, Oxford University Press, Oxford

Woodcock, G. (1962) *Anarchism*, World Publishing Company, Cleveland, OH

Worster, D. (1977) *Nature's Economy: the Roots of Ecology*, Sierra Club Books, San Francisco

2 SYMBIOSIS AND MUTUALISM: CRISP CONCEPTS AND SOGGY SEMANTICS

D.H. Lewis

The Semantic Background

As in the appreciation of beauty, interpretations of technical terms and their definitions depend on the eye of the beholder. The contributors to, and many readers of, this book will certainly know that much time and space has previously been taken up by semantic discussion of the words, *symbiosis* and *mutualism*. I would like to hope that what follows will be the last word but am under no illusions that this will be the case! However, I begin with the assurance that I am *not* going to indulge in an historical witch-hunt into previous discussions. For those who prefer the meaning of symbiosis *to contrast with* that of parasitism, Lewin (1982) and some contributors to this volume are recent champions. For those who wish symbiosis *to encompass* parasitism, no less a person than de Bary himself — the originator of the term in 1879 — is their man. Read (1970), Starr (1975), Macinnis (1976) and Cooke (1977) are among recent advocates of this view. I am firmly in the latter camp (Smith, Muscatine and Lewis 1969; Lewis 1973a, 1974) and will merely comment to those in the former that, because their view is often repeated in textbooks, it does not mean that it is correct! Citations of earlier discussions may be found in the references already quoted in Chapter 1.

Outcomes Versus Mechanisms

Although my interests in symbiosis (*sensu lato*) as a plant physiologist concern *mechanisms* of nutrient exchange between interacting species, I shall, in this chapter, restrict attention to the *outcomes* of *direct* interactions between two species. Mechanisms are not relevant to the concepts to be discussed. Indirect interactions, not to be discussed here, can be considered as the outcomes, in terms of fitness, of symbioses between pairs, or groups of pairs, of organisms (see Vandermeer, Hazlett and Rathcke; and Lane, this volume).

The Symbiotic Continuum

In the same way that formal biological systematics is a two phase process: (a) the recognition of taxa and (b) the naming of them, so, in considerations of the interactions between taxa, concepts have first to be delimited and then an acceptable nomenclature devised. Odum (1953 and later editions) and Read (1970) examined the range of interactions between *populations* of two species on a +, 0, – basis, i.e. + where growth of the population was increased as a result of interaction, – where it was decreased and 0 where it was not affected. It is more appropriate from an evolutionary standpoint to view the combinations of these three possibilities in relation to *individuals* (see Templeton and Gilbert, this volume). It follows that, in terms of the potential fitness of the interacting individuals (Roughgarden 1975; Margulis 1981; Law and Lewis 1983, Law and the already mentioned other contributors to this volume), the six different combinations of –, 0 and + form a continuum from the doubly detrimental (– / –) to the doubly enhancing (+ / +). With a slightly modified terminology from that of Odum (1953) and Read (1970), Table 2.1 shows the segments of this continuum. 'Agonism' is preferred to 'antagonism' as the term for the (– / +) interaction. In this way, antagonism can be used as a collective noun for competitive, amensalistic and agonistic interactions,

Table 2.1: Segments of the Symbiotic Continuum Defined in Terms of Potential Fitness of the Two Associated Organisms

– / –	Competition	
– / 0	Amensalism	
– / +	Agonism	– Potential fitness decreased
		0 Potential fitness not affected
0/0	Neutralism	
0/ +	Commensalism	+ Potential fitness increased
+ / +	Mutualism	

that is, those in which the fitness of one or more interactants is reduced. Within the agonistic segment, a somewhat hazy distinction may be made between parasitic and predatory interactions. With respect to fitness and population biology, there is little difference between the outcomes of these alternatives. Essentially, predators are phagotrophs which employ some active capturing and ingestive processes, whereas parasites are absorptive osmotrophs on or in another organism (see Lewis 1974). The difference is one of mechanism, not of outcome. Some predators, for example, nematode-trapping fungi and carnivorous plants, bridge the gap in that they possess active or passive trapping mechanisms but lack ingestive systems.

Even those with a broad view of symbiosis do not usually include competition (– / –), amensalism (– /0) and neutralism (0/0) within their concept (see Read 1970; Macinnis 1976) although, within the definition of de Bary, a living together of dissimilarly-named organisms, there is no reason why the concept should not be extended to include them. This also applies to the inclusion of interactions within a species (Starr 1975). However, if this is done, symbiotic interactions become universal, the concept loses its intra-biological discrimination and can merely be contrasted with *asymbiosis* — interactions with the non-living (abiotic and dead) environment (cf. Starr 1975). This is what should happen to the concept of symbiosis. In this way, from a nomenclatural point of view, the discriminatory value of the word *symbiosis* is downgraded by its very broadness. It can be withdrawn from the semantic melée involving mutualism, parasitism, etc. but persist as the term for interactions in the widest possible sense. With this much debated concept safely on the sidelines, attention can instead become focused more sharply on different aspects of biological interactions. Reproductive fitness and population dynamics, which gave rise to the continuum above, are only two of many such aspects, albeit the two to be explored in most depth in this book. Alternative but complementary views of symbiosis (*sensu lato*) are considered in the next section.

Other Continua

The most obvious overall approach to the study of the individual segments of the symbiotic continuum is via the traditional divisions of biology. These are the horizontal divisions of the multi-layered biological cake of Odum (1953), vertical wedges of which represent different taxa, ultimately individual species. The horizontal slices of the biological cake transcend these taxonomic boundaries. In addition to taxonomy itself, there are slices concerned with structure (morphology, anatomy, cytology) and function (physiology and its sub-divisions, biochemistry, biophysics and biomathematics). The interplay between these, which involves development and culminates in reproduction, leads to those horizontal slices concerned with continuity — genetics and evolution. (See also, the continuum based on mode of integration discussed below; Table 2.2.) From what has been said above, no individual of any species remains in isolation from individuals of the same or other species for very long and so, in the natural world, *symbioses* substitute for *species* very quickly (Lewis 1973b). The bottom layer of the biological cake is ecology. Its terms of reference are the interactions within and between individuals of particular species and symbioses, and between these and the environment. As noted by many, environmental effects can readily shift the outcome of a particular symbiosis along the symbiotic spectrum between its competitive and mutualistic extremes.

A more detailed and heuristic approach to symbiology has been elaborated by Starr (1975). What follows is an adaptation of his ideas to complement the concept of symbiosis advocated above, which is probably close to, if not identical with, his. Starr's scheme, itself an elaboration of an unpublished one by himself and H. Heise outlined in Starr and Chatterjee (1972), employed a set of nine (A-I) continua which were based on a variety of criteria and by combined use of which the multiplicity of organismic interrelations can be described. As he noted, the descriptions employed for any given criterion 'can be specified quantitatively or qualitatively or both — depending on the fact-base and on the purpose behind the statement'. The segments of the continua can then be combined in a multi-dimensional network to describe the nature of interactions with the appropriate detail commensurate with the objectives of the description. The scheme's heuristic value lies in the amount of both qualitative and quantitative ignorance about specific symbioses which application of it will reveal. The modified scheme is outlined in Table 2.2 and should be read in conjunction with the paper of Starr (1975) which gives the rationale for its construction. Differences in emphasis or concept from Starr's original scheme are explained in the notes to Table 2.2 and below.

In his meaning of symbiosis, Starr (1975) required the association to be 'significant to the well-being (or unwell-being or both) of a least one of the associants'. This concept was elaborated in his continuum H based on the criteria of benefit or harm. In his discussion of this continuum, he listed a few mechanisms by which harm and benefit could be derived. As noted earlier, I am here solely concerned with outcome in terms of fitness. From this viewpoint, Starr's benefit and harm continuum evolves into the one already conceived by Odum. This has been redefined in relation to individuals in Table 2.1 and becomes continuum 1 of Table 2.2. In other words, the valuational continuum of Heise and Starr (see Starr and Chatterjee 1972) can be lengthened to include the spectrum from competition to mutualism. In Starr's terms, this spectrum extends from associations significant to the unwell-being of both associates to those significant to the well-being of both, where 'well-being' and 'unwell-being' are defined in terms of the relative fitness when symbiotic compared with that when not.

As noted above, I stress that the concept of symbiosis used here merely requires two organisms potentially to have some effect on each other's fitness. It requires neither *close association*, a physical feature, (cf. Boucher, James and Keeler 1982) nor *prolonged association*, a temporal attribute (cf. my earlier views in 1973a, 1974 and criticism of these by Starr 1975). These two characteristics, together with the relative sizes of the interacting organisms, form the basis of the next three continua in Table 2.2. The fifth continuum concerning specificity is self-explanatory but the sixth, concerned with nutrition, requires some elaboration.

In his discussion of his continuum F, equivalent to my continuum 6, Starr comments on some differences between his views and mine at that time. The revision of the concept of symbiosis here removes these differences with respect to this continuum and my views now coincide with his. I have, however, incorporated 'saprotrophy' into the continuum, albeit in parentheses. Saprotrophy is, by definition, an asymbiotic activity when asymbiosis itself is used as defined above. However, in necrotrophic interactions in which the fed-upon organism is killed and the killer continues to feed upon it after its death, the continued relationship is essentially an integral part of the nutritional continuum. This situation does not occur only in cases of agonistic symbioses, both predatory and parasitic, but also in those mutualistic interactions in

Table 2.2: An Heuristic Scheme, Based on Starr (1975), for Specifying Symbiotic Interactions Between Organisms (or Organelles)

Criterion	Continuum	Equivalent Continuum in Starr (1975)
1. Fitness[a]	Competition ↔ Amensalism ↔ Agonism ↔ Neutralism ↔ Commensalism- Mutualism	H
2. Duration[b]	Transient ↔ Prolonged ↔ Permanent	C
3. Relative Size[c]	Anisosymbiotic ↔ Isosymbiotic	B
4. Physical Contact[d]	Exhabitational (physical contact incidental) ↔ Exhabitational (physical contact essential) ↔ Inhabitational (organismic, partial complete) ↔ Inhabitational (organellar)	A
5. Specificity	Non-specific ↔ Specific	G
6. Nutrition[e]	(Saprotrophic) ↔ Necrotrophic ↔ Biotrophic	F
7. Interdependence		
(i) Degree	Facultative ↔ Obligate	D
(ii) Mode[f]	At biological level: Organelle ↔ Cell ↔ Tissue ↔ Organ ↔ Organism ↔ Population	E
	At molecular and sub-molecular level: Biophysical ↔ Biochemical	
8. Integration		
(i) Degree	Independent ↔ Integrated	I
(ii) Modes	Structure (Morphology, Anatomy, Cytology)	
	Function (Physiology, Biochemistry, Biophysics)	
	Development and Reproduction (Interplay of structure and function)	
	Continuity (Genetics and Evolution)	
	Ecology (Interplay of all above features)	

Notes:

a: See text and Table 1.

b: This continuum, here defined in very qualitative terms including 'prolonged' not used by Starr, should be elaborated quantitatively in relation to the duration of the life cycle of the interacting organisms.

c: This continuum merely describes symbionts as different or equal in size. Micro-, macro- and iso-symbionts are useful terms to describe these relationships.

d: The organellar level is included in continuum 4 and 7 (ii) to cover situations such as persistance of functional chloroplasts derived from algae in sacoglossan molluscs etc. and, if required, the symbiotic origins of eukaryotic organelles (Margulis 1981).

e: See text for rationale for including 'saprotrophy', an asymbiotic activity, in this continuum.

f: The interpretation here of Criterion 7 (ii), mode of interdependence, differs somewhat from that of Starr's equivalent Continuum E. That of 7 (ii) aims to allow the mode of interdependence to be specified at any appropriate level from the sub-molecular (e.g. ionic inter-relationships) to the population. Starr's sub-divisions were set out as follows.

1. Dependent physically
2. Dependent chemically (including bio-organic-chemically)
3. Dependent organismically (e.g. physiologically, nutritionally, regulatively, metabolically, genetically, anatomically, behaviourally
4. Dependent societally.

See text for rationale for this arrangement.

which specific organisms culture others as food (for example, ants and fungus gardens, ruminants and their micro-organisms, humans and crops). Also, during pollination, feeding on nectar may be regarded as a saprotrophic trait of the mutualism whereas feeding on pollen, a living 'tissue', is necrotrophic. Denizens of digestive tracts may also be regarded nutritionally as saprotrophic symbionts.

Both continuum 7, based on the criterion of interdependence, and continuum 8, based on integration, have been split into two aspects, degree and mode. In both cases as shown in Table 2.2, those based on degree are self-explanatory whereas those based on mode are more complex and require further explanation. In order that the scheme for continuum 7 (ii) (mode of interdependence) in Table 2.2 can be directly compared with Starr's equivalent, his scheme is reproduced in note f of Table 2.2. Additional mechanistic features which perhaps could be covered by this kind of continuum are nutrition, reproduction, locomotion and behaviour (including cleaning symbioses, protection against abiotic environment, protection against biotic environment, etc.).

The essence of continuum 8 (ii) is the formation, at its integrated end in the various manners listed, or 'third entities', that is, structures, functions, etc. which are more than, and different from, those which the participants are capable of as individuals. The various manners of integration listed derive from the horizontal slices of the 'biological cake' mentioned above.

Translations of Other Terminologies

A major recommendation above is that the concept of symbiosis be extended to include the whole spectrum of interactions from the doubly detrimental to the doubly beneficial, with assessment of detriment and benefit in terms of potential fitness. As noted by Templeton and Gilbert (this volume), this potential may not be realised because of the intervention, between formation of a reproductive propagule and its development into a reproductive adult, of many factors which contribute to mortality of one or both partners. In this way, as all mutualisms are included as a sub-set of the superset, symbiosis (Table 2.1), the phrase, non-symbiotic mutualism as used for example by Boucher *et al*. (1982) and Keeler (this volume), is a contradiction of the terminology advocated here. The problem arises because they define symbiosis as 'the living together of two organisms *in close association*' and non-symbiotic mutualisms as 'those in which the two species are *physically unconnected*'. In the scheme outlined in Table 2.2, the physical nature of associations is described by continuum 4, a separate exercise from assessment of potential fitness of the associates within the confines of continuum 1. Pollination of angiosperms by most animals which, in their terms, are non-symbiotic mutualisms can be described by the first four continua of Table 2.2 as anisosymbiotic, exhabitational, mutualistic symbioses involving transient, but essential, physical contact. Different examples of pollinating interactions can be specified further by the remaining four continua. Pollination of figs by fig wasps involves *prolonged* physical contact. Other examples of their so-called non-symbiotic mutualisms, for example, animal dispersal of seed, fungal spores and other propagules; anemone-dwelling fishes, ant-plant interactions; burrow-sharing gobies and shrimps; anemones on crab shells; cleaning mutualisms; mixed species aggregations, etc. can all be readily specified by the continua of Table 2.2, that is, *within symbiosis*.

Essentially, definitions based on single criteria are less liable to misinterpretation than those which involve more than one, since it is all too easy for greater emphasis to be given to one than to others by different interpreters, or by the same interpreter on different occasions. Semantically and conceptually alert readers will already have noticed my fault in this respect on at least one occasion — deliberately to make the point about use of varying criteria! In the paragraph above on the semantic background, 'parasitism' was first used in its agonistic sense, that is, it was defined by reference to a valuational criterion. Later, in the opening paragraph concerning the symbiotic continuum, I used 'parasitic' to contrast with 'predatory', a distinction based on mechanism of nutrition. This problem of two criteria for the concept of parasitism has been discussed by Hall (1974) who advocates a definition based on nutrition. His term for my first use of parasitism would be 'pathogenism' although disease, rather than specifically reduced fitness, was noted as the nature or the harmful outcome of the association by him.

I began by writing that the above was unlikely to complete discussion of these problems. My solution relies heavily on Starr (1975). It is therefore appropriate to end as he did. 'Given the nature of intellectual dialogue and personal development, this classificatory scheme does not purport to be the last word — mine or other's — on this subject. What is asserted here (including the meaning given

to the term 'symbiosis') can best be viewed as being my last word —
just for now!'

Acknowledgements

I am most grateful to Dr Richard Law for stimulating discussions
during the writing of this chapter.

References

de Bary, A. (1879) *Die Erscheinung der Symbiose*, Verlag von Karl J. Trübner,
    Strassburg
Boucher, D.H. (1985) 'The Idea of Mutualism, Past and Future', this volume, Chap-
    ter 1
Boucher, D.H., S. James and K.H. Keeler (1982) 'The Ecology of Mutualism',
    Annual Review of Ecology and Systematics, *13*, 315–47
Cooke, R. (1977) *The Biology of Symbiotic Fungi*, Wiley, London
Hall, R. (1974) 'Pathogenism and Parasitism as Concepts of Symbiotic Relation-
    ships', *Phytopathology*, *64*, 576–77
Keeler, K.H. (1985) 'Cost-Benefit Models of Mutualism', this volume, Chapter 5
Law, R. (1985) 'Evolution in a Mutualistic Environment', this volume, Chapter 7
Law, R. and D.H. Lewis (1983) 'Biotic Environments and the Maintenance of
    Sex — Some Evidence from Mutualistic Symbioses', *Biological Journal of the
    Linnean Society*, *20*, 249–76
Lewin, R.A. (1982) 'Symbiosis and Parasitism — Definitions and Evaluations', *Bio-
    Science*, *32*, 254–9
Lewis, D.H. (1973a) 'Concepts in Fungal Nutrition and the Origin of Biotröphy',
    Biological Reviews, *48*, 261–78
Lewis, D.H. (1973b) 'The Relevance of Symbiosis to Taxonomy and Ecology, With
    Particular Reference to Mutualistic Symbioses and the Exploitation of Marginal
    Habitats' in V.H. Heywood (ed.), *Taxonomy and Ecology*, Academic Press,
    London, pp. 151–72
Lewis, D.H. (1974) 'Micro-organisms and Plants: the Evolution of Parasitism and
    Mutualism', *Symposia of the Society for General Microbiology*, *24*, 367–92
Macinnis, A.J. (1976) 'How Parasites Find Hosts: Some Thoughts on the Inception
    of Host-Parasite Integration' in C.R. Kennedy (ed.), *Ecological Aspects of
    Parasitology*, North Holland Publishing Company, Amsterdam, pp. 3–20
Margulis, L. (1981) *Symbiosis in Cell Evolution*, Freeman, San Francisco
Odum, E.P. (1953) *Fundamentals of Ecology*, W.B. Saunders, Philadelphia
Read, C.P. (1970) *Parasitism and Symbiology*, Ronald Press Company, New York
Roughgarden, J. (1975) 'Evolution of a Marine Symbiosis — A Simple Cost-Benefit
    Model', *Ecology*, *56*, 1201–8
Smith, D.L. Muscatine and D. Lewis (1969) 'Carbohydrate Movement From
    Autotrophs to Heterotrophs in Parasitic and Mutualistic Symbiosis', *Biological
    Reviews*, *44*, 17–90
Starr, M.P. (1975) 'A Generalized Scheme for Classifying Organismic Associations',
    Symposia of the Society for Experimental Biology, *29*, 1–20
Starr, M.P. and A.K. Chatterjee (1972) 'The Genus *Erwinia*: Enterobacteria

Pathogenic to Plants and Animals', *Annual Review of Microbiology*, *25*, 649–78
Templeton, A.R. and L.E. Gilbert (1985) 'Population Genetics and the Coevolution of Mutualism' this volume, Chapter 6
Vandermeer, J., B. Hazlett and B. Rathcke (1985) 'Indirect Facilitation and Mutualism', this volume, Chapter 14

3 THE NATURAL HISTORY OF MUTUALISMS

D.H. Janzen

Mutualisms are the most omnipresent of any organism-to-organism interaction. All terrestrial higher plants, vertebrates, and arthropods are involved in one diffuse mutualism and many are involved in several. The allospecific mutualisms of the terrestrial world can be fairly cleanly placed in five groups: harvest mutualisms (gut flora and fauna, root rhizosphere occupants, lichens, decomposers in unit resources such as carcasses, epiphyllae and epiphytes, ants feeding plants), pollination mutualisms, seed dispersal mutualisms, protective mutualisms (ants and sometimes other insects that protect plants or Homoptera), and human agriculture/animal husbandry. Here I ignore the latter category as its traits are generated largely through the replacement of genetic fitness by the desires of humans, the most mutualistic of all organisms. I restrict this essay to terrestrial systems because I have no personal familiarity with aquatic ones. I disregard the conspecific mutualisms, as they generally involve parent-offspring and other kin selection subjects amply treated elsewhere.

The natural history of allospecific mutualisms has long been a subject of description and analysis under the rubric of pollination, seed dispersal, gut floras, etc. It is my opinion that any specific mutualism should continue to be studied as one of these processes. However, I have been asked to comment on the natural history of mutualism. I will do so, even though mutualism does not have a natural history above or beyond that already categorized under pollination, protection of plants by ants, rumen ecology, etc.

The natural history of organisms is what they do and are in nature. Natural history of mutualisms is therefore what mutualisms do and are in nature. But what are mutualisms? A mutualism is an interaction between individual organisms in which the realized or potential genetic fitness of each participant is raised by the actions of the other. The participants are called mutualists. Since a species has no trait that is analogous to the genetic fitness of an individual, mutualism cannot be defined with reference to species. Almost all allospecific mutualisms involve feeding of one mutualist on the other. The interaction is therefore from a member of one trophic level to a member of another, and the ease of recognizing the event as a mutualism is asymmetrical; there is almost always no doubt that the consumer is getting something that raises its fitness while it is the effect on the consumed that requires close scrutiny. Recognition of the effect is made doubly difficult by the fact that in most diffuse mutualisms the benefit to the individual comes about as a quite serendipitous outcome of the biology of the consumer, rather than something that is selected for by the mutualism. Seeds as contaminants of gut contents, pollen stuck on pollinator bodies, soil pH modification by rhizosphere bacteria, etc. are usually by-products of consumption rather than traits selected for through enhancement of consumer fitness.

There is a second major area of confusion in analyses of particular mutualisms. Be the mutualism diffuse or one-on-one, the standard first step in analyses of a mutualism is to separate the interactants in nature and see what happens. If a one-on-one interaction (for example, *Catasetum maculatum* orchid pollinated by male *Eulaema tropica* bees, *Barteria fistulosa* treelet being protected by *Pachysima aethiops* ants), the outcome is generally a severe and conspicuous decline in the fitness of the subjects of the experiment and a dramatic change in their ecological parameters as well. However, if it is a diffuse mutualism, the outcome is likely to be much more obscure. First, the removal of one or more mutualists may be followed by their replacement by other members of the mutualist coterie. The fitness of the remaining mutalists may even increase, if those removed happen to have been superior competitors and inferior mutualists. Second, the removal of a diffuse mutualist may be expressed as only a very small apparent change in the fitness of its partner (for example, number of seeds dispersed to good sites drops 5 per cent) but have a large ecological impact on the partner, with subsequent selection for a suite of traits that seemingly have little to do with the mutualism. For example, if the 5 per cent of the seeds that are no longer dispersed has been those that were directed at wet river banks because the dispersal agent removed was a denizen of that habitat (the remaining 95 per cent of the seed shadow landing on slopes and ridges), the selection for vegetative phenotypes capable of survival on wet river banks is deleted and selection can be more finely tuned to a vegetative phenotype for dry slopes and ridges.

A third major area of confusion in analyses of mutualisms is that by definition, it is situation- and outcome-dependent. A mutualist today may be a parasite of the mutualism tomorrow. A commensal may become a mutualist in some particular circumstance, and then drop out entirely or revert to being a commensal as other animals competitively push it about. The coinages of mutualism, usually food and behaviour, are coinages in many other kinds of interactions. A protective mutualism will be impossible to demonstrate when the plant is growing in a habitat impoverished in the relevant herbivores. The participants in mutualisms are not species or Latin binomials, but rather individuals. A naïve juvenile seed disperser may be more or less of a mutualist than is its experienced parent, depending on what kind of a seed shadow yields the highest fitness. A leaf-cutter ant colony is a mutualist to one large plant (its fungus), parasite to many others, and commensal to yet others.

I will illustrate the above introductory comments with a few natural history statements about the grey squirrel and oak trees in eastern US. A squirrel that collects an acorn is a seed predator on that acorn if it kills it. The acorn is a parasite of the squirrel (and of the mutualism, if you wish) if the squirrel buries the acorn and does not later recover it. The squirrel is never involved in a mutualism with the acorn. The squirrel is a mutualist to (with) the parent oak tree if it eats some acorns and disperses others (and does not recover them) *and* the consequence is that the fitness of that oak is higher than it would have been in the absence of that squirrel's acts. The array of squirrels that interacts with a particular oak tree (a portion of the oak's disperser coterie) has a diffuse mutualism with that oak (and likely with that oak's population), as does the entire disperser coterie of that oak (and that oak's population).

The squirrel's central position as an apparent oak mutualist may be very old or may be a recent event brought about by the extinction of the passenger pigeon and the gross readjustment of animal-plant interactions by the destruction of virtually all natural circumstances in the eastern US. The Pleistocene megafaunal extinctions severely altered the squirrels' interaction with oaks by greatly increasing the number of available acorns and their duration on the ground; whether this affected the mutualism cannot be determined without intense study of squirrel response to low and high acorn availability and knowledge of squirrel density in the presence of intense acorn harvest by other animals. We know so little of the sources of oak recruitment in natural forests (contemporary or pre-Columbian) that, strictly speaking, it is even inappropriate to assume that a squirrel or a squirrel population has a mutualism with oaks. The obvious experiment is certainly in order and worth doing. There are numerous natural stands of many species of oaks in many parts of the world that have no squirrel or squirrel analogue; would the introduction of squirrels to these stands raise the fitness of these oaks?

Might it even be that the normal way for oaks to get around (disperse individuals) is through pollen flow to those individuals that just happen to be growing on the edge of the canopy gap where a large adult has senesced? Acorns would then be viewed as iterative probes for the existence of adjacent safe sites by a blind organism. Suppressed seedlings, wherever they are, are just waiting for a safe site to appear. The acorn-parent (functional female) might well have a higher fitness, in the contemporary big-herbivore-free forest, if her acorns mostly died of competition or made suppressed seedlings in her immediate vicinity rather than have 90 per cent plus of them eaten by squirrels in return for a few metres movement. Over evolutionary time, one certainly cannot know if a tiny seed shadow and a huge pollen shadow necessarily give a lower fitness than a slightly larger seed shadow and a smaller pollen shadow (that is, in the absence of acorn predators like the squirrel, make fewer and larger acorns and a bigger pollen shadow).

The categories of diffuse and one-on-one mutualisms cut across the categories of facultative and obligatory mutualisms. An organism can be involved in any combination of these four. However, in general the obligatory mutualisms involve one-on-one relationships or a set of very closely related (and often interchangeable) species that almost act as one in their interactions with a similarly close set of partners. Close observation may be adequate to label organisms as diffuse or one-on-one mutualists, but experiments are necessary to distinguish facultative and obligatory mutualisms. Even with experiments, the outcome may be fuzzy. A particular fruit crop may be only a small fraction of a disperser's annual intake; deprive it of that food, and it may just eat other apparently 'compensating' foods, yet be on a slow population decline that will eventually put its density too close to the minimal population density to persist in that habitat. Is it a facultative or obligatory mutualist with that plant? Ant-acacias can survive without their ants in certain habitat types (even in some natural ones); are they then not obligate mutualists? Even figs occur in some habitats where they probably do not have fig-wasps.

Is there anything left to say about mutualisms, something that would not be said by working on seed dispersal or gut floras or extra-floral nectaries? Virtually every paper on 'coevolution', one of the common outcomes of mutualism, goes on and on about mutualistic systems. Mutualism is not a complex subject and is easily explored through the application of common sense and natural history knowledge. On the other hand, there is seed dispersal biology that is conspicuously not mutualistic (for example, wind and water dispersal of seeds, dispersal of burrs by large mammals) and all pollination is not done by mutualists. The authors of this volume apparently think that there is something to say, but I wonder if we are not beating a dead horse. First, mutualism has been thought to death; what we need are solid descriptions of how organisms actually interact, experiments with what happens when a potential mutualist is removed. And these descriptions and experiments already have categorical homes: seed dispersal, pollination biology, ant-plants, etc. Second, humans have displayed so very thoroughly their humanity by destroying some major portion of virtually all the arena in which the participants in terrestrial mutualisms evolved and can be recognized as such. I work in Costa Rica, a tropical habitat that can still make some claim to approximate some of the conditions in which its mutualisms came about. I cannot name a single mutualism in the forest that can be comfortably viewed as operating as it did in the pre-Columbian environment in which its members evolved. Uncontestably we are studying a system that is not in evolutionary equilibrium for human-generated as well as natural reasons. The degree that this matters is yet to be the subject of major research programmes.

To introduce the natural history of mutualisms, I will briefly look at the following six questions for the four naturally occurring groups of terrestrial allospecific mutualists.

(1) What are they and who participates in them?
(2) What are the coinages, the goods and services paid and obtained, in mutualisms?
(3) How are participants lost and gained; what determines coterie sizes?
(4) What are the genetic programmes of mutualisms?
(5) Who are the parasites of mutualisms?
(6) What would the world be like without mutualisms?

In this chapter I adopt an unconventional format; no references are included in the text, but a bibliography is given at the end. It contains a substantial array of the kind of literature that one should be familiar with in working with mutualisms. My goal here is to foment an attitude about mutualisms rather than directly to elucidate or document examples of them.

What are They and Who Participates in Them?

Virtually all mutualisms are seed dispersal, pollination, resource harvest (including food processing), and/or protection by ants and a few other carnivorous insects. For the vast majority of species, the mutualism is diffuse and apparently facultative; the removal of one participant does not result in immediate extinction of its partner. While diffuse mutualisms are present throughout trophic webs and touch on practically all members of the habitat, the obligatory one-on-one mutualisms range from trivial and isolated from the other organisms in the habitat (for example, neotropical ant-acacias) to having a potentially very large effect if deleted (for example, the primary mycorrhizal fungus of oaks in an oak forest).

Seed Dispersal Mutualisms

Most plants move twice in their lives — once as a pollen grain and once as a seed (though each of those movements may have several stages). In the pollen stage, the only survivors are among those that move to a conspecific stigma; the target is more identifiable in morphology and in concept than is a seed's target. A seed's fitness can only be realized if it can germinate at a point in space and time that has traits that match its genotype; that exact point where a conspecific is surviving as an adult is likely to lack those traits. The seed is in search of safe sites, and directly under the parent (or another conspecific) is unlikely to be a safe site for many life forms.

Animals move often and animals often kill what they swallow. Animals big enough to swallow a seed and animals with cause to sequester a seed appear in hindsight to be obvious evolutionary and/or ecological candidates for a seed dispersal mutualism with plants. On the animal side, the associated evolutionary changes or ecological requisites are sloppiness of ingestion, digestion, and sequestration coupled with acceptance of certain plant products as food. On the plant side, the changes or requisites are seed adornment and presentation, coupled with some seed protection.

A seed dispersal mutualism is an interaction through which the seed ends up in a better place than it would have, had the interaction not occurred. The plant half of this definition is, however, complex; there is a mutualist and a parasite on the plant side. As in the squirrel-oak example above, there is a mutualism from the viewpoint of the parent; here numerous seed deaths may be associated with an unambiguous mutualism if offspring are the coinage to the animal mutualist (see later section as well). There is also a parasitism from the viewpoint of the seed; however, if the seed was digested, there was no parasitism of the disperser, only seed predation by the potential disperser. This semantic and biological tangle can be at least partly eliminated by backing up in developmental time and including the dispersal agent in the definition of a safe site. That is, the safe site may start at the parent tree (at either the stigma or the fruit) and have a physical beginning as well as a terminus. Seeds in safe sites do not have 100 per cent survival to germination, but just a relatively high per cent survival. Mortality caused by dispersal agents can be part of the per cent mortality operating on a safe site as well as on an unsafe one.

There is a second semantic and biological tangle embedded in seed dispersal mutualisms, one that causes no end of trouble in comparisons among systems and researchers. The word 'seed' (or 'diaspore') is convenient in that it represents a physical diploid entity and we all think we know what we mean when we use it. However, it is a bag made of maternal tissue around a mix of the diploid offspring and the endosperm (the latter is a genetically asymmetrical composite of maternal and paternal genomes which seems to be involved in mediating or regulating the physiological interaction between offspring and maternal parent). It is customary to think of the fruit (or other genetically identical maternal baits) as under maternal control in the evolution of dispersal and the 'seed' as an organism different from the maternal parent. In fact, the seed is an aggregate of three things and certainly the seed coat is as much under maternal influence as are the fruit and other bait or protective structures (nut walls, tough pericarps, seasonally dormant fruits, fruit crop size, etc.).

Almost all seed dispersal mutualisms are diffuse and non-obligatory for any one participant, in that numerous species and individuals make up the disperser coterie of a given plant or plant population and they also depend on many other sources of food.

This pattern is in great part generated by the fact that were any plant to have the continuous fruit production necessary to support a full-time vertebrate obligate mutualist dispersal agent, the seed killers and fruit parasites would build up on its seed crops and eliminate them. Second, there is a large array of animals persisting on other food types that can be temporarily lured away from their (generally well-defended foods) with high quality and cheaply obtained (and processed) disperser rewards. Third, any mutant that gets onto an evolutionary track of supporting a seed disperser coterie year-round is going to have to produce a very large amount of nutrient material. Fourth, since seasonal dormancy appears to be unavailable to tropical vertebrates, an obligate monophagous dispersal mutualist cannot be active only during a short period of fruit maturation. Looked at the other way round, nearly all seed dispersers disperse the seeds of a number of species of plants. The reasons are contained in the above listing.

What kinds of seed dispersal are not mutualisms? From the viewpoint of the seed, almost all seed dispersal is a parasite of the mutualism between disperser and parent plant. Additionally, by definition, inanimate dispersal of seeds is not a mutualism. Wind and water have no fitness. Explosive capsules are plants that move themselves. Burrs stuck on horse legs do not benefit the horse. Hard red *Erythrina* seeds swallowed by a fruit pigeon and defecated entire do not benefit the pigeon. The squirrel does not benefit from the acorn that it buries and never recovers. An ant-acacia whose fruits are eaten by a bird that carefully spits out each seed below the parent ant-acacia does not benefit.

There is also a potential confusion over proximate/ultimate causes. Accept that a large tree evolved a large nutrient-rich fruit with hard pill-shaped seeds, and these large fruits come in large crops evolutionarily designed to be eaten by large herbivores that are relatively gentle on the seeds and defecate them to some degree in safe sites. In this scenario, wild pigs that get a few of the fruits and grind up 99.999 per cent of the seeds along with them are seed predators, parasites of the mutualism. The fitness of the tree is decreased by the peccaries. Now extinguish the large herbivores. After the tree population has ecologically come to some new equilibrium density, much reduced in habitat-coverage, there may still be habitats where it is sympatric with peccaries. The wild pigs still grind up 99.999 per cent of the seeds in the fruits they eat, but they now get to process nearly all of the fruit crop (what they don't get rots below the parent tree or is processed by seed-killing mice that do not disperse the seeds they miss). The price for pig dispersal of a few seeds is enormous, but the interaction is still a mutualism. The fitness of the tree is increased by the pigs. However, such a situation is a prime candidate for the evolution of traits to either render the pig a high-quality mutualist (through the invention of such a thing as a smaller and non-spitable extremely hard seed) or attract some other animal that will be more gentle on the seeds (through the invention of such a thing as a sweeter persistent fruit that would attract monkeys or birds). In fact, what we got was the introduction of Spanish horses and cattle which partly replaced the extinct herbivorous megafauna, followed by the elimination of the habitat and subsequent extinction of trees, peccaries, monkeys and anybody else of interest. The contemporary ecologist or evolutionary biologist studying seed dispersal mutualisms is confronted with the horrible problem of having to figure out where the snapshots lie within the movie described above.

Plants involved in dispersal mutualisms are to be found from the Arctic (berries eaten by arctic foxes, bears, wolves, ptarmigan, etc.) to all other habitats. Bird berries occur well up into the paramo of tropical mountains. Cacti, desert legumes and desert gourds all feed large herbivores. However, a more ecological and less taxonomic list suggests that seed dispersal mutualisms are not distributed evenly over the earth's surface. Mangrove swamps generally lack them, probably because water is a very appropriate and directed dispersal agent for mangrove seeds and because it would be difficult to present the large mangrove seeds to an earth-bound frugivore on the muddy substrate washed daily by the tides. Perhaps it is more than the absence of an edible layer of ground-level herbaceous plants that is the cause of the absence of vertebrate herbivores in mangrove swamp understorey. At first glance grasslands seem to lack them, but there are many ant and rodent species in grasslands that are major parts of seed dispersal coteries (along with the wind). Further, dispersal of grass and broadleaf herb seeds through seed consumption by large herbivores is probably an unrecognized mutualism. At the other extreme, almost every reproducing angiosperm in forest understoreys is involved in a seed dispersal mutualism involving a fleshy and bait-adorned fruit (except for a few burrs and explosive capsules). The conspicuous incompetence of the understorey breezes at getting seeds to safe sites in forest understorey is probably the primary cause. At the shrub and treelet level in forest, the vast majority of the mutualists with animals bear fruits containing bird-dispersed seeds, but at ground level, ants, birds, bears, rodents and big herbivores are involved. Moving up into the crown of the forest, there appears a substantial proportion of wind-dispersed samaras, propellers, gliding planes, parachutes, etc. whether the forest is tropical or extra-tropical. It is also from this level that numerous species produce large-seeded fruits that are behaviourally presented to earth-bound dispersal agents by falling to the ground. The avocado, rock hard when shed from the tree to fall 20–40 m, is a familiar example; it 'ripens' (turns soft and edible) after being shed and it is easy to see why. It is clear that the different proportions of the dispersal types mentioned above and generated by the properties of the mutualists available, the habitat, the best kinds of seed shadows, phylogenetic inertia, etc.

What kinds of plants are involved? Conifers are only rarely involved in dispersal mutualisms, and those that are have conspicuous departures in other aspects of their biology from that of the mast-seeding conifers in huge monospecific stands. Grasses that have fleshy fruits are generally restricted to tropical forest understoreys. The angiosperm Dipterocarpaceae have invented wind-dispersed mast-seeding northern conifers over again in the southeast Asian lowland rainforest, and the whole family stays away from seed dispersal mutualisms. However, in general, seed dispersal mutualisms are scattered throughout the woody broadleafed plant families.

The animal side of seed-dispersal mutualisms is hard to characterize. Almost all terrestrial and volant families of vertebrates have members that are seed dispersers in some kind of an apparent mutualism, with the glaring exceptions of most of the insectivorous bats, almost all shrews and edentates, most lizards and all snakes and amphibians. The taxonomic distinction between Carnivora and the taxonomically herbivorous groups of mammals is ecologically blurry because cats, dogs, hyaenas, bears, mustelids, procyonids and viverrids all eat fruits and disperse seeds. While their seed consumption by volume may be much less than that of monkeys and large herbivores, their short and fast gut systems and their less grinding molars result in a high proportion of surviving seeds. Also, they are likely to defecate in quite different places and patterns than are the other major groups of seed swallowers, potentially adding other dimensions to seed shadows.

Frugivorous vertebrates are found wherever there are fruits for them to eat (and seed-caching vertebrates are found wherever there are seeds to cache, except perhaps in the far Arctic and in the uppermost vegetation of tropical mountains). Their relative abundance not only follows the geographic trends of where there are fruits to eat, but probably generates the trends as well. The extra-tropical absence of bat-dispersed seeds is probably due to the difficulty a fruit bat will have in making a living on a continuously low quality food in climates with winters, rather than there being something in winter-possessing areas that is inimical to summertime fruits that a bat would eat. Likewise, many species of large herbivores are unlikely to be able to live entirely on fallen fruits; I suspect that their geographic availability to be mutualists to large trees is in great part controlled by the amounts and kinds of the herbage or browse that they need. The opportunity for complex interactions is very great. Large stands of oaks might well have sustained forest bison acting as seed predators on oaks but being important mutualists to honey locust as seed dispersers (with the honey locust never being sufficiently abundant to support the bison). The passenger pigeon clearly played the same kind of game with acorns on the one hand (though it might also have been a disperser for them) and forest berries on the other.

Pollination Mutualisms

A paternal plant spreads a thin and particulate sheet of itself over the habitat. The adaptive significance of this behaviour and morphology appears to be that a few of the partlets land at points where they can join up with tissues of another plant to produce new individuals. The formation of these new individuals is adaptive because (1) some are in better sites than the parent, (2) some have a better genetic programme than does the parent, (3) some carry on the parental lineage as the parent senesces or meets an untimely fate, and/or (4) there are more of them than the parent. In this one move there are portions of the process that animals conduct with many moves (though seed dispersal again conducts these processes a bit later in the ontogeny of the organism). The process of concern here is the generation of the pollen shadow with the highest fitness, given a plant's conspicuous inability to directly determine the location of the points on which its pollen grains should land. Inanimate generation of pollen shadow conspicuously differs from pollen shadows generated by animals in the following ways:

(1) Almost all inanimate-generated pollen shadow are wind generated.

(2) You cannot satiate the wind.

(3) While you can predict and perhaps assay the amount and direction of the wind, it can only be controlled by the timing of when you use it and the morphology of what interacts with it.

(4) The wind is blind and cannot smell.

(5) Almost all pollen in a wind-generated pollen shadow lands in the lethal ocean between the minute islands represented by conspecific stigmas.

Almost by definition a functional pollen shadow generated by animals involves at least some mutualism, since animals are quite unlikely to go to a flower unless they get some reward (ignoring the few cases of floral deception); if the pollen shadow is functional, the plant has a higher fitness than if it had no pollen shadow. I shall ignore cleistogamy, which is analogous to self-generated seed shadows.

While a substantial number of pollinator mutualisms are diffuse and non-obligatory in both directions, a larger absolute number and proportion of them are obligatory with one or a very few species of mutualists than is the case with dispersal mutualisms. This is probably due to the fact that insects are small (thereby acquiring relatively small rewards per visit) and some insects can be dormant through major parts of the year in the tropics as well as out of them.

Pollination biology is both the generation of some best pollen shadow by the paternal parent and the reception of some best distribution of incoming pollen by the maternal parent. However, much of the maternal moulding of the genetics of her clutch is done at the physiological level, from the stigma to the abortion of fruits and seeds. Much of the paternal moulding of its chances or pollination has to be done with the configuration in time and space of the pollen shadow. The traits of the pollen vectors are therefore differentially important to the paternal and maternal sides of the genome and the mutualism need not be symmetrically distributed among the plant sexes. Likewise seed dispersal and pollination are not words with analogous definitions; seed dispersal is the creation of a seed shadow, and contains no implication about the fate or fitness of the seeds. Pollination is the event of landing on a conspecific's stigma, and the word implies that the pollen grain on a stigma has a greater chance of survival than does one that does not land there; there is no English word in pollination biology analogues to seed dispersal, unless one wishes to invent the phrase pollen dispersal (and pollen shadow).

Plants involved in pollination mutualisms occur in all terrestrial habitats. They are least prominent among grasses and large extra-tropical trees, both being groups of plants whose crowns are maximally exposed to the wind and tend to live in stands of low species-richness. One is left with the distinct impression that if you grow shoulder-to-shoulder with conspecifics in a windy habitat (or microhabitat), wind pollination is more likely to evolve than is some kind of a pollination mutualism. Pollination mutualisms are most prominent among plants that usually grow in mixed species-rich stands in relatively wind-free habitats. Excluding species of plants of long-term disturbance sites (riverbanks, steep slopes), many tropical species-rich habitats contain only animal-pollinated plants. The huge family Leguminosae has only one species that is unambiguously wind-pollinated and two others that might be; all three exceptions live in very seasonal tropical habitats and are canopy members in dense stands. There is one widespread exception to the generality that tropical woody plants are usually animal-pollinated; bamboos, which grow (grew) in enormous nearly monospecific stands are apparently wind-pollinted (although their visitation by pollen-collecting *Apis* confuses the matter). There are two major exceptions to the generalization that plants that grow in large pure stands in the tropics are usually wind-pollinated; dipterocarps, members of a low species-richness canopy (but some dipterocarp forests have a non-dipterocarp species-rich subcanopy), are apparently pollinated by thrips. However, these thrips feed on vegetative parts of the plants when flowers are not available. These tiny insects, with presumably low pollen dispersal abilities per thrip, may be viewed almost as animate wind. The other exceptions are those species of mangroves that are pollinated by animals from nearby species-rich forests.

While pollination mutualisms are essentially omnipresent in broadleafed tropical forests, and in many extratropical habitats as well (excluding grassland habitats), only a small part of the forest's animals are involved in the mutualism (another significant fraction is a parasite of the mutualism, see below). Bees, wasps, moths, butterflies, flies and some beetles, along with hummingbirds, some other birds, some bats, and quite rarely rodents, marsupials and primates are at least 99 per cent of the mutualists in pollination mutualisms. Virtually all these mutualists are volant or very arboreal; they may be regular flower visitors because they are volant, may be volant in order to be regular flower visitors, or both.

Digestive Mutualisms

Just as animals can be viewed as frills around a set of gonads, they can also be viewed as frills around a compost heap. Virtually all animals have a gut which is partly to nearly totally (ruminants) filled with a complex coterie of other organisms (microbes, yeast, nematodes, protozoans, etc.). While in theory it might not be absolutely necessary for an animal to have such an array to process the food (or consume it and have the gut mutualists' corpses serve as food), animals clearly do it. At present the gut flora and fauna cannot be divided into mutualists and parasites of the mutualists, but it is clear that as a group these organisms are certainly diffuse and often obligatory mutualists. The truly large puzzle is why the animal itself does not evolve the capacity for doing its own digestion and simply dumps its food through a thorough acid or other sterilizer before it hits the absorptive surfaces. The answer probably lies in the fact that the gut coterie has very diverse ability, does its work for a low cost (sometimes), has a high rate of change of ability when exposed to new food, and may even be able to do some things that higher animals have never invented. Some animals have their compost heap on the outside, such as the fungus-growing leafcutter ants and termites.

Plants wear their guts on the outside but have practically the same kind of diffuse and often obligatory mutualism with bacteria, nematodes, fungi and other litter decomposers. The core question is the same for them as it is for animals; why not do the digestion yourself rather than pay someone else to do it? Just as with animals, it may well be cheaper to hire someone than to do it yourself, as long as there are different life forms with different desires and abilities. In other words, in contrast to bees and flowers, the gut micro-flora was certainly not invented *de novo* as a mutualism, but rather the presence of particular kinds of potential mutualists has led to their being hired into a mutualistic relationship. The relationship also extends up out of the ground to the nitrogen-fixing ephiphyllae (and sometimes epiphytes) on the aerial parts of the plant. It even may take the form of ants feeding the plant in return for domatia and/or food.

The pattern is essentially the same in plants and animals. The organism takes in (or grows to where it is near) a potential food item that is in a form that cannot directly assimilated, is probably contaminated with a large non-mutualist set of microbes and fungi, and variously contains toxic or otherwise detrimental materials (for example, cellulose is indigestible to many animals, being bulk that reduces the amount of food per bite swallowed unless the microbial flora can degrade it). The micro-flora then becomes part of inactivating the noxious incoming living organisms, renders some of the potential nutrients assimilable, and disarms some of the nasty compounds. It may even make a number of molecules that are not present at all in the original food item (for example, vitamins by gut bacteria, nitrogenous compounds by root-nodule bacteria or leaf epiphyllae). The mix is then discarded by the animal or plant, after a certain fraction has been removed. One of the extremes is represented by the observation that the faeces of a millipede feeding on dead fallen leaves may have a higher nutrient content than does the food that it eats (but the nutrient content would be even higher if the millipede did not take its cut).

It is now generally appreciated that animals have lots of waste to defecate because they take in so much unusable material — however, if they had to rely entirely on their abilities, they might well evolve to where they would take in less, but more selectively (and be rarer in the habitat as a consequence). Plants are thought of as having little waste to excrete because they take in little more than exacly what they want from the soup in the rhizosphere and from their more directly mutualist mycorrhizal fungi and ants. However, the intestinal wall is subject to a much greater amount and variety of odd molecules that later have to be filtered out of the blood or chopped up by the liver, than is any root. That is to say, if plants ate the kinds of organic things that animals eat, and especially herbivorous animals, they would have to have a liver and kidneys. If there were no mutualist coteries available to higher animals or plants, both would have to tolerate the risk of taking in an array of nasty molecules, and have to have quite spectacular personal defences against either their uptake or their action once inside.

In harvest mutualisms there is a process not generally present in either seed dispersal or pollination mutualisms. The microbes, etc. have little mobility themselves and therefore the process of transmission from one generation of higher organisms to the next becomes very important. This is evident when a leaf-cutter ant queen carries a bit of her mother colony's fungus garden with her to start her own garden, or a newly moulted termite feeds directly on a sib's faeces to obtain a gut flora inoculum. Some insects even transmit their gut symbionts transovarially. It appears that plants carry out transmission somewhat passively, in that where there are species-specific mutualisms, the only seedlings that have much chance of surviving are those that land on soils already contaminated by mutualist-infected root systems of a conspecific. However, the lectin glues that bind the nodule-forming nitrogen-fixing bacteria to plant roots may be viewed as a form of active inoculum capture.

Harvest mutualisms differ strongly from seed dispersal and pollination mutualisms in that in digestion, one organism is generally interacting with a geographically restricted suite of mutualists, and may well contain a whole population of them. This opens the way for the evolution of a variety of traits of the animal or higher plant that are unambiguously functional in the mutualism and probably were evolved specifically for it. The multi-chambered digestive tract of a ruminant very likely was not evolved for some other function and then just happened to be colonized by a complex microbe and protozoan array. The leakage of higher plant metabolites into the rhizospheres of roots might be just a leaky bucket, but I doubt it. It is quite possible that the acid bath given to food by most non-ruminants is meant to sterilize the food before it gets to the intestinal array as well as to begin the actual digestion of certain nutrient types; such sterilization would be highly functional in avoiding competitive disruption of the intestinal flora. The anaerobic conditions in the intestine may have been evolved more to close off that habitat to outsiders than as an accident of animal morphology. Likewise, for most large trees their true ecological roots (absorptive organs) are mycorrhizal fungi; the patterns of exploration and division of small roots, and the places they go in the soil, may be determined more by the biological actions of the fungi than the plant.

In like manner, each of the digestive mutualists within an animal is interacting with at least one organism in common, leading to mutualistic interactions between and among the gut mutualists as well as to strongly competitive ones. Since all the mutualists obtain all of their sustenance at the site (gut), they are also more likely to be mutualists than incidental parasites of the system. The mutualist gut coterie has a vested interest, with some chance of realization, in getting rid of the parasites; this chance is less available in pollination and seed dispersal. In the case of plants, however, not only may roots lie very close to one another and therefore share the rhizosphere, but one individual mycorrhizal fungus may even colonize the roots of several different plants.

It is striking that gut digestive coteries are almost entirely very small organisms with a high population turnover rate. They are also organisms with a world-wide reputation for having quite different biochemical abilities from their containers. The multicellular animals (nematodes, tapeworms, etc.) are probably parasites of both mutualism and container animal except in those few cases where a particular organ (for example, caecum) is regularily occupied by large monospecific stands of these animals. It is also striking that while the microbes, yeasts, and protozoa are often closely related when compared across container species, they often appear very distantly related to other microbes, yeasts and protozoa that are not members of gut coteries. This dichotomy is much less the case with the rhizosphere microbial flora, though roots are such an omni-present part of soil that it seems almost impossible to speak of soil microbes as having a life independent of root rhizosphere.

The mycorrhizal fungi that appear to be such an integral part of mineral harvest by large plants are taxonomically as well as ecologically a relatively distinct set of fungi from the large array of free living and pathogenic species. The epiphyllae (including the bacterial flora) that fix nitrogen on leaf surfaces and may even have other kinds of harvest mutualisms with plants are quite a specialized flora. The ants that deposit nutrient-rich materials in cavities of plants, with later absorption of the nutrients by the plant, tend to be species that are found only in their association.

Protection Mutualisms

Perhaps the most simple-mindedly spectacular and least complex are the protection mutualisms, with the obligate plant-ant protection systems at the head of the list. Of all the mutualisms, they are also the most amenable to experimental field analysis, primarily because all the mutualists are big enough to see and move around, the challenges (herbivores, are easy to observe and mimic, and the organisms involved are relatively sessile with respect to their partners. Most of the actual cases can be put in one of two categories: ant-plants, whereby an ant colony lives in some hollow (or hollowable) part of the plant and is directly or indirectly fed by the plant, and extra-floral nectary-bearing plants, whereby ants and a few other insects visit the nectaries but also forage and live elsewhere. In the former case, the protection ranges from extremely intense and thorough to barely measurable. Such protection cases may also blend into mutualisms in which the ants are feeding the plant as well. In the case of extra-floral nectaries, the protection is always somewhat diffuse and more a by-product of foraging in the area of the nectaries or protecting them than it is aggressive behaviour that evolved in direct response to the fitness or the plant bearing the nectaries. Here, the animal mutualists are always a diffuse and non-obligatory array and the multi-specific array of nectary-bearing plants usually constitutes a diffuse array of plants as well.

There remain two major stumbling blocks in this area of research. On the one hand, more experiments are needed in the field to determine if particular ant-plant interactions are feeding mutualisms or not. There is also the question of from what sort of ant-plant interaction the mutualism evolved. On the other hand, there is the nagging question of whether the numerous parasitoids of insects' larvae that visit extra-floral nectaries along with non-obligatory mutualist ants and other predators are in fact mutualists or just parasites of the interaction. This question will be resolved only when we know the foraging actions of the offspring of the parasitoids that have parasitized the herbivores that they encountered because they were visiting extra-floral nectaries. Perhaps needless to say, if the herbivores are sessile for a generation, such parasitoids could have a strong impact on their herbivory.

A major conceptual problem in protection mutualisms (as in dispersal and pollination) is that the herbivore threat changes in intensity and pattern over time and geographic area. For a given degree of mutualism, the protection will be overkill in one circumstance and insufficient in another. If relevant herbivores are absent, the system may seem to be, or actually be non-adaptive. The outcome is that the system may seem not very well tuned to any particular habitat or time. The threat of herbivory can vary enormously in space and time, and therefore this mutualism is probably the one that should be least in evolutionary equilibrium of all of them. Worse, protective mutualisms need to be maintained in times of low herbivore challenge just so that they will be available when the herbivores do appear; at these times of low herbivore challenge, the system may seem particularly poorly adjusted to the real world.

Obligate protective mutualists are always ants living in the plant that is being protected. The plants are almost all drawn from six families: Leguminosae, Passifloraceae, Polygonaceae, Moraceae, Verbenaceae, and Euphorbiaceae. There is no particular reason to think that these families were especially prone to this way of life; each has gone at it in its own way.

Diffuse protective mutualisms are sprinkled all over the plant families. Some families have many (for example, Leguminosae) while others have virtually none even though they are very species-rich (for example, grasses). The ants are drawn from almost all subfamilies, and genera with arboreal members seem most likely to be involved. However, when one leaves the ants behind, problem arise. There seems to be no doubt that aculeate (stinging) wasps and egg-parasitizing parasitic wasps can be part of a diffuse protective mutualism. However, a much better understanding of the biology of other nectary visitors and their hosts is needed before we can know if they are other than parasites of the mutualism.

What is the Coinage of Mutualism?

In almost all mutualisms, food is what the plant pays as final coinage in mutualisms with other organisms. However, many other behavioural and morphological plant traits may be viewed as subsidiary coinage of great, if indirect, importance. Since, the core of a mutualism is the payment of what is fitness-cheap to you for what is fitness-dear to you, it is commonplace for the coinage of mutualisms to be a resource that one organism is specialized at making or harvesting while the other cannot make or harvest it (at anything approaching a reasonable price or with a reasonable phenotype). Such a view stresses that all mutualists pay something for what they get. How do you get something for nothing? By making the bottom line fitness. Mutualists never pay each other in the same physical coinage at the same time. I pay you three iron nails for one pig. We both win. For this to occur we have to assume that each of us has a different cultural phenotype so that our respective cultural fitnesses can each be raised by this mutually asymmetrical trade. If there is any one non-definitional universal trait of allospecific mutualisms, it is that the partners are always very distantly related. This is because each mutualist is offering, serendipitously or otherwise, an activity or structure quite unattainable by the other.

Whether such a pair of reciprocally asymmetrical abilities was a precursor for the initiation of a given mutualism or was evolved as the mutualism became established is often unknowable. Were the first ant-acacias exceptionally poorly chemically defended and therefore ants foraging on their surfaces were exceptionally fitness-enhancing; were the first ant-acacias well-protected chemically but evolved chemically bland phenotypes as their chemical defences became redundant to the patrolling ants; were the first ant-acacias plants of habitats where ordinary acacia chemistry was an inferior defence compared with ants; or were they some combination of these three? Whichever is the case, the mutant acacia that made somewhat more hollowable larger thorns, more productive extra-floral nectaries, more edible leaflet tips (Beltian bodies), and more evergreen foliage, had a higher fitness. These four items, all based fairly directly on photosynthates (comparatively cheap in fully insolated habitats) and highly feasible for a conventional acacia phenotype to produce, become the plant's coinage in the mutualism. Likewise were the first acacia-ants exceptionally aggressive, did they evolve patrolling and cleaning aggressiveness through its enhancement of their fitness through a healthier acacia, did they live in a habitat where a carnivorous diet was especially unrealiable; or was it some combination of these three (and more)? Whichever is the case, the mutant ant colony that was somewhat more vicious, thorough and herbivorous had a higher fitness. These items, potentially cheap in a food-rich habitat, then became the ant's coinage in the mutualism.

Over time, the coinage of both plants and ant becomes modified to give higher yields with less silver tied up in no-interest-earning coins, more efficient minting machines, fewer defaults on loans, more accurate assessments of future cash flow, etc. But the bottom lines is not the absolute value of the coinage *vis à vis* any particular physiological production process or strategic consideration but rather the fitness of the genotype making the coins.

A major proportion of an organism's resources may be paid out in mutualism-inducing coinage. This proportion becomes most important if you are an outsider wanting to divert some of that resource budget to your own ends. On the other hand, an animal may be a very important mutualist in, for example, the consumption of a plant's fruits and dispersal of its seeds, yet interact with that plant over a minute fraction of its annual time budget and obtain only a tiny fraction of its daily resource intake from those fruits. Likewise, the payout by a carnivore to the vitamin-synthesizing bacteria in its gut may be too small to measure yet have a huge depressing impact on the carnivore's fitness if not paid.

There are three aspects to coinage. It can be cheap (e.g. cellulose used to make ant-acacia thorns), you can make it cheap (e.g. evolve fat storage patterns and sleeping behaviour that keeps your rumen flora warm in winter), or you can pay out a lot for a big return (e.g. have 90 per cent of your seed crop eaten by dispersers in return for getting a few seeds to very good germination sites). The big problem with human research is that while we can often measure the ecological properties of the coinage (and specially the coinages that we use in our own mutualisms), the fitness values of these coinages are much harder to measure and conceptualize. We really have no idea how the fitness of an oak tree changes with acorn edibility, acorn size (and thus number of offspring per crop), and supra-annual fruit crop periodicity (and thus sensation of dispersers and seed predators); however, we might easily measure how each change affects the number of acorns eaten by a mastodont, passenger pigeon, American Indian or squirrel. Perhaps the largest obstacle to translating ecological events into fitness value is our lack of knowledge of the natural history linkages between any specific ecological event and eventual representation of the relevant genotypes n generations later. Almost no field studies are directed at this class of question.

Once a structure, tissue or behaviour becomes coinage to a mutualism, two quite different processes generate upper boundaries to the degree to which it may be evolutionarily modified to be a higher quality coinage for the donor or receiver. First, the material or behaviour may well serve other functions mutually incompatible with the increased use or modification as coinage. If herb foliage is the bait (fruit) in a case where a large herbivore disperses small seeds consumed along with the foliage, only a certain amount or direction of relaxation of the chemical defences of the foliage can occur before the load of detrimental herbivores begins to increase due to the relaxation of defences. Second, the mutualist may well become satiated. Ever-increasing sugar load in fruits will not improve the quality of the seed shadow past a certain point, and may even decrease it by attracting the wrong animals or too many of the right animals.

The coinage that initiated a mutualism may not necessarily remain with that interaction over evolutionary time. Escalation (further evolution) of the interaction, once established, may involve structures that would not be initially of sufficient importance to have led to the bond. The high vitamin content of Beltian bodies was probably a coinage that came into play after acacia-ants lived with the plant for many selective changes, and the loss in function of petiolar nectaries on medium- to old-aged leaves of African acacias occupied by *Crematogaster* ants may represent the loss of a trait that may have first attracted the ants.

It is commonplace for coinage to serve numerous functions outside of the mutualism; evolution should both initially choose and converge toward molecules, tissues and structures that simultaneously serve many functions for the organism, with mutualisms being one of those functions. The coinage is often a structure or behaviour impossible or infeasible for the phenotype of one member of the mutualism to produce or perform. Some of the coinage in this case may actually cost nothing. For example, in large vertebrates deep-core body heat may largely be there anyway. However, one of the reasons for maintaining a *constant* high body heat — be a mammal — is to produce a higher quality gut flora than would be achieved with just the fluctuating gut temperature produced through normal activity. That is to say, mammals may have evolved in part as gut incubators.

Seed Dispersal Mutualisms

Plants pay maternal tissues, behaviour and strategic opportunity, and offspring for dispersal. Animals pay behaviour, strategic opportunity, morphological traits, gut space, and some other things associated with feeding for the nutrient rewards of seed dispersal mutualisms.

Maternal Possessions. The most common and easily conceptualized reward in seed dispersal mutualisms is nutrient-rich (including water) tissues immediately surrounding the seeds. These tissues are what humans commonly term 'fruit' (but include such things as tomatoes, cucumbers, squashes, avocadoes and carob beans). The fruit tissue often appear to have been evolutionarily designed to match the desires of a particular type of seed disperser (for example, tough, fibrous, dry, sweet, tannin- and protein-rich fruits for big herbivores). The animal often appears to have been evolutionarily designed to match the particular fruits (for example, long-armed brachiating and highly frugivorous spider monkeys feeding on sweet juicy fruits that match the desires of a highly diverse dispersal coterie). All kinds of combinations of the above occur as well.

Viewed over all vascular plants, the fruit tissue reward are extremely varied. Water, carbohydrates, fats, and proteins are all represented along with vitamins, stimulants (perhaps), minerals and other dietary needs. However, an edible fruit and its chemistry have many other functions; these make it nearly impossible to determine the sole function, if any, of a given fruit constituent. A fruit protects the developing seeds and often photosynthesizes; some of the materials in a ripe fruit are remnants of this activity, left in the ripe fruit rather than removed during the ripening process (for example, the congealed, permanently inactive tannin globs in vacuoles in ripe banana and persimmon pulp). A ripe fruit discourages the wrong vertebrates from eating it (and thereby dispersing the seeds to the wrong place, or a worse place than would the 'right' members of the dispersal coterie). It also has to have traits to deter insects and microbes that would destroy the fruit's attractiveness to the right members of the dispersal coterie. Finally, numerous traits of fruits are functional primarily as signals and not of material use to the seed disperser. While ideal chemical traits might serve both an attractant and protective role, many probably do not; their function can be determined only with the kinds of close examination generally not given to wild fruits. The high carotenoid content of yellow fruits ranging from oil palm fruits to persimmons can only be understood in the context of the exact animals that do or could interact with these fruits.

For analysis purposes, there is a particularly annoying trait of fruits. The protective and reward tissues are layered. The chemical analysis that grinds up all the fruit (non-seed) tissues generally obtains a soup totally different than that eaten or digested by any organism in the field. If the disperser's reward is a thin oily arillate pulp around each seed, then chemical analysis of the 2 cm thick fruit wall around the cluster of seeds is quite irrelevant to the biology of the disperser (though not to the seed predators like squirrels that may try to cut through it). Chemical analyses that grind up the seeds along with the fruit are even more absurd from a biologist's viewpoint.

A maternal parent plant pays for more than fruit tissues for the seed dispersal mutualism. Seasonal timing, synchrony with other conspecific and allospecific foods, synchrony with the fruit crop and other ripening patterns, green fruit dormancy, location of fruits within the crown, supra-annual timing, age at first maternal reproduction, dioecy, and other behavioural traits are all evolutionarily moulded by seed dispersal mutualisms. Each of these traits has a cost both in direct output and in strategic considerations. Additionally, just as fruit tissues are evolutionarily moulded by more than just the dietary desires of the seed disperser(s), these other traits are moulded by such things as herbivore avoidance, avoidance of predators on immature seeds, detrimental weather, allocation of resources to vegetative activities, etc.

There is a strong temptation to expect the plant's coinage in seed dispersal mutualisms to be in items obtained cheaply by the plant but dear to the animal. But this is an area perhaps impossible to unravel. Are many small fruits rich in water and sugars (and not much else) because these are cheap for the plant or because they are particularly desirable to birds, for example? And have they become particularly desirable to certain birds because fruits have been so reliably present for so long that birds have evolutionarily lost the traits that would allow them to obtain these compounds easily from other sources? Are the oil-fibre-protein-rich large fruits of the Lauraceae (avocado family) that way because oil and protein is easily made by members of this family, because if you want a big seed carried you need a different kind of bait than for a small seed, because the dispersal agents are slow to get to the fruits so you need a fruit that persists a long time once ripe, because you want to exclude those animals that primarily want sugar and water from your fruits, etc? One may perhaps some day be able to distinguish among these possibilities, but it will come about through extensive and detailed understanding of each of the options rather than through some 'test' of a single correlative or experimental model. This is an area of science where one simply has to want to know for its own sake rather than to verify some model or generality about such systems.

Offspring. Payment of offspring for mutualistic seed dispersal is generally hard for humans to intuit. We have been reared on the double standard that children are high on the list of whom to put in the lifeboat when the Titanic sinks, but children may also be the most appropriate cannonfodder (if we cannot locate enough young adults at the bottom of the power pyramid to put in their place). There are many cases in nature where the highest fitness is achieved for a parent by haphazardly, randomly or explicitly using offspring as the coinage in a mutualism. The squirrel-oak example is well known; in the Neotropics, agoutis and large seeds provide a parallel case.

From the viewpoint of the maternal parent, an offspring is simply a piece of her tissue (like fruit pulp) that happens to have got itself cemented onto a piece of paternal tissue (which in turn may raise that piece's fitness, just as does the yolk or cotyledons tacked onto a zygote). The programmed death of offspring as mutualism coinage is conspicuously a case where the inclusive fitness of the parent-offspring group is the frame for the evolution of the mutualism. In this particular case, the surviving individuals not only merely survive, but have their fitness raised by the death of their sibs or offspring.

Given human propensity for feeling that things that can be more directly translated into fitness values (for example, seeds) are less likely to be given away than are vegetative structures such as fruit pulp, one has the temptation to ask why such things as oaks, hickories, pecans, etc. have not evolved a fruit tissue type that is either a complete replacement for the seed as squirrel-reward, or a partial replacement (as in fruit around a hard palm nut that is buried by an agouti after eating the pulp)? The answer may be hardly more than 'it turned out that way'. However, one can also recognize that the northern hard 'nuts' can lie on the ground for months before being dispersed without losing their attractiveness (perhaps a valuable trait for a seed that is a member of very large and predator-satiating seed crops). If the seeds were covered with a high quality but perishable bait, perhaps the animals would just get fat on that and never bother to bury any in the first place.

The zygote in a seed pays a generally unappreciated price for dispersal by animals. The seed has to be swallowed rather than spat out, pass molar mills and gizzards unharmed, not block digestive systems passageways, and do other size- and specific-gravity-related activities. This places limits on the shapes *and* volumes of seeds, and on the amount that is seed coat versus seed content. In effect, the fact that the seed is going to have to pass through an animal is one of the limits on the size of the bag lunch carried by the seed. The optimal size of a *Enterolobium* seed might well be the size of an avocado pit, if only the number of seeds in the crop and the seedling reserves were to be taken into account. However, such a seed would never be swallowed intact by anything except perhaps a mammoth.

What do Animals Pay? The movement that the plant obtains in the seed dispersal mutualism is largely *not* what the animal pays. The movement is usually a by-product of the animal moving for all those reasons that animals move, including going to cache the seed in a place where the animal can later recover it. Animals most directly pay such things as time spent to spit out seeds, tooth wear and fractures on hard seeds, gut space occupied by indigestible materials, and perhaps the risk of lumen blockage by excess seeds. More indirectly, seed dispersal mutualists evolve a variety of morphological structures (for example, ridged palates, large gapes, dextrous gizzards) and behaviours (for example, traplining ability, short intestines, short intestinal passage rates, cud chewing) which are involved both with frugivory and some aspects of processing the seeds (other than simply spitting them out below the parent plant). Likewise indirectly, seed dispersal mutualists tend to have reproductive timing, migration timing and direction, developmental patterns and other timing regimes that are conspicuously adjusted to the patterns of fruit availability. Fruit availability is also adjusted to the animals. Finally, the animals have also ecologically elected to deal with those fruits that match the animal's timing regime dictated by other processes or selective pressures.

Perhaps the only area potential confusion in this subject is contained in the question 'do plant support their dispersal agents?', where 'support' is taken to mean that the plant or plant population has traits whose sole function is keeping the dispersal agents alive (or present) until there are again seeds to be dispersed. I know of no case that can be argued this way that is not better argued as the animal harvesting the resources of value to it and in the process dispersing seeds. It is, however, a puzzle that plants do not seem to have evolved the behaviour of producing an array of fruits of a size appropriate to attract a particular dispersal agent or coterie of dispersal agents, with only some of the fruits containing (expensive?) seeds. However, the frequently occurring evolutionary reduction in seed number in multi-seeded fruits, without concomitant reduction of fruit mass, may be a step in this direction. A relevant process may be that animals show enough negative response to seediness that the inclusion of seedless fruits in a crop might well result in the animals first eating all the seedless fruits, or destructively discarding many seedy fruits while in search of seedless fruits.

Pollination Mutualisms

Plants pay at least pollen, floral glandular secretions, floral structure (and flower-bearing structures), and phenological behaviour for pollination mutualisms. Animals pay the structures and behaviours of gathering and depending on floral resources, the risk of being on flowers, and the contamination of food and body with pollen.

Pollen. While it is tempting to think of pollen as offspring, and therefore the harvest of pollen by pollination mutualists as qualitatively different from the harvest of other plant parts, in fact pollen is simply paternal tissues. It is as expendable as is any other plant part, and its payment as a reward to animals represents no genetically special circumstance. The death of pollen through consumption by a bee larva is no different than the death of pollen through falling off a hummingbird head before that hummingbird's head brushes a stigma. It is also no different than death of a cell eaten by a caterpillar. Whether pollen is cheaper than other currencies in either fitness units or energy units simply cannot be examined with the data at hand on fitness or production cost. Pollen has the peculiar property that while each individual grain is a living organism, it has zero fitness (and therefore is genetically dead) unless it teams up with another plant. In this respect it differs strongly from seeds and spores (but is the epitome of a conspecific mutualism).

Whether, or to what degree, pollen becomes the primary coinage in a mutualism seems to be largely based on the nutritional needs of the particular pollinator (it is the primary coinage only with certain bees, birds, beetles and butterflies). It has the property that it can be used as sole coinage only with perfect flowers, and among bees, only females will be interested in it. Since it has to be living tissues and has to do other things than be coinage (or at least some of it does), it cannot be evolutionarily moulded into as many different forms as can be nectar (however, nectar's lack of diversity may be more due to similarity of external selective forces than physiological or evolutionary limits to production). It has the risk property that once pollen has become something other than a contaminant to the pollinator, the pollinator is not only likely to harvest it from the flower, but becomes good at harvesting it from itself, thereby evolutionarily turning itself into a parasite of the mutualism (as in many tropical stingless bees). On the other hand, pollen has the property of being much less digestible to animals in general than is nectar (though this digestibility is itself an evolved trait of nectar). The use of pollen as coinage therefore may be viewed as just one more way of excluding the variety of animals that are potentially available to harvest floral resources but not be pollinators.

When pollen becomes coinage, it is immediately subject to three potentially incompatible suites of selective forces. All pollen is subject to the two different suites of selective forces associated with transport through contamination on the one hand and survival to fertilization on the other hand. However, if pollen is also food, larger amounts with certain edibilities are also selected. This third suite of selective forces is what generates extreme specialization such as in *Cassia* flowers. They have one kind of tubular anther containing the fertile pollen that contaminates the bee and another kind of tubular anther containing the sterile pollen that it collects as food. The inedibility of pollen is suggested by the general failure of grass pollens to be good foods for solitary bees (though it might well be possible to evolve a bee that can do very well on wind-dispersed pollen, and some species of social bees collect grass pollens on occasion). Pollen inedibility is also suggested by the general disinterest of bees in collecting the pollen from flowers that are normally pollinated by birds. Pollen grain size, stickiness, and other standard pollen traits should certainly be affected by whether the pollen is food to mutualist, as opposed to only being a contaminant and/or eaten by parasites of the mutualism.

While the actual content of floral secretions has received much attention, there are two areas of major ignorance. First, what are the costs of nectar and metabolic scheduling *vis à vis* the plant as a whole? Second, to what degree is the pattern of nectar flow to be viewed as a general trait whose function is manipulation and avoidance of animals, and to what degree is it a simple outcome of physiological rates and steps determined by other plant demands? I opt for the view that like all traits important in interactions with the outside world, its actual manifestation is a combination of the two.

Floral rewards have remained in the area of pollen and nectar, with only the occasional quite trivial case of oils or resins being evolutionarily substituted. Malpighiaceae, with their oil-producing flowers, are a very good example of how, by manipulation of nectar traits, virtually all animals except the particular pollinating bees can be excluded. The consequences of a plant community made up entirely of such flowers would, however, be most dramatic; the enormous set of generalist pollinators and parasites of the pollination mutualism would be missing, as would all the things that they in turn do to the other organisms.

Sugary floral glandular secretions have another trait that is very relevant to the plant reproductive behaviour of engaging in intense brief bouts of sexual activity interspersed with much longer bouts of vegetative activity. When the plant does come into flower, if it is not closely tied to a specialist with excellent synchronization abilities, it has to rely on pollinators that are being supported by a set of the plants of the habitat at large. It has to use a currency that is compatible with that of the other plants.

The convergence of floral nectars on fruit traits (both are sweet and juicy, with various amounts of included extras) is probably a simple result of the fact that all animals need water, and sugars are very close to universal resource units in biochemical systems. Is there any reason to believe that the situation in the Oligocene was any different than at present, or that 100 million years from now it will be any different? No.

Floral Structures. Just as the function of a fruit is to be appropriately edible to certain animals and keep the vast majority of other organisms from eating it, along with protecting the seeds, a flower has been selected to be attractive and available to certain organisms and ignored, undiscovered or impenetrable to a very large number of other animals. In addition to nectar, the floral coinage in the mutualism is visual and odour signals, landing platforms, nectar guards, structures excluding this or that sort of animal, and even defensive chemistry. It is no accident that the female inflorescences of plants like marijuana and pyrenthrum have the highest per gram content of pharmacologically active ingredients of any part of the plant.

It is conspicuous that when a plant is monecious or dioecious (as opposed to having hermaphroditic flowers), it makes many more male flowers than it does female flowers. Is this because female flowers are more expensive to make than are male flowers, or because there is something about receiving a part of various pollen shadows that is different from generating a pollen shadow? It seems reasonable to guess that the number of female flowers is determined by such things as how many fruits the plant can afford to set or will set, the number of seeds per fruit, how much opportunity for selective abortive culling the plant wishes to have, how many of the fruits and/or seeds will be taken by animals before they mature, etc. The number of male flowers should, on the other hand, be set by how large and intense a pollen shadow gives the highest fitness, cost per flower, how animals repond to various proportions of flowers to flower size, the best phenological pattern, etc.

In other words, the traits of the floral portion of the mutualism should be moulded by both the traits of the animals and by the plant's problems associated with unit cost and fine tuning the genetic composition of the plant's clutch through abortion. Some of the coinage of the flower as a unit are therefore things of little material cost, such as whether to have a large number of small flowers or a small number of large flowers. The other coinages are those of the tissues themselves. These coinages should be largely determined by the desires of the animals, which may themselves be in small to large part determined by previous ecological and evolutionary interactions with flowers.

Before leaving this area, I should mention that traditional plant study renders the flower a quite separate structure from the remainder of the plant, and thus its price is viewed as somehow separate from that of the 'vegetative' structure. However, there are numerous morphological structures of plants quite aside from the flower itself that must be counted as coinage in the pollination mutualism. All of the storage facility that is selected for by the need to make flowers at pulsed intervals, support structures that put flowers in positions associated with particular kinds of visitation by animals, crown structures that do the same, and the hormonal system that determines flowering times, nectar flow, etc. are all parts of the cost of the mutualism, even though they are only in part the actual coinage.

Phenology. It is traditional to think of flowering phenology as by and large driven by or cued by climatic circumstances. However, the question of interest here is to what degree has the evolutionary choice of a given environmental cue for flowering been determined by the pressures of a pollination mutualism, as opposed to such things as internal allocation decisions (for example, flowering is unlikely to occur at the instant of leafing out, irrespective of what a nice pollinator relationship it might generate), competitor considerations (for example, heavy flowering at a time when your competitors are capable of intense vegetative growth is more risky than flowering at a time when they are dormant), herbivore considerations (for example, a flower crop at a time in the tropical rainy season when somewhat generalist herbivores are very abundant or specialists are evolutionarily very present), weather considerations (for example, it's too dry or too rainy to have a functional flower even if the pollinators will arrive), etc. To the degree that the phenology is guided by pollinator actions, the timing is part of the coinage of the system; the costs of such coinage will be both strategic and material, and we know virtually nothing of either.

The functional significance of phenology in the pollinator mutualism has the unhappy property that it generally cannot be observed at any one plant or time of year. Nearly the impossible is needed; we need a picture of how the various timings of flowering by different individuals and species push and pull at the foraging decisions of the individual pollinators. What would happen to a mutant of a dry-season flowerer, one that flowered in the middle of the rainy season? What would happen to a population of them, introduced as seeds, that flowered in the rainy season? What would happen to the mutant if it flowered in the rainy season *and* weather cues stopped a pollinator-sharing allospecific from flowering in that rainy season? Not only do we need to ask what sort, number, and quality of seed set does the mutant make, but what sort of pollen shadow, did it generate?

Contamination with Pollen. While it is an essential part of the pollination mutualism, it is not at all clear what price an animal pays for carrying around the pollen that contaminates it. Certainly pollen can be disruptive to the animal (for example, apparent selection for orchid pollinia that are coloured so as not to be visible to the hummingbird on whose bill they are glued). It seems possible that the first use of pollen as food by proto-bee larvae might have been that cleaned off the body by the parent wasp that was provisioning the nest with other foods — such cleaning, and that often seen on flowers, implies that pollen grains are detrimental. However, there is an alternative hypothesis — that the pollen grains are cleaned off as a by-product of cleaning that is functional in removing dust that would scratch cuticles and fungal spores that might germinate.

Much pollen that is later to be eaten by the flower visitor is carried in special structures or on specific parts of the body; it never ends up on a stigma. However, there are some social bees that are both sloppy about moving the pollen into the special pollen bearing areas and have pollen bearing areas such that pollen can be removed from them by the stigma. In short, the better the floral visitor is at sequestering the pollen, the less likely it is to be a mutualist (except for fig-wasps, which very thoroughly sequester pollen, but then very carefully dole it out again to the fig's stigmas).

It seems unlikely that contaminant pollen is a significant weight or drag to a flying insect. However, if every pollen grain with which a flower visitor came in contact, was to stick to it until removed by a stigma, many flower visitors would very quickly be flying (?) balls of pollen weighing much more than their original weight. I suspect that part of the pollinator's contribution to the mutualism is to have evolved an external coating that is *moderately* unlikely to pick up pollen grains. Those that have nearly uncontaminatable exteriors would be parasites of the mutualism. However, only in obligatory mutualisms such as fig-wasps and figs, or yucca-moth and yuccas, do there appear traits that are unambiguously evolved by the animal through their selective value to the plant. There is simply no way to select for, for example, branched hairs on a wasp because it then becomes a better pollinator to many species of plants. Branched hairs because they carry more pollen home to the proto-bee's nest, yes; because the proto-bee is a better pollinator, no. However, plants have evolved a multitude of pollen traits — sculpture, size, amount of clustering, accessory threads and glues — that increase the likelihood that a pollen grain will be picked up as a contaminant by certain flower visitors. One can certainly expect selection for contamination, but it will be a very one-sided kind of selection.

Flower visitors do clean themselves of pollen and other contaminant particles. The degree to which they pursue this anti-coinage, so to speak, can surely be influenced by the traits of the pollen grains themselves. Pollen grain size, colour, stickiness (especially when between two sliding and facing cuticular surfaces) and weight may all influence the assiduousness of a flower visitor's cleaning activities.

Risk of Being on Flowers. Flowers are generally conspicuous. Equally conspicuously, insectivorous birds are very rarely if ever observed gleaning flower heads for insects. A high per cent of the non-vertebrate diurnal visitors on flowers are stinging Hymenoptera or their mimics, or extremely fast (flies and some beetles and butterflies). While sugar and water may be an exceptionally fine resource to flying Hymenoptera, many of them can also sit on an ostentatious landing platform with much less risk of carnivory than can many other groups of insects that might sit there. This may be at least part of why the enormous number of cryptic and edible nocturnal moths do not become diurnal flower visitors (as well as being active at night). In short, the pollinator pays a price for nectar and pollen other than an energetic one, and part of its coinage in the mutualism is the morphology, physiology and behaviour to survive on the ostentatious face of a flower. Flowers also have their carnivorous parasites of the mutualism — spiders and assassin bugs (Reduviidae, Phymatidae). These animals take a steady but very small toll of the visitors to many species of flowers. Again, avoiding these animals is another price that the pollinator mutualist must pay. However, all the world is full of arthropod carnivores, and it is doubtful if flower visitors are at any higher risk from these animals than is the ordinary caterpillar or other non-flower visitor.

Structures and Behaviours for Gathering Floral Resources. It is no secret that Hymenoptera, Lepidoptera, bats, hummingbirds, and a variety of other animals have evolved a variety of behaviours, morphologies and physiologies in direct response to the traits of floral resources. Bees are the epitome of this evolution and many can be said to be hardly more than a wasp that is gene-spliced onto a pollen grain.

Just as with seed dispersal, the movements of the pollinators do not have as their function the pollination of flowers or the creation of a pollen shadow. Rather, the animals move in order to do other things, and to go to other plants. The question central to such movements is why don't the animals just sit at the flower crop and fill themselves up with resource, digest it, fill up, digest . .?

(1) Competitors take so much resource that you move on in search of missed flowers, or flowers that have not been visited for some time.
(2) Floral resources are not complete food so you must move to other areas to get other foods (or other resources).
(3) You are so harassed at a given plant by males or aggressive competitors that you move on to escape them.
(4) The plant regulates nectar flow between and within days to develop all three of the above processes to where the pollen shadow and pollen input gives the highest fitness.

Harvest Mutualisms

You have to feed your mutualist. Plants pay carbohydrates to their mycorrhizal fungal associates and the fungi pay minerals to the plants. Plants also exude organic compounds from their roots, and these chemicals apparently result in a rhizosphere bacterial world that renders minerals and other compounds more easily extractable from the soil. The bacteria are paying minerals for their food, though it is extremely difficult to determine if the situation is analogous to seeds and pollen carried as contaminants (the bacteria doing nothing to aid and abet this process other than what they do any time they find a resource) or analogous to fig-wasps (the bacteria do things for the plant that are of no direct value to the bacteria). Likewise, it does not seem to be known if the bacteria of the rhizosphere should be viewed as a distinctive subset of the soil flora that exist only in this context (something bordering on an obligate mutualism) or is it just that they are soil flora that become much more abundant at the sites where roots are putting out food for them. It is, however, clear that the mycorrhizal fungi are a distinctive set of fungi, often with no independent life of their own. Some even use root exudates as their cue to germinate (from a spore) and grow.

It appears that plants generally pay carbohydrates to harvest mutualists; they should be relatively cheap to produce by the plant with its crown in the full sun. It would not be surprising to find that the degree of mycorrhizal association becomes ever less intense as sunlight's products become dearer, just as absence of mycorrhizal association seems to be associated with times or habitats of great soil nutrient richness. Mutualists of higher plants, on the other hand, seem to be best at paying either in direct mineral coinage or in modification of mineral uptake rates.

Digestive mutualists of animals appear to pay in a more varied coinage, from vitamins and the bodies of protozoa to direct detoxification of plant poisons eaten by the animal to enzymatically degrading food materials. It is impossible to guess to what degree they just happened to make these things and therefore ended up not being selected out by the system, and to what degree they were established flora that evolved these abilities because their host mutualist then had a higher ecological fitness. The animals themselves pay many things: from body heat to digestive passage to remixing rates to base-rich salivas to patterns of food consumption. Animal fussiness over what is eaten (and its microbe content) may be as much determined by avoiding damage to its gut microflora as to traits directly inimical to the animal. Again, it is very difficult to separate traits that just happen to be there for other functions from those of direct function for the mutualism; is the stomach's acid bath a food sterilizer or a food digestor, or both?

Protective Mutualisms

In the case of obligate ant-plants, those systems where the ant colony obligatorily lives and feeds on the plant, the usual coinage from the plant is domatia, carbohydrate-rich fluids as nectaries or food bodies and/or ant-tended Homoptera, protein-lipid rich food bodies made by the plant or Homoptera carcasses, and continued production of these resources throughout the year (as well as very early in seedling life). The coinage from the animal is aggressive cleaning and attack behaviour, patrolling the plant during the 24-hour cycle or at least substantially longer than their non-plant-ant congeners, and all the behaviours associated with living within the confines of a plant and feeding on plant products. Little change may be required in the later case, since the plant products often resemble quite closely a mixed diet of insects and nectar from extra-floral nectaries on non-obligatory mutualism plants.

In the case of ants and other insects visiting extra-floral nectaries, it appears that the plant simply provides nectaries and nectary products and the animal harvests them. By foraging in the area of harvest, or by aggressively protecting the nectary against other insects, the mutualist then incidentally makes life more difficult for the herbivores than it would be if there were no nectaries attracting carnivores. In short, the mutualist appears to have made little or no change in behaviour to raise the fitness of the plant, and the plant has only made the change of producing nectary products. However, there is little doubt that an ecological consequence is that there is generally a substantial increase in the overall set of resources available to carnivores in a habitat rich in nectary-bearing plants. In addition to making nectaries and nectar, plants for which this system works well may reduce their chemical defenses against herbivores. This is, however, not so much a coinage as a consequence of the mutualism.

How are Participants Lost and Gained: What Determines Coterie Size?

Acquisition and loss of partners in diffuse mutualisms must be a commonplace in ecological and evolutionary time. One may think of a given mutualist as being 'handed' from coterie to coterie as it moves across geographic space and evolutionary time. When a potential diffuse mutualist first arrives, and all species first arrive somewhere, they encounter established diffuse mutualisms in which they can participate (except on some depauperate islands; for example, native bee-pollinated plants are conspicuously absent from many Caribbean islands, but practically all have complexes of sphinx moth-pollinated plants).

In obligatory mutualisms with coteries of low species richness — and indeed they may be obligatory simply because they have low species richness in their coteries — the new arrival may find itself in a severe competitive interaction with the residents; likewise, the loss of an obligatory mutualist may result in strong competitive release among the remaining mutualists or in some severe depression in the density of the mutualists, or both.

In envisioning the evolution of mutualism, it is probably important to make use of the fact that many of the plant traits — fruit characteristics, ant domatia, extra-floral nectaries, flower colours, etc. — are the sorts of traits that could easily be acquired as supergenes, or their equivalents, through interspecific introgression; non-mutualists may turn into mutualists without speciation occurring and nonsense traits with no apparent functional antecedents (for example, foliar nectaries) might not have to be independently invented nearly as often as is suggested by the diversity of plant taxa in which they are found. On the animal side of this coin, it is relevant that practically the entire animal coinage in the mutualism is made up of traits that are functional and used in other parts of animal biology. Digestive tracts were not invented for seed dispersal.

Just as we are fond of counting up the number of species involved in other kinds of ecological interactions, mutualisms offer fertile ground for enumerators. There are numerous conspicuous patterns and asymmetries. Symmetrically obligate mutualisms tend to have roughly the same number of species on each side of the interaction — about ten ant-acacias in Central America and about ten acacia-ants (about two to four of each in any major geographic area); approximately one species of pollinator fig-wasp per fig species; one to two species of large leaf-cutter ants and one to two species of their associated fungi at any one site. Asymmetrically obligate mutualisms tend to have asymmetrically species-rich coteries. One orchid bee or hummingbird may be the sole pollinator or an orchid or *Heliconia* species, but each of those animals will be part of the visitor or pollinator coteries of ten to hundreds of other species of plants.

The size of diffuse mutualism coteries reflects the overall species richness of the habitat for the particular groups of organisms involved, though in some cases — for example, the dicot plants and their bees in southwestern US desert — one enormous and species-rich diffuse mutualism (with occasional more obligatory and/or one-on-one nodes) may account for virtually all the members of the taxonomic groups involved. Throughout mutualisms, there is every reason to assume that the normal competitive and mutualistic inter-actions occur among members of the coteries, just as within and between any other guild (for example, detritivores, leaf-eating caterpillars, etc.). Yes, the other side of the mutualism wants its mutualists, but it wants them in certain patterns, amounts and forms. Just any old ant is not evolutionarily or ecological welcome into the ant-acacia mutualism nor is just any old fruit-swallower welcome into the species-rich disperser coterie of a large tropical tree with juicy fruits. The mostest is very unlikely to be the bestest in seed dispersal, pollination, digestion, etc. coteries.

At the other end of the list, mutualist coteries may range from zero to other low numbers in ecological time and space. It is easy to imagine that an absentee member of a mutualist coterie will be rapidly replaced by competitive release, immigration or evolution. Additionally, low species-richness diffuse mutualism coteries should have a tendency to grow because addition of new members may well lead to geographic and ecological range expansions, which in turn should lead to contact with more potential new members of the coterie. Along the same lines, a short-term diffuse mutualist (a tree that is in fruit for a few weeks, a large plant that is in flower for a few days) is very unlikely to drift to having a mutualistic coterie of low species richness, simply because it cannot be the sole support of its mutualists and because one or few species have a low chance of being reliably present. The conspicuous exceptions to this general-ization — figs and orchids pollinated by euglossine bees — are unrepresentative plants in numerous ways that maintain their excep-tional status.

On the very local and contemporary scale, mutualists are acquired through quite ordinary processes. When a tree comes into fruit, it does it in a presentation pattern that is easily interpreted as part of the process to generate seed shadow. When an ant acacia makes its first thorn, and even before, it is found by a searching queen acacia-ant. A nursing mammal probably acquires much of its gut flora from the nipples located conveniently for faecal contamination: to the rear and below. Virtually every flower and fruit harvester has some kind of search pattern to locate resources new to the season and habitat that year.

To illustrate some of the above processes, I offer a scenario of how a wind-dispersed perennial legume may turn into a plant dispersed by large mammals. *Pithecellobium platylobum* has its large flat seeds attached to the flat dehiscent fruit valves which are blown a maximum of a few tens of metres by the wind. This fruit is a likely model of the ancestor of the fruit of *Pithecellobium saman*, whose indehiscent seed-rich sweet, dry and fibrous fruits are eaten by large herbivores and whose hard seeds are defecated days later. During the evolutionary change from wind dispersal to mammal dispersal, the plant should acquire a diffuse coterie of seed dispersal mutualists. The coterie should have the following properties:

(1) Over the large geographic range of the plant (for example, *Pithecellobium saman* ranges from Mexico to tropical South America) the fruits will be eaten by many species and life forms of animals, many of which will disperse them to the best sites.

(2) At any one site, a specific *P. saman* population and individual will have its seed shadow generated by a number of species of mutualists *and* seed predators.

(3) The seed shadow so generated will yield different densities, demographies and distributional patterns in different habitats because (a) different habitats impose different mortality patterns on a given seed shadow and (b) the dispersal coterie (hence seed shadow) is different in different habitats.

(4) The members of the dispersal coterie will interact with many plants other than *P. saman* during most of the year (and in some cases all year if *P. saman* fails to fruit).

(5) While *P. saman* is bearing ripe fruits, its seed dispersal mutualists (and seed predators) will feed on both *P. saman* fruits and other foods; it is the movement to and from these other foods that is largely responsible for the particular configuration of the composite seed shadow that they will generate.

(6) The *P. saman* dispersal agents are conspicuously those types of animals that would eat a fallen pod of *P. platylobum* simply as one more piece of fibrous browse, but could be attracted by a mutant *P. platylobum* with sweet pods.

(7) Many of the mutualists will be obtaining seed contents — acting as seed predators — along with the strictly maternal fruit tissues from the ripe fruit. Even those animals whose absence would raise the fitness of an individual tree may still contribute to the form of the aggregate seed shadow — such an animal may be a mutualist from the viewpoint of some seeds and a seed predator from the viewpoint of the parent.

(8) The degree to which a given individual or species contributes to a *P. saman* or *P. saman* population's seed shadow is a function of both (a) the availability and desirability of other foods and (b) the activities of other *P. saman* fruit and seed eaters.

(9) Many members of the fruit-eating coterie of *P. saman* are not mutualists but the cause may be as much a function of (8) above as of their species-specific or even individual-specific traits.

Not only does it acquire a coterie with the above-listed traits during this evolutionary change, but many other aspects of the plant besides its detailed fruit chemistry will have a quite different ecology. Now it will be better to drop its fruits at a particular time during the dry season rather than any time during the dry season. Now its seed shadow will cover many more kinds of habitats than would be reached by a wind-generated seed shadow (selected for quite different life forms of the adult plant). Now seed crop size influences the fate of individual seeds within it, both in where they are dispersed to and their chance of being dispersed. Now the cost per fruit goes up and the seed crop size must decline if other parameters do not change. Now dry weather means conditions inimical to fruit-rotting organisms. Now dispersal patterns change depending on the composition of the allospecific plant array in which the adult is embedded. All these changes and many more may be associated with a change as dramatic as evolutionarily moving from a small scandent shrub on creek banks (*P. platylobum*) to a huge tree in forest and grasslands (*P. saman*). We cannot know to what degree the fruit is the tail that wags the dog, but in this game quite unexpected things are possible. Just because the tree is huge and the fruit is small, there is no reason to think that either consistently plays a much larger part than the other in evolutionary adjustments.

Assuming *P. saman* as we know it, to be many millions of years in age, there must have been two major circumstances under which the coterie of its seed dispersal mutualists has changed. First, over geological time, the tree's seed dispersal coterie has varied due to extinctions and invasions. When extinctions occur, the remaining members of the seed dispersal coterie are not only likely to eat more of the fruit and/or seeds than before, they are likely to generate a different seed shadow. In addition, species that previously were of little importance or species that did not even eat *P. saman* fruits may now have a major role in generating a high quality seed shadow. How much the ecological distribution of *P. saman* changes in time and space as a consequence will depend on all of the nine considerations listed above.

However, each time it changes there will be a new suite of selective pressures brought to bear on the relationships of the plant with *all* aspects of the environment, not just the dispersal event. For example, if there is a disperser coterie that generates a seed shadow spread over many different habitats, there will be selection for a different vegetative phenotype than if much of the coterie is eliminated and the new seed shadow is tighly focused on riparian bottomland.

Second, as the tree progressively invaded more of Centra̅. America, Mexico or South America from wherever it first evolved, it encountered different arrays of dispersers (even if its primary method of movement was via the guts of some long-moving or newly invading disperser, so that at least one old member of its disperser coterie was present).

The question then becomes, why didn't *P. saman* quickly become a series of local species, each adapted to a local set of dispersers (and habitat conditions)? The answer is at least in part that the array of animals common to most lowland tropical habitats is (was) about the same. The array of interest to *P. saman* would consist of several ruminants, several perissodactyls (or ground sloths with similar gut behaviour), several large rodents, a proboscidian, a primate or two, a suid or two, and perhaps an herbivorous reptile or bird. I expect selection, over millions of years, to hit on a fruit and fruiting phenology that would be quite robust to the deletion or replacement of any particular member of the coterie. A second answer is probably that with the seed shadow being generated by a large array of animals, the deletion of any of them will lead to only a small effect. The third answer is that the different mutualists themselves (dispersers) want about the same thing from the interaction. In other words, an Eocene rhino may well want a stomach full of a *P. saman*-like fruit at about the same time of year as did a Pleistocene gomphothere. Both may well have wanted about the same proportion of tannin to fibre to sugar in that fruit, and both might have been willing to walk about as far to get it.

At this point I encounter a long standing question among botanical inquiries. Why are flowers and fruits (reproductive parts) of plants so varied on a fine scale yet conservative in basic plan? This very conservatism is the basis of a plant taxonomy that has proven very robust as more and more information on relatedness accumulates. A previously unconsidered answer to this question is that the selective pressures by mutualists on fruits, for example, may be quite insensitive to basic structural traits and instead their variation in fitness is associated with such things as sugar and vitamin content, toughness of pericarp, seed number per ovary, fruiting phenology, etc. All of these traits are notably absent from the fruit traits characterizing plant families and subfamilies. Vegetative structures, characteristically of little use in family-level botany, are commonly of direct functional significance to photosynthesis, desiccation, wind resistance, browsers, etc. A camel may take no note of whether there is still a suture present in the wall of an indehiscent *Acacia* fruit, but it is very hard to think up a leaf trait that cannot (does not) influence its function. Likewise, a switch from bee to hummingbird pollination requires little or no change in the kinds of traits important in family-level classification.

Additions and deletions to seed dispersal coteries over geographic ranges and over evolutionary time illustrate quite well what may be expected of diffuse mutualisms. If there is a set of 23 species of birds feeding on the fruits of a tree at one point in geographic space and evolutionary time, and if four of them are mutalists, these four are likely to be replaced in time and space. When one is deleted, the system is ecologically and evolutionarily begging for one of the non-mutualists to become a mutualist. This is because in a mutualist-depleted system one of the previously non-mutualists has a high chance of raising the plant's fitness with comparatively inferior seed dispersal. If indeed this is the case, there is immediately selection for the plant genotype to drift in a direction that favours the actions of the new mutualist. Such a drift will in turn favour any mutant of the new mutualist that favours the plant as food. The situation is ideal for continual pulling of the new mutualist into the interaction, and vice versa for the plant. The selective forces should abate as once again the combined services of the mutualist coterie do not raise the fitness of the parent plant with ever-increasing rewards per seed, per fruit, per crop, etc. All of the above applies equally to geographic movements and to the original evolution of the system.

While it is easy to think of a plant as being passed from mutualist to mutualist in evolutionary and ecological time (and space), it is a bit harder to think of mutualists as being passed from plant to plant. However, it must occur. Were we to introduce the agouti (*Dasyprocta punctata*) to Africa, a habitat free of agouti seed dispersal analogues, we would be mimicking what must have happened each time the agouti moved into a new neotropical habitat. When it arrives in Africa, it will take on seed predation and seed dispersal activities with plants that have never seen such a beast. It will quickly become a mutualist to some and even displace some current mutualists. It will quite literally have been passed from Neotropical Sapotaceae, for example, to African Sapotaceae. Frugivorous seed-dispersing birds, migrating from Canada to Mexico, may be thought of as being passed from plant to plant all along the way.

The essence of diffuse seed dispersal mutualisms is that many forces are acting on the interaction to make it such that it contains parts interchangeable with the remainder of the world. To be sure, if an essential part of a perennial herb's seed shadow lands on treefalls deep in the forest, the forest bison cannot be replaced with a horse. On the other hand, if the forest bison is extinguished without replacement, so may be that portion of the herb population that lived in deep forest tree falls; but the portion that lived along grassland-forest edges may also become more abundant owing to more horse and grassland bison activity by the same process that eliminated the forest bison.

The essence of convergence in seed-dispersal mutualisms is that the fruit designed to be eaten by a number of species of animals is likely to be desirable to many more if exposed to them. Likewise, the seed size, hardness, shape, etc. to be optimally dispersed by a multispecies coterie is likely to also work well with a variety of other animals. The fig is at the apex of such an argument.

It is when seed-dispersal coteries become focused on single species or life forms of animals, and vice versa, that the parts become progressively less exchangeable with the other members of the habitat. Seeds of plants only dispersed by bats are often imbedded in fruit pulp of conspicuously low interest to monkeys, rodents and other arboreal vertebrates; arboreal mammals take little or no interest in the orange or yellow juicy fruits of mistletoes, fruits avidly eaten by birds. The woody fruits of many legumes that are dispersed by large herbivores seem to be little or no interest to birds and primates in Costa Rica, though in Africa this does not appear to be the case. Elephant-dispersed large seeds often sit on the forest floor in their steroid-rich fruit pulp for months in forests purged of elephants but still rich in other large and small mammals.

It is easy to conclude that as the traits of the mutualism become less generally compatible with the biologies of most relevant organisms in the habitat, the more disruptive will be the addition or deletion of a member of the mutualist coterie. However, such a trend is diluted in effect by the fact that almost all seed-dispersal mutualists feed on many species of plants and on many fruit life forms; it is very rare that a seed-disperser is so tied to a particular plant that selection can proceed to where that plant is dispersed solely by that one disperser. Most briefly, seeds are generally large and require large animals for dispersal; large animals are usually vertebrates, and unlike some of the very specialized insects tied tightly to flowers, vertebrates need so much food so much of the year that they cannot become closely tied to the fruit crop of a particular plant species.

The Genetic Programme

Mutualism — just as competition, predation and other isms — is a property of the interaction, not the organism. A bird finds a fruit and swallows it. The enclosed seeds survive the trip and are defecated in a good site because the bird happened to be there. In less outcome-dependent terminology, the bird encountered food and ate it. Its genetic programme said 'eat', not 'mutualize'. There is no gene for mutualism, even in the tightly obligatory ones like ant-acacias and acacia-ants. In dispersal, fruit traits are selected because plants with them have more of their seeds land in good places. Whether it was the wind, water or a bird that moved the seeds is unknowable to the plant genome. Sugar arranged one way gives you a gliding wing; sugar arranged another way gives you a flapping wing. Which is best depends on the habitat and a host of other things.

Organisms do not seek (or avoid) mutualisms any more than they seek to avoid competition. Organisms ecologically or evolutionarily forage where they get resources, where they don't get bashed over the head. That the resources are available because of the absence of a competitor or good weather is irrelevant. That one doesn't get hurt while foraging is what matters, not whether it was a competitor or a falling tree that did it. The same applies to mutualisms.

It is tempting to wonder if the reliance on mutualists by many organisms is due to some sort of difficulty in one organism having the genetic programming for the many abilities represented by a mutualist coterie. I think not. Rather, the adaptive peaks on which most organisms sit have some very steep sides; and mutualistic interactions appear when some other organism can make the slope more gentle. Yes, I suppose one could invent a plant phenotype that walked about planting its seeds in little piles of fertilizer in forest tree falls, but it certainly involves many fewer steps and resources to surround the seed with a bit of nutrient tissue and thereby splice a forest bison to your genome. In a very real sense, a defensive acacia-ant colony is simply the biosynthetic outcome of the autocatalytic reaction when the plant maintains nectaries, domatia, food bodies and evergreen foliage in the same physical space. The vast majority of plants programme for exactly the same result with the genetic code for alkaloids, tannins, drought resistance, branch patterns, etc. That the proto-ant-acacia flipped over to a new mode of programming was probably not due to the programme for ordinary defences becoming impossibly complex, large, etc. Rather, the acacia-ant colony is, for example, one of many ways that acacias have acquired a new set of traits that result in them moving into new habitats (ant-acacias in general occupy substantially moister habitats than do their assumed ancestral mimosoid legumes).

A long-standing question in biology is why genotypes don't congeal. And of course many fractions of them do — those involved in processes that can well do without the disruptions of mutations and new combinations, such as photosynthesis and cell respiration. In a real sense, a mutualist, with its genetic programme out of reach of genetic events in its partner, is a piece of congealed genotype. The traits of the suite of fruit-eating and seed-dispersing birds that have serviced Central American trees for millions of years have probably had almost no aggregate change for that time. Until the Pleistocene extinctions of large herbivores, the same may be said for the large herbivore seed-dispersers. A fig tree may go through all sort of genetic changes to adjust to climatic and biotic changes, without its pollinator fig-wasp being subjected to any of the genetic perturbations that were necessary for there to be material on which selection could act for these changes.

Such a situation is made graphic by the statement that a non-acacia can acquire the entire genome of an acacia-ant simply by acquiring the ant-acacia genome through inter-specific introgression. Likewise, a plant may totally change its pollen sources by a single gene change in flower colour, a change that may cause it to flip from being pollinated by hummingbirds to being pollinated by bees. The programme for the change in pollen shadow sources lies not in the flower but in the programme for activities of the quite different hummingbird and bee pollinator coteries.

With harvest mutualisms there is an added complexity. If digestive microflora can trade genetic information with ease, in one sense we may have to view the gut or rhizosphere as occupied by one enormous genetic programme with nodes at this or that point of biochemical activity. Coupling this with a very high turnover rate for the physical holders of the genetic programme, the opportunity for rapidly adjusting the biochemical abilities of the digestive coterie to the heterogeneities in the substrate should be truly awesome.

It is important to recognize that many mutualisms may come about through quite serendipitous ecological juxtaposition of just the right phenotypes, followed by only very slight evolutionary change. They may be the right phenotypes because they were previously involved in a similar interaction, or just by luck. It is much more reasonable to guess that the large-mammal seed dispersal system came about through these large herbivores eating wind-dispersed fruits as simply plant matter, than to think that some mutation flipped a small juicy bird berry over to a large woody fruit in response to large herbivores licking up fallen fruits on the forest floor (though this *could* in theory be an evolutionary route). It is not a surprise that all protective plant-ants belong to arboreal genera, the members of which live in hollow twigs and capture insects, and are often members of diffuse protective mutualisms in their own right. Viewed this way, the evolution of even very complex mutualisms seems not difficult evolutionarily; the intermediate steps seem simple, logical and ecologically reasonable (in contrast with such absurdities as vertebrate eyes, turtle limbs attached inside the rib cage and chloroplasts). Extra-floral nectaries are an exception. However, even extra-floral nectaries are easy to understand if one will grant that they were invented in the angiosperm flower and then evolutionarily moved to other parts of the plant, rather than vice versa.

Parasites of Mutualisms

Given the definition of mutualism, mutualists should drift back and forth between mutualist and parasite status in ecological and evolutionary time and space. The plants should rarely be parasites of the mutualism, for the simple reason that they almost always feed their mutualists. It is rather hard to convince an animal it is being fed when it is not. When an orchid is pollinated by a bee attempting to mate with it, it is being a parasite of the bee, not a parasite of a mutualism (that the male bee is a member of a variety of mutualisms with other species of flowers is probably serendipitous). However, when a hard red seed is ingested and defecated undigested, or a flower is nectarless but visited by nectar-seeking pollinators, a true parasitism of the mutualism by a plant has occurred.

In any particular case, identifying a parasite of a mutualism may be very difficult indeed. Who is to say that the few seeds carried off by some apparent parasite of the bird-fruit diffuse mutualism are not a highly significant portion of the seed shadow? There is little substance to the argument that if those seeds were important, the system would have evolutionarily proceeded to put most of the seed crop into that particular bird. Perhaps it will, or perhaps it cannot because of properties of the bird (for example, perhaps it is a rare species for reasons quite unrelated to its fruit consumption). Or perhaps it would be prevented by ecological actions of the disperser coterie being ousted in the evolutionary process.

When an organism is unambiguously a parasite of the mutualism — as in the non-pollinating wasps in figs, many gut invertebrates, and ants that occupy ant-acacias but do not protect them — it is largely a competitor with the mutualists and has a strong effect on both sides of the mutualism. It is the elimination of these parasites that undoubtedly drives a large number of the distinctive fine detail traits of fruits and other mutualism coinages. The trick is to make coinage that is highly desirable to one organism and rejected (or blocked) to another. This would appear to be technically more difficult than simply making a protected structure, but part of the impasse is resolved by the mutualists themselves evolving peculiar traits. The morphology and behaviour of pollinating fig-wasps allows access to flowers that are quite inaccessible to the majority of organisms that would quite easily consume their chemically undefended parts if they could get to them without puncturing the bitter and latex-rich green fig wall. The short intestinal gut tract of frugivorous bats probably relies very little on a digestive mutualism, and therefore bat fruits can be laced with antibiotics and other compounds, rendering them considerably less interesting to other animals. Acacia-ants are just as vicious to organisms attempting to harvest the highly edible food produced for them by the plant as they are to herbivores eating the leaves directly. However, it is absolutely unambiguous that in considering the evolution of the traits of coinage in mutualisms, keeping out the potential parasites of the system may be as important or more so than is letting in the potential mutualists.

What Would the World Be Like Without Mutualisms?

Certainly inter-specific interactions would be less complex if it were not for mutualisms allowing mutualists to 'have' traits that are incompatible with their primary phenotype. Another manifestation of the same phenomenon is that mutualism-free plants and animals would have narrower ecological ranges in terms of habitats and microhabitats occupied. Many plants have the population pattern they do within and among habitats because of the shape and intensity of their seed shadow (this not to say that the distribution pattern will be correlated with the shape and intensity of the seed shadow). If all plants had only the kinds of small and ovoid seed shadows normally generated by wind and explosive capsules, or the linear ones produced by water (barring mammal-dispersed burrs here), many kinds of plant distributions would be missing since a seed shadow would only touch a habitat if that habitat was within a quite short distance of a habitat already containing parent plants. With a horse-dispered seed, reproductive adult trees may be found just about anywhere; with a wind or water distributed seed, the plant would have a very local distribution with there being widely scattered individuals in no habitat. What such an event would do to the biologies of animals that normally live in species-rich forests is anybody's guess, but it is tempting to wonder if the gross differences between the plant habitat structure and animal community in southeast Asian dipterocarp (wind dispersed) rainforests, and those of the rest of the species-rich lowland tropics, may be due at least in part to the dominance of wind-dispersal in the tallest trees in the habitat (as well as the crummy soils, the dipterocarp mast seeding, etc.). The analogous case applies to animal diets. If omnivorous seed-dispersal mutualist birds had the fruit deleted from their diets, many would disappear from certain habitats, have to have quite different migration behaviours, and obtain much larger amounts of food from other sources. If animals had to rely solely on their own digestive abilities, they would be presumably much less of a threat to plants than they are. Plants would have more resources for competition and reproduction, both because of less direct loss and less expenditures for defence against herbivores. The herbivores themselves would be either less efficient (than now) generalists or more tightly restricted specialists; the rumen-less deer might well evolve into a sage-brush specialist, but that might well be all it could eat. Plants might well win many coevolutionary races with herbivores.

The elimination of pollination mutualists would wreak havoc with many plant population structures. They would go extinct at much lower inter-plant distances than at present, with the consequence that more species would be edaphic specialists and occur in denser stands. Within habitat species-richness would decline and between habitat species-richness increase (compared to within habitat), while overall species-richness would drop substantially owing to the impossibility of being a plant customarily found at a long distance from its conspecifics. It may not be only competition and weather that result in virtually all truly reproductive conifers occurring in stands shoulder to shoulder with conspecifics. A pine species whose other biology resulted in it being represented by widely scattered individuals would simply go extinct from lack of pollination, quite apart from its fate *via à vis* herbivores, weather, etc. If tropical plants only existed in strongly monospecific stands, and therefore had to expend little resources on flowering, they might well be fiercer competitors as individuals, starting from bigger seeds and devoting more of their resources to competitive interactions, longevity, or anti-herbivory.

Similar consequences might well occur in the animals. In a world without nectar-producing flowers, the thousands of species of nectar-harvesting moths would be reduced to those that can harvest enough resources as larvae and water as adults to carry out all adult functions. Adults would be short-lived and large, and caterpillars longer-lived and probably better-protected. Bat lifeform-richness would be substantially reduced, and those bats that have an omnivorous diet including floral parts would either become more frugivorous and carnivorous, or disappear. Hummingbirds are probably so trivial in the tropical food web (excluding their activities as pollinators) that they would scarcely be missed.

Plants would be much more restricted to 'good' soils and be a resource base reduced in species richness, biomass and replacement rate following herbivory. If certain minerals and nitrogen-rich compounds were no longer available from mycorrhizae and nitrogen-fixing bacteria, the options for defence chemicals and growth patterns would be greatly reduced. Decomposition rates would also be slowed, since the gut flora and rhizosphere are prime inocula when an animal or root dies. This in turn would slow the return of nutrients to the soil cycle, and mean that more carcass material would be available to higher animals in the decomposition process.

The elimination of obligate ant-plants would be hardly missed — a few major groups of ants would go down the drain; vice-versa, the elimination of obligate plant-ants would again be just the loss of a few species of plants and a relatively small amount of biomass except in a very few neotropical secondary successional systems.

The elimination of diffuse protective mutualisms would on the other hand have a major disruptive effect on such things as the density and species-richness of ants (decline), density and species-richness of ant-tended Homoptera (increase), and the array of parasitic Hymenoptera (gross decline in certain groups). There would presumably be an increase in plant investment in other kinds of protections, and presumably in some circumstances these would be less effective. This should be coupled with a concomitant increase in herbivore density because of the lowered carnivore resources of the kind that support carnivores while they are searching for prey and hosts, and waiting in between.

All of the above changes might be identified as a response to the loss of mutualists. However, the loss of the mutualism would mean the loss of very substantial resources to the parasites of the mutualism. Where these parasites are very specific (such as the ants that live in ant-acacias unoccupied by protective ants) their absence would hardly be noticed. However, the parasites of large and diffuse mutualisms are often prominant parts of the habitat. Many birds feed on fruits without contributing to the seed shadow, and the fruit consumed may be as important in their diets as is the fruit eaten by mutualists. An enormous number of species of insects would drop out of the habitat if the flowers they visit (and do not pollinate) were removed. I suspect that the removal of plant exudates from root-occupied soil would create such a different pattern of nutrient input from higher plants that the ordinary soil microflora would take on a very different aspect. Even more indirectly, there is a large fauna of fruit-eating insects that would disappear.

In short, a world without the four common mutualisms would be less species-rich, less ecologically and life-form diverse, richer in extreme specialists and incompetent generalists, and have population patterns based more on monospecific stands (lower within-habitat diversity). Competition would be more intense among the survivors, and the survivors would expend more resources on it than they do now. Seed and flower predators would play an even larger role than at present in moulding the reproductive biology of plants. It might well be that such a view is a way to speculate on why pre-Cretaceous populations and habitats had their novel biological structures.

Select Bibliography

Twenty-five years ago, familiarization with the literature on mutualism was a matter of picking through ecological, behavioural, physiological and evolutionary studies in search of a smattering of useful facts about the organisms of interest. Today, there is an ever-deepening snowfall of papers directly examining this or that mutualism or the ecology/evolutionary biology of this or that mutualistic organism. Where does the newcomer to this field begin? Since mutualisms are everywhere, the answer is 'everywhere'. A solid understanding of, say, lichens and their algae, is essential to understanding of exploring, for example, birds and fruits. However, the wide-ranging reader will quickly encounter a morass of terms, dicta and impressions coined and used by different authors and with different organisms. The only way to avoid cynical yet lethal repulsion from this seemingly undisciplined approach to science is to have your own favourite reference point. Evaluate and notice what is there in the context of the particular mutualistic system you are personally familiar with, a system that exists independently of what people say of it. If I state that a microbial flora is a universal mutualistic system yet you work on an animal that lacks one, then instead of discarding the questions that derive from my statement, ask how your organism must differ in other traits if it truly lacks microbial mutualists.

My choice of papers for this bibliography was dictated by the desire for overview and to draw attention to particular case studies. I have omitted many papers of equal stature to those cited, with the assumption that the concerned reader will use both the current journals and the references of the papers cited to locate more depth than is offered by my sketchy bibliography. This bibliography was not constructed for the likes of those that have written the other chapters in this book, but rather for the newcomer to this area.

Bibliography

Admadjian, V. (1970) 'The Lichen Symbiosis: its Origin and Evolution', *Evolutionary Biology*, *4*, 163–84

Aker, C.L. (1982) 'Regulation of Flower, Fruit and Seed Production by a Monocarpic Perennial, *Yucca whipplei'*, *Journal of Ecology*, *70*, 357–72

Armstrong, J.A., J.M. Powell and A.J. Richards (eds.) (1982), *Pollination and Evolution*, Royal Botanic Gardens, Sydney, Australia

Atsatt, P.R. and D.J. O'Dowd (1978) 'Mutual Aid among Plants', *Horticulture*, *56*, 22–31

Baker, H.G., P.A. Opler and I. Baker (1978) 'A Comparison of the Amino Acid Complements of Floral and Extrafloral Nectaries', *Botanical Gazette*, *139*, 322–32

Batra, S.W.T. (1984) 'Solitary Bees', *Scientific American*, *250(2)*, 120–7

Bawa, K.S. (1980) 'Evolution of Dioecy in Flowering Plants', *Annual Review of Ecology and Systematics*, *11*, 15–59

Bawa, K.S. (1982) 'Outcrossing and the Incidence of Dioecism in Island Floras', *American Naturalist*, *119*, 866–71

Beach, J.H. (1982) 'Beetle Pollination of *Cyclanthus bipartitus* (Cyclanthaceae)', *American Journal of Botany*, *69*, 1074–81

Beattie, A.J. and D.C. Culver (1982) 'Inhumation: How Ants and Other Invertebrates Help Seeds', *Nature*, *297*, 627

Beattie, A.J. and D.C. Culver (1983) 'The Nest Chemistry of Two Seed-Dispersing Ant Species', *Oecologia*, *56*, 99–103

Bentley, B.L. (1977) 'Extrafloral Nectaries and Protection by Pugnacious Bodyguards', *Annual Review of Ecology and Systematics*, *8*, 407–27

Bentley, B.L. (1981) 'Ants, Extrafloral Nectaries and the Vine Life-Forms: an Interaction', *Tropical Ecology*, *22*, 127–33

Bequaert, J. (1922) 'Ants in Their Diverse Relations to the Plant World', *Bulletin of the American Museum of Natural History*, *45*, 333–583

Bernhardt, P., R.B. Knox and D.M. Calder (1980) 'Floral Biology and Self-Incompatibilities in Some Australian Mistletoes of the Genus *Amyema* (Loranthaceae)', *Australian Journal of Botany*, *28*, 437–51

Berryman, A.A. (1972) 'Resistance of Conifers to Invasion by Bark Beetle-Fungus Associations', *BioScience*, *22*, 598–602

Bertin, R.I. (1982) 'Paternity and Fruit Production in Trumpet Creeper (*Campsis radicans*)', *American Naturalist*, *119*, 694–709

Bertin, R.I. (1982) 'The Ruby-Throated Hummingbird and its Major Food Plant: Ranges, Flowering Phenology, and Migration', *Canadian Journal of Zoology*, *60*, 210–19

Bertin, R.I. (1982) 'Floral Biology, Hummingbird Pollination and Fruit Production of Trumpet Creeper, *Campsis radicans* (Bignoniaceae)', *American Journal of Botany*, *69*, 122–34

Bertin, R.I. and M.F. Wilson (1980) 'Effectiveness of Diurnal and Nocturnal

Pollination of Two Milkweeds', *Canadian Journal of Botany, 58*, 1744–6

Boggs, C.L., J.T. Smiley and L.E. Gilbert (1981) 'Patterns of Pollen Exploitation by *Heliconius* Butterflies', *Oecologia, 48*, 284–9

Bonaccorso, F.J., W.E. Glanz and C.M. Sanford (1980) 'Feeding Assemblages of Mammals at Fruiting *Dipteryx panamensis* (Papilionaceae) Trees in Panama: Seed Predation, Dispersal, and Parasitism', *Revista de Biologia Tropical, 28*, 61–72

Bond, G. (1963) 'The Root Nodules of Non-Leguminous Angiosperms', *Symposia of The Society for General Microbiology, 13*, 72–91

Bossema, I. (1979) 'Jays and Oaks: An Eco-Ethological Study of a Symbiosis', *Behavior, 70*, 1–117

Boucher, D.H., S. Jones and K.H. Keeler (1982) 'The Ecology of Mutualism', *Annual Review of Ecology and Systematics, 13*, 315–47

Bowen, G.D. (1978) 'Dysfunction and Shortfalls in Symbiotic responses', *Plant Disease, 3*, 231–56

Bowen, G.D. and C. Theodorou (1973) 'Growth of Ectomycorrhizal Fungi Around Seeds and Roots', *Ectomycorrhizae. Their Ecology and Physiology*, G.C. Marks and T.T. Kozlowski, Academic Press, New York, pp. 107–50

Bowen G.D. and A.D. Rovira (1976) 'Microbial Colonization of Plant Roots', *Annual Review of Phytopathology, 14*, 121–44

Brown, J.H. and A. Kodric-Brown (1979) 'Convergence, Competition, and Mimicry in a Temperate Community of Hummingbird-Pollinated Flowers', *Ecology, 60*, 1022–35

Burns, R.C. and R.W.F. Hardy (1975) *Nitrogen Fixation in Bacteria and Higher Plants*, Springer-Verlag, New York

Cahalane, V.N. (1942) 'Caching and Recovery of Food by the Western Fox Squirrel', *Journal of Wildlife Management, 6*, 338–52

Carothers, J.H. (1982) 'Effects of Trophic Morphology and Behavior on Foraging Rates of Three Hawaiian Honeycreepers',*Oecologia, 55*, 157–9

Connell, J.H. (1980) 'Diversity and the Coevolution of Competitors, or the Ghost of Competition Past', *Oikos, 35*, 131–8

Cornell, H.V. (1983) 'The Secondary Chemistry and Complex Morphology of Galls Formed by the Cynipinae (Hymenoptera): Why and How?', *American Midland Naturalist, 110*, 225–34

Cortes, J.E. (1982) 'Nectar Feeding by European Passerines on Introduced Tropical Flowers at Gibraltar', *Alectoris, 4*, 26–9

Cox, P.A. (1982) 'Vertebrate Pollination and the Maintenance of Dioecism in *Freycinetia*', *American Naturalist, 120*, 65–80

Cox P.A. (1983) 'Search Theory, Random Motion, and the Convergent Evolution of Pollen and Spore Morphology in Aquatic Plants', *American Naturalist, 121*, 9–31

Cox, P.A. (1983) 'Extinction of the Hawaiian Avifauna Resulted in a Change of Pollinators for the Ieie, *Freycinetia arborea*', *Oikos, 41*, 195–9

Cruden, R.W. and S. Miller-Ward (1981) 'Pollen-Ovule Ratio, Pollen Size, and the Ratio of Stigmatic Area to the Pollen Bearing Area of the Pollinator: an Hypothesis', *Evolution, 35*, 964–74

Dafni, A. (1983) 'Pollination of *Orchis caspia* — a Nectarless Plant Which Deceives the Pollinators of Nectariferous Species from Other Plant Families', *Journal of Ecology, 71*, 467–74

Davidson, D.W. and S.R. Morton (1981) 'Competition for Dispersal in Ant-Dispersed Plants', *Science, 213*, 1259–61

Denslow, J.S. and T.C. Moermond (1982) 'The Effect of Accessibility on Rates of Fruit Removal From Tropical Shrubs: an Experimental Study', *Oecologia, 54*, 170–6

Dommergues, Y.R. and S.V. Krupa (eds.) (1978) *Interactions Between Non-*

Pathogenic Soil Microorganisms and Plants, Elsevier Scientific Publishing Company, Amsterdam, 475pp.

Eastop, V.F. (1981) 'Coevolution of Plants and Insects', *The Evolving Biosphere*, P.L. Forey (ed.), Cambridge Univ. Press, England, pp. 179–90

Emmons, L.H., A. Gautier-Hion and G. Dubost (1983) 'Community Structure of the Frugivorous-Folivorous Forest Mammals of Gabon', *Journal of Zoology, London*, 199, 209–22

Feinsinger, P. and L.A. Swarm (1982) ' "Ecological Release", Seasonal Variation in Food Supply, and the Hummingbird *Amazilia tobaci* on Trinidad and Tobago', *Ecology*, 63, 1574–87

Fox, J.F. (1982) 'Adaptation of Grey Squirrel Behavior to Autumn Germination by White Oak Acorns', *Evolution*, 36, 800–9

Fritz, R.S. (1983) 'Ant Protection of a Host Plant's Defoliator: Consequence of an Ant-Membracid Mutualism', *Ecology*, 14, 789–97

Futuyma, D. and M. Slatkin (1983) *Coevolution*, Blackwells Sci. Pub., Oxford, England

Givnish, T.J. (1980) 'Ecological Constraints on the Evolution of Breeding Systems in Seed Plants: Dioecy and Dispersal in Gymnosperms', *Evolution*, 34, 959–72

Givnish, T.J. (1982) 'Outcrossing Versus Ecological Constraints in the Evolution of Dioecy', *American Naturalist*, 119, 849–65

Gross, R.S. and P.A. Werner (1983) 'Relationships among Flowering Phenology, Insect Visitors, and Seed-Set of Individuals: Experimental Studies on Four Co-Occuring Species of Goldenrod (*Solidago:* Compositae)', *Ecological Monographs*, 53, 95–117

Hale, M.E. (1974) *The Biology of Lichens*, E. Arnold, London

Handel, S.N., S.B. Fisch and G.E. Schatz (1981) 'Ants Disperse a Majority of Herbs in a Mesic Forest Community in New York State', *Bulletin of the Torrey Botanical Club*, 108, 430–7

Harley J.L. and S.E. Smith (1983) *Mycorrhizal Symbiosis*, Academic Press, London

Heithaus, E.R. (1982) 'Coevolution Between Bats and Plants', *Ecology of Bats*, T.H. Kunz (ed.), Plenum Press, New York, pp. 327–67

Heithaus, E.R., D.C. Culver and A.J. Beattie (1980) 'Model of some Ant-Plant Mutualisms', *American Naturalist*, 116, 347–61

Herrera, C.M. (1981) 'Are Tropical Fruits More Rewarding to Dispersers than Temperate Ones?', *American Naturalist*, 118, 896–907

Herrera, C.M. (1982) 'Defence of Ripe Fruits from Pests: Its Significance in Relation to Plant-Disperser Interactions', *American Naturalist*, 120, 218–41

Herrera, C.M. (1982) 'Seasonal Variation in the Quality of Fruits and Diffuse Coevolution Between Plants and Avian Dispersers', *Ecology*, 63, 773–85

Herrera, C.M. (1984) 'Determinants of Plant-Animal Coevolution: the Case of Mutualistic Dispersal of Seeds by Vertebrates', *Oikos*, (in press)

Herrera, C.M. and P. Jordano (1981) '*Prunus mahaleb* and Birds: the High Efficiency Seed Dispersal System of a Temperate Tree', *Ecological Monographs*, 51, 203–18

Herrera, C.M., J. Herrera and X. Espadaler (1984) 'Nectar Thievery by Ants from Southern Spanish Insect-Pollinated Flowers', *Insectes Sociaux*, (in press)

Hocking, B. (1970) 'Insect Associations with the Swollen Thorn Acacias', *Transactions of the Royal Entomological Society of London*, 122, 211–55

Howe, H.F. (1983) 'Annual Variation in a Neotropical Seed-Dispersal System', *The Tropical Rainforest: Ecology and Management*, S.L. Sutton, T. Whitmore and A.C. Charwick (eds.), Blackwell, Oxford, pp. 211–27

Howe, H.F. (1984) 'Constraints on the Evolution of Mutualism', *American Naturalist*, 123, 764–77

Howe, H.F. and G.A. Vande Kerckhove (1979) 'Fecundity and Seed Dispersal by Birds of a Tropical Tree', *Ecology, 60,* 180–9

Howe H.F. and J. Smallwood (1982) 'Ecology of Seed Dispersal', *Annual Review of Ecology and Systematics, 13,* 201–28

Hutchins, H.E. and R.M. Lanner (1982) 'The Central Role of Clark's Nutcracker in the Dispersal and Establishment of Whitebark Pine', *Oecologia, 55,* 192–201

Huxley, C. (1980) 'Symbiosis Between Ants and Epiphytes', *Biological Review, 55,* 321–40

Inouye, D.W. (1980) 'The Terminology of Floral Larceny', *Ecology, 61,* 1251–3

Inouye, D.H. and O.R. Taylor (1979) 'A Temperate Region Plant-Ant-Seed Predator System: Consequences of Extra-Floral-Nectar Secretion by *Helianthella quinquenervis', Ecology, 60,* 1–7

Janos, D.P. (1980) 'Mycorrhizae Influence Tropical Succession', *Biotropica Supplement, Tropical Succession,* pp. 56–64

Janos, D.P. (1983) 'Tropical Mycorrhizae, Nutrient Cycles and Plant Growth', *Tropical Rainforest: Ecology and Management,* S.L. Sutton, T.C. Whitmore and A.C. Chadwick (eds.), Blackwell Sci. Public. Oxford, pp. 327–45

Janson, C.H. (1983) 'Adaptation of Fruit Morphology to Dispersal Agents in a Neotropical Forest', *Science, 219,* 187–89

Janzen, D.H. (1966) 'Coevolution of Mutualism Between Ants and Acacias in Central America', *Evolution, 20,* 249–75

Janzen, D.H. (1967) 'Fire, Vegetation Structure, and the Ant x Acacia Interaction in Central America', *Ecology, 48,* 26–35

Janzen, D.H. (1971) 'Seed Predation by Animals', *Annual Review of Ecology and Systematics, 2,* 465–92

Janzen, D.H. (1972) 'Protection of *Barteria* (Passifloraceae) by *Pachysima* (Pseudomyrmecinae) in a Negerian Rain Forest', *Ecology, 53,* 885–92

Janzen, D.H. (1973) 'Dissolution of Mutualism Between *Cecropia* and *Azteca* Ants', *Biotropica, 5,* 15–28

Janzen, D.H. (1973) 'Evolution of Polygynous Obligate Acacia-Ant in Western Mexico', *Journal of Animal Ecology, 42,* 727, 750

Janzen, D.H. (1974) 'Swollen-Thorn Acacias of Central America', *Smithsonian Contributions to Botany, 13,* 1–131

Janzen, D.H. (1974) 'Epiphytic Myrmecophytes in Sarawak: Mutualism Through the Feeding of Plants by Ants', *Biotropica, 6,* 237–59

Janzen, D.H. (1975) '*Pseudomyrmex nigropilosa*: a Parasite of a Mutualism', *Science 188,* 936–7

Janzen, D.H., C.A. Miller, J. Hackforth-Jones, C.M. Pond, K. Hooper and D.P. Janos (1976) 'Two Costa Rican Bat-Generated Seed Shadow of *Andira inermis* (Leguminosae)', *Ecology, 56,* 1068–75

Janzen, D.H. and D. McKey (1977) '*Musanga cecropiodes* is a *Cecropia* Without its Ants', *Biotropica, 9,* 57

Janzen, D.H. (1979) 'How Many Babies do Figs Pay for Babies?', *Biotropica, 11,* 48–50

Janzen, D.H. (1979) 'How Many Parents do the Wasps from a Fig Have?', *Biotropica, 11,* 127–9

Janzen, D.H. (1979) 'Why Food Rots', *Natural History, 88,* 60–6

Janzen, D.H. (1979) 'How to be a Fig', *Annual Review of Ecology and Systematics, 10,* 13–51

Janzen, D.H. (1980) 'When is it Coevolution', *Evolution, 34,* 611–12

Janzen, D.H. (1981) 'Bee Arrival at Costa Rican Female *Catasetum* Orchid Inflorescences, and a Hypothesis on Euglossine Population Structure', *Oikos, 36,* 177–83

Janzen, D.H. (1981) 'Differential Visitation of *Catasetum* Orchid Male and Female Flowers', *Biotropica, 13 (supplement)*, 77

Janzen, D.H. (1982) 'Removal of Seeds from Horse Dung by Tropical Rodents: Influence of Habitat and Amount of Dung', *Ecology, 63*, 1887–900

Janzen, D.H. and P.S. Martin (1982) 'Neotropical Anachronisms: the Fruits the Gomphotheres Ate', *Science, 215*, 19–27

Janzen, D.H. (1983) 'Dispersal of Seeds by Vertebrate Guts', *Coevolution*, D.J. Futuyma and M. Slatkin (eds.) Sinauer Associates, Sunderland, Massachusetts, pp. 232–62

Janzen, D.H. (1983) 'Seed and Pollen Dispersal by Animals: Convergence in the Ecology of Contamination and Sloppy Harvest', *Biological Journal of the Linnean Society, 20*, 103–13

Janzen, D.H. (1984) 'Dispersal of Small Seeds by Big Herbivores: Foliage is the Fruit', *American Naturalist, 123*, 338–53

Jeffries, R.A., A.D. Bradshaw and P.D. Putwain (1981) 'Growth, Nitrogen Accumulation and Nitrogen Transfer by Legume Species Established on Mine Spoils', *Journal of Applied Ecology, 18*, 945–56

Jones, C.E. and R.J. Little (eds.), *Handbook of Experimental Pollination Biology*, Scientific and Academic Editions, New York, 558pp.

Jones, C.G. (1984) 'Microorganisms as Mediator of Plant Resource Exploitation by Insect Herbivores', *A New Ecology: Novel Approaches to Interactive Systems*, P.W. Price, C.N. Slobodchikoff and W.S. Gaud (eds.), J. Wiley and Sons, New York, pp. 53–99

Jordan, W.P. and F.R. Rickson (1971) 'Cyanophyte Cephalodia in the Lichen Genus *Nephroma*', *American Journal of Botany, 58*, 562–8

Jordano, P. (1982) 'Migrant Birds are the Main Seed Dispersers of Blackberries in Southern Spain', *Oikos, 38*, 183–93

Keast, A. (1958) 'The Influence of Ecology on Variation in the Mistletoe-Bird (*Dicaeum hirundinaceum*)', *The Emu, 58*, 195–206

Keeler, K.H. (1981) 'Function of *Mentzelia nuda* (Loasaceae) Postfloral Nectaries in Seed Defense', *American Journal of Botany, 68*, 295–9

Kephart, S.R. (1983) 'The Partitioning of Pollinators Among Three Species of *Asclepias*', *Ecology, 64*, 120–33

Kessler, K.J. (1966) 'Growth and Development of Mycorrhizae of Sugar Maple (*Acer saccharum* March.)', *Canadian Journal of Botany, 44*, 1413–25

Kevan, P.G. and H.G. Baker (1983) 'Insects as Flower Visitors and Pollinators', *Annual Review of Entomology, 28*, 407–53

Kleinfeldt, S.E. (1978) 'Ant-Gardens: the Interaction of *Codonanthe crassifolia* (Gesneriaceae) and *Crematogaster longispina* (Formicidae)', *59*, 449–56

Kraus, B. (1983) 'A Test of the Optimal-Density Model for Seed Scatterhoarding', *Ecology, 64*, 608–10

Kukor, J.J. and M.M. Martin (1983) 'Acquisition of Digestive Enzymes by Siricid Woodwasps from their Fungal Symbiont', *Science, 220*, 1161–3

Lack, A.J. (1982) 'The Ecology of Flowers of Chalk Grassland and Their Insect Pollinators', *Journal of Ecology, 70*, 773–90

Lanner, R.M. (1982) 'Adaptations of Whitebark Pine for Seed Dispersal by Clark's Nutcracker', *Canadian Journal of Forest Research, 12*, 391–402

Law, R. and D.H. Lewis (1983) 'Biotic Environments and the Maintenance of Sex — Some Evidence from Mutualistic Symbioses', *Biological Journal of the Linnean Society, 20*, 249–76

Leighton, M. and D.R. Leighton (1982) 'The Relationship of Size of Feeding Aggregate to Size of Food Patch: Howler Monkeys (*Allouatta palliata*) Feeding in *Trichilia cipo* Fruit Trees on Barro Colorado Island', *Biotropica, 14*, 81–90

Leighton, M. and D.R. Leighton (1983) 'Vertebrate Response to Fruiting Seasonality Within a Bornean Rain Forest', *Tropical Rain Forest: Ecology and Resource Management*, S.L. Sutton, T.C. Whitmore and A.C. Chadwick (eds.), Blackwell Scientific Publ., Oxford, pp. 181–96

Livingston, R.L. and A.A. Berryman (1972) 'Fungus Transport Structures in the Fir Engraver, *Scolytus ventralis*, (Coleoptera: Scolytidae)', *Canadian Entomologist*, *104*, 1793–800

Malloch, D.W., K.A. Pirozunski and P.H. Raven (1980) 'Ecological and Evolutionary Significance of Mycorrhizal Symbioses in Vascular Plants (a review)', *Proceedings of the National Academy of Science*, 77, 2113–18

Manasse, R.S. and H.F. Howe (1983) 'Competition among Tropical Trees for Dispersal Agents: Influences of Neighbors', *Oecologia*, *59*, 185–90

Marshall, J.J. and F.R. Rickson (1973) 'Characterization of the a-D-glucan from the Plastids of *Cecropia peltata* as a Glycogen-Type Polysaccharide', *Carbohydrate Research*, *28*, 31–7

May, R. (1982) 'Mutualistic Interactions among Species', *Nature*, *269*, 803–4

McDade, L.A. and S. Kinsman (1980) 'The Impact of Floral Parasitism in Two Neotropical Hummingbird-Pollinated Plant Species', *Evolution*, *34*, 944–58

McDiarmid, R.W., R.E. Ricklefs and M.S. Foster (1977) 'Dispersal of *Stemmadenia donnell-smithii* (Apocynaceae) by Birds', *Biotropica*, *9*, 9–25

McKey, D. (1975) 'The Ecology of Coevolved Seed Dispersal Systems', *Coevolution of Animals and Plants*, L.E. Gilbert and P.H. Raven (eds.), Univ, Texas Press, Austin, Texas, pp. 159–91

Meeuse, A.D.J. (1978) 'Nectarial Secretion, Floral Evolution, and the Pollination Syndrome in Early Angiosperms', *Proceed Kon. Nederl. Akad. Wetensch., Series C*, *81*, 300–26

Meeuse, A.D.J. (1979) 'Why were the Early Angiosperms so Successful: A Morphological, Ecological and Phylogenetic Approach', *Proceed. Kon. Nederl. Akad, Watensch.*, series C, *82*, 343–69

Messina, F.J. (1981) 'Plant Protection as a Consequence of an Ant-Membracid Mutualism: Interactions on Goldenrod (*Solidago* sp.)', *Ecology*, *62*, 1433–40

Michener, C.D. (1974) *The Social Behavior of the Bees: A Comparative Study*, Harvard Univ. Press (Cambridge), 404 pp.

Milton, K. (1981) 'Distribution Patterns of Tropical Plant Foods as an Evolutionary Stimulus to Primate Mental Development', *American Anthropologist*, *83*, 534–48

Milton, K., D.M. Windsor, D.W. Morrison and M.A. Estribi (1982) 'Fruiting Phenologies of Two Neotropical *Ficus* Species', *Ecology*, *63*, 752–62

Moermond, T.C. and J.S. Denslow (1983) 'Fruit Choice in Neotropical Birds: Effects of Fruit Type and Accessibility on Selectivity', *Journal of Animal Ecology*, *52*, 407–20

Newman, E.I. (1978) 'Root Microorganisms: Their Significance in the Ecosystem', *Biological Review*, *53*, 511–54

Nilsson, L.A. (1981) 'The Pollination Ecology of *Listera ovata* (Orchidaceae)', *Nordic Journal of Botany*, *1*, 461–480

O'Dowd, D.J. (1982) 'Pearl Bodies as Ant Food: An Ecological Role for Some Leaf Emergences of Tropical Plants', *Biotropica*, *14*, 40–9

O'Dowd, D.J. and M.E. Hay (1980) 'Mutualism Between Harvester Ants and a Desert Ephemeral: Seed Escape From Rodent', *Ecology*, *61*, 531–40

O'Dowd, D.J. and E.A. Catchpole (1983) 'Ants and Extrafloral Nectaries: no Evidence for Plant Protection in *Helichrysum* sp.—Ant Interactions', *Oecologia*,

Osmaston, H.A. (1965) 'Pollen and Seed Dispersal, in *Chlorophora excelsa* and other Moraceae, and in *Parkia filicoidea* (Mimosaceae), with Special Reference to the Role of the Fruit Bat, *Eidolon helvum*', *Commonwealth Forestry Review*, *44*, 96–103

Petelle, M. (1981) 'More Mutualisms Between Consumers and Plants', *Oikos, 38*, 125-7

Pickett, C.H. and W.D. Clark (1979) 'The Function of Extrafloral Nectaries in *Opuntia acanthocarpa* (Cactaceae)', *American Journal of Botany, 66*, 618-25

Pijl, L. Van Der (1966) 'Ecological Aspects of Fruit Evolution: A Functional Study of Dispersal Organs. I.-III.', *Proceed. Kon. Ned. Acad. Wet. (C), 69*, 597-640

Plowright, R.C. and T.M. Laverty (1984) 'The Ecology and Sociobiology of Bumble Bees', *Annual Review of Entomology, 29*, 175-99

Pratt, T.K. and E.W. Stiles (1983) 'How Long Fruit-Eating Birds Stay in the Plants Where They Feed: Implications for Seed Dispersal', *American Naturalist, 122*, 789-805

Rausher, M.D. and N.L. Fowler (1979) 'Intersexual Aggression and Nectar Defense in *Chauliognathus distinguendus* (Coleoptera: Cantharidae)', *Biotropica, 11*, 96-100

Real, L., J. Ott and E. Silverfine (1982) 'On the Tradeoff Between the Mean and the Variance in Foraging: Effect of Spatial Distribution and Color Preference', *Ecology, 63*, 1617-23

Regal, P.J. (1977) 'Ecology and Evolution of Flowering Plant Dominance', *Science, 196*, 622-9

Rice, B. and M. Westoby (1982) 'Heteroecious Rusts as Agents of Interference Competition', *Evolutionary Theory, 6*, 43-52

Rickson, F.R. (1971) 'Glycogen Plastids in Mullerian Body Cells of *Cecropia peltata* — a Higher Green Plant', *Science, 173*, 344-7

Rickson, F.R. (1975) 'The Ultrastructure of *Acacia cornigera* L. Beltian Body Tissue', *American Journal of Botany, 62*, 913-22

Rickson, F.R. (1976) 'Anatomical Development of the Leaf Trichilium and Mullerian Bodies of *Cecropia peltata* L.', *American Journal of Botany, 63*, 1266-71

Rickson, F.R. (1977) 'Progressive Loss of Ant-Related Traits of *Cecropia peltata* on Selected Caribbean Islands', *American Journal of Botany, 64*, 585-92

Rickson, F.R. (1979) 'Ultrastructural Development of the Beetle Food Tissue of *Calycanthus* Flowers', *American Journal of Botany, 66*, 80-6

Rickson, F.R. (1979) 'Absorption of Animal Tissue Breakdown Products into a Plant Stem — the Feeding of a Plant by Ants', *American Journal of Botany, 66*, 87-90

Risch, S.J. and D. Boucher (1978) 'What Ecologists Look For', *Bulletin of the Ecological Society of America, 57(3)*, 8-9

Risch, S.J. and F.R. Rickson (1981) 'Mutualism in Which Ants Must be Present Before Plants Produce Food Bodies', *Nature, 291*, 149-50

Roubik, D.W. (1982) 'The Ecological Impact of Nectar-Robbing Bees and Pollinating Hummingbirds on a Tropical Shrub', *Ecology, 63*, 354-60

Roubik, D.W., J.D. Ackerman, C. Copenhaver and B.H. Smith (1982) 'Stratum, Tree, and Flower Selection by Tropical Bees: Implications for the Reproductive Biology of Outcrossing *Cochlospermum vitifolium* in Panama', *Ecology, 63*, 712-20

Roubik, D.W. and S.L. Buchmann (1983) 'Nectar Selection by *Melipona* and *Apis mellifera* (Hymenoptera: Apidae) and the Ecology of Nectar Intake by a Bee Colony in a Tropical Forest', *Oecologia*

Roubik, D.W. and M. Aluja (1983) 'Flight Ranges of *Melipona* and *Trigona* in Tropical Forest', *Journal of the Kansas Entomological Society, 56*, 217-22

Salomonson, M.G. (1978) 'Adaptations for Animal Dispersal of One-Seed Juniper Seeds', *Oecologia, 32*, 333-9

Savageau, M.A. (1983) '*Escherichia coli* Habitats, Cell Type, and Molecular Mechanisms of Gene Control', *American Naturalist, 122*, 732-44

Schaffer, W.M., D.W. Zeh, S.L. Buchmann, S. Kleinhans, M.V. Schaffer and J. Antrim (1983) 'Competition for Nectar Between Introduced Honey Bees and Native North American Bees and Ants', *Ecology*, 64, 564–77

Schemske, D.W. (1982) 'Ecological Correlates of a Neotropical Mutualism: Ant Assemblages at *Costus* Extrafloral Nectaries', *Ecology*, 63, 923–41

Schemske, D.W. (1983) 'Limits to Specialization and Coevolution in Plant-Animal Mutualisms', *Coevolution*, M.H. Nitecki (ed.), Univ. Chicago Press, Chicago, pp. 67–109

Scoot, J.K. and R. Black (1981) 'Selective Predation by White-Tailed Black Cockatoos on Fruit of *Banksia attenuata* Containing the Seed-eating Weevil *Alphitopis nivea*', *Australian Wildlife Research*, 8, 421–30

Seeley, T.D. and R.H. Seeley (1982) 'Colony Defense Strategies of the Honeybees in Thailand', *Ecological Monographs*, 52, 43–63

Shettleworth, S.J. (1983) 'Memory in Food-Hoarding Birds', *Scientific American*, 248, 102–10

Smith, N.G. (1980) 'Some Evolutionary, Ecological and Behavioral Correlates of Communal Nesting by Birds with Wasps or Bees', *Proceedings of the 17th International Ornithological Congress*, pp. 1199–205

Snow, D.W. (1981) 'Coevolution of Birds and Plants', *The Evolving Biosphere*, P.L. Forey (ed.), Cambridge Univ. Press, England, pp. 169–78

Sorenson, A.E. (1983) 'Taste Aversion and Frugivore Preference', *Oecologia*, 56, 117–20

Sork, V.L. (1983) 'Mammalian Seed Dispersal of Pignut Hickory During Three Fruiting Seasons', *Ecology*, 64, 1049–56

Sprent, J.I. (1979) *The Biology of Nitrogen-Fixing Organisms*, McGraw-Hill, London

Stapanian, M.A. (1982) 'Model for Fruiting Display: Seed Dispersal by Birds for Muberry Trees', *Ecology*, 63, 1432–43

Stephenson, A.G. (198c) 'Flower and Fruit Abortion: Proximate Causes and Ultimate Functions', *Annual Review of Ecology and Systematics*, 12, 253–79

Stephenson, A.G. (198c) 'Toxic Nectar Deters Nectar Thieves of *Catalpa speciosa*', *American Midland Naturalist*, 105, 381–3

Stephenson, A.G. (1982) 'The Role Of the Extrafloral Nectaries of *Catalpa speciosa* in Limiting Herbivory and Increasing Fruit Production', *Ecology*, 63, 663–9

Stephenson, A.G. (1982) 'Iridoid Glucosides in the Nectar of *Catalpa speciosa* are Unpalatable to Nectar Thieves', *Journal of Chemical Ecology*, 8, 1025–34

Stephenson, A.G. (1982) 'When Does Outcrossing Occur in a Mass-Flowering Plant?', *Evolution*, 36, 762–7

Stevens, G.C. (1983) '*Atta cephalotes* (Zompopas, Leaf-Cutting Ants)', *Costa Rican Natural History*, D.H. Janzen (ed.), Univ. Chicago Press, Chicago, pp. 688–91

Tempel, A.S. (1983) 'Bracken Fern (*Pteridium aquilinum*) and Nectar-Feeding Ants: a Nonmutualistic Interaction' *Ecology*, 64, 1411–22

Thompson, J.N. (1981) 'Elaiosomes and Fleshy Fruits: Phenology and Selection Pressures for Ant-Dispersed Seeds', *American Naturalist*, 117, 104–8

Thompson, J.N. (1981) 'Reversed Animal-Plant Interactions: the Evolution of Insectivorous and Ant-Fed Plants', *Biological Journal of the Linnean Society*, 16, 147–55

Thompson, J.N. (1982) *Interaction and Coevolution*, John Wiley and Sons Inc., New York, New York, 179 pp.

Thompson, J.N., and M.F. Willson (1978) 'Disturbance and the Dispersal of Fleshy Fruits', *Science*, 200, 1161–3

Thompson, J.N., W.P. Maddison and R.C. Plowright (1982) 'Behavior of Bumble Bee Pollinators of *Aralia hispida* Vent. (Araliaceae)', *Oecologia*, 54, 326–36

Tilman, D. (1978) 'Cherries, Ants, and Tent Caterpillar; Timing of Nectar Production in Relation to Susceptibility of Caterpillar to Ant Predation', *Ecology, 59,* 686–92

Tomback, D.F. (1982) 'Dispersal of Whitebark Pine Seeds by Clark's Nutcracker: a Mutualism Hypothesis', *Journal of Animal Ecology, 51,* 451–67

Tomback, D.F. (1983) 'Nutcrackers and Pines: Coevolution or Coadaptation?', *Coevolution,* M.H. Nitecki (ed.), Univ. Chicago Press, Chicago, pp. 179–223

Troyer, K. (1982) 'Transfer of Fermentative Microbes Between Generations in a Herbivorous Lizard', *Science, 216,* 540–2

Van Soest, P.J. (1982) *Nutritional Ecology of the Ruminant,* O and B Books Inc., Corvallis, Oregon, 374 pp.

Vander Wall, S.B. and R.P. Balda (1981) 'Ecology and Evolution of Food-Storage Behavior in Conifer-Seed-Caching Corvids', *Zeitschrift fur Tierpsychologie, 56,* 217–42

Vogt, K.A. C.C. Grier, C.E. Meier and R.L. Edmonds (1982) 'Mycorrhizal Role in Net Primary Production and Nutrient Cycling in *Abies amabilis* Ecosystems in Western Washington', *Ecology, 63,* 370–80

Waddington, K.D. (1981) 'Factors Influencing Pollen Flow in Bumblebee-Pollinated *Delphinium virescens' Oikos*; *37,* 153, 159

Waddington, K.D., T. Allen and B. Heinrich (1981) 'Floral Preferences of Bumble-bees (*Bombus edwardsii*) in Relation to Intermittent Versus Continuous Rewards', *Animal Behavior, 29,* 779–84

Waterman, P.G., C.N. Mbi, D.B. McKey and J.S. Gartlan (1980) 'African Rain-forest Vegetation and Rumen Microbes: Phenolic Compounds and Nutrients as Correlates of Digestibility', *Oecologia, 47,* 22–33

Watmough, R.H. (1983) 'Mortality, Sex Ratio and Fecundity in Natural Populations of Large Carpenter Bees (*Xylocopa* sp.)', *Journal of Animal Ecology, 52,* 111–25

Weis, A.E. (1982) 'Use of a Symbiotic Fungus by the Gall Maker *Asteromyia carbonifera* to Inhibit Attack by the Parasitoid *Torymus capite',* *Ecology, 63,* 1602–5

Wheelwright, N.T. (1983) 'Fruits and the Ecology of Resplendent Quetzals', *The Auk, 100,* 286–301

Wheelwright, N.T. and G.H. Orians (1982) 'Seed Dispersal by Animals: Contrasts With Pollen Dispersal, Problems of Terminology and Constraints on Coevolution', *American Naturalist, 119,* 402–13

Wiens, J.A. (1983) 'Competition or Peaceful Coexistence', *Natural History, 92(3),* 30–4

Wille, A. (1983) 'Biology of the Stingless Bees', *Annual Review of Entomology, 28,* 41–64

Willson, M.F. and M.N. Melampy (1982) 'The Effect of Bicolored Fruit Displays on Fruit Removal by Avian Frugivores', *Oikos, 41,* 27–31

Willson, M.F. and J.N. Thompson (1982) 'Phenology and Ecology of Color in Bird-Dispersed Fruits, or Why Some Fruits are Red When They are "Green" ', *Canadian Journal of Botany, 60,* 701–13

Willson, M.F. and N. Burley; *Mate Choice in Plants,* Princeton Univ. Press, Princeton, New Jersey, 251 pp.

Wilson, D.S. (1983) 'The Effect of Population Structure on the Evolution of Mutualism: A Field Test Involving Burying Beetles and Their Phoretic Mites', *American Naturalist, 121,* 851–70

Wood, T.K. (1982) 'Selective Factors Associated with the Evolution of Membracid Sociality', *The Biology of Social Insects,* M.D. Breed, C.D. Michener and H.E. Evans (eds.), Westview Press, Boulder, pp. 175–9

Wyatt, R. (1980) 'The Impact of Nectar-Robbing Ants on the Pollination System of *Asclepias curassavica',* *Bulletin of the Torrey Botanical Club, 107,* 24–8

Young, A.M. (1982) 'Effects of Shade Cover and Availability of Midge Breeding

Sites on Pollinating Midge Populations and Fruit Set in Two Cocoa Farms',
Journal of Applied Ecology, *19*, 47–63
Young, A.M. (1983) 'Seasonal Differences in Abundance and Distribution of
Cocoa-Pollinating Midges in Relation to Flowering and Fruit Set Between Shaded
and Sunny Habitats of the La Lola Cocoa Farm in Costa Rica', *Journal of
Applied Ecology*, *20*, 801–31

4 COST: BENEFIT MODELS OF MUTUALISM

Kathleen H. Keeler

Cost: benefit models are an amorphous class of models. Evolutionary ones are usually phrased in terms of fitness: fitness 'costs' must be less than fitness 'benefits', otherwise the organisms dies, the population goes extinct, or the interaction dissolves. There does not seem to be a unified approach, other than the cost and benefits are estimated and compared (Roughgarden 1975; Covich 1976; Huey and Slatkin 1976; Pyke *et al.* 1977; Stapanian and Smith 1978; Kodric-Brown and Brown 1979; Magnusen *et al.* 1979; Holldobler and Lumsden 1980; Keeler 1981; Rubenstein 1981).

My use of cost: benefit models is to define the conditions under which mutualism is favoured. In extant mutualisms, benefits exceed costs, since it is axiomatic in evolutionary biology that activities or interactions that decrease fitness do not persist (although a particular trait could be in a state of transition or maintained by phylogenetic intertia). Cost: benefit models are used here as a mechanism for predicting or explaining the existence of mutualism, and perhaps more importantly, its absence. Because I have not analysed the interplay of multiple interactions between two individuals, all the models can be seen as describing the 'genetic unit of selective response' of Templeton and Gilbert (Chapter 5).

Applying a cost: benefit model to mutualism requires envisioning a population polymorphic for mutualism (and, ideally, only for that character), which contains three classes of individuals: (1) successful mutualists who assist another organism and receive assistance, (2) unsuccessful mutualists, who provide the mutualistic service of reward but (for any number of reasons) receive nothing in return, and (3) non-mutualists who make no investment in mutualism and receive no return (note that this type of model excludes mutualism maintained by interdemic selection, in which non-mutualists receive the greatest benefit from the mutualism because the partner does not distinguish them from mutualistic partners (Wilson 1983)). For a population to be mutualistic, the fitness of the successful mutualists must be greater than that of either the non-mutualists or the unsuccessful mutualists and furthermore, the total fitness of all mutualists, successful or not, must exceed those who do not try. If it does not, the trait or behaviour will be lost from the population, and with it the interaction. Some mutualists have developed greater intimacy and/or obligacy, usually both, so that the unsuccessful mutualist class may be missing (in an obligate mutualism it has a fitness of zero and thus shouldn't exist; in a symbiotic mutualism it simply does not exist as a mutualist). For these it may be necessary to look first at selection for the origin of the mutualism and then at its subsequent evolution. Where the interaction is coevolved and coevolving, a pair of interrelated equations are necessary and the situation is complex.

The chief problem includes (1) estimating costs and benefits (2) putting all costs and benefits into a framework that can be compared (3) making the cost and benefits measurable so that falsifiable predictions can be tested, and (4) writing simultaneous equations when the responses of the two species are coevolved.

Cost: benefit models of mutualism have been written for sea anemone-damselfish interactions (Roughgarden 1975) and ant-plant interactions at extrafloral nectaries (Keeler 1981). The analyses of co-operation by Axelrod and Hamilton (1981) and Maynard Smith (1982) determining evolutionary stable strategies which result in co-operation also present a type of cost: benefit analysis (Reichert and Hammerstein 1983). In this section I derive cost: benefit models for pollination, myrmecochory, mycorrhizal fungi, fungus-gardening ants and mixed feeding flocks of birds. I then consider the common and unique factors of the models, problems of cost: benefit approaches, and the findings of selective models in general.

Pollination

Mutualistic pollination is a two-species interaction involving the exchange of food for transfer of gametes. There are two non-mutualistic alternatives for a plant producing gametes by meiosis: self-fertilization (within flower gamete transfer) and wind-pollination (abiotic outcrossing). Since the angiosperms are believed to have been ancestrally animal-pollinated (Kevan and Baker 1983 and included references), they can be seen as being derived from mutualists. While not symbiotic, pollination mutualism is often obligate: many animal-pollinated plants are dependent upon pollinators for seed set (obligate outcrossers). Although Feinsinger (1983) reviewed coevolution in pollinator-plant systems and concluded it is infrequent or diffuse, certainly some pollinator-plant communities are coevolved (Carlquist 1965; Stiles 1975).

The chief problem with pollination as a mutualism is that the benefit conferred by cross-pollination is not readily apparent. Consequently it is difficult to model and measure. The conventional wisdom, upon which I cannot improve, is that the benefit is from outcrossing and overshadows the obvious expenses and risks of mutualistic pollination, such as timing flowering to match pollinator activity, losses to pollen and nectar thieves, and producing sufficient rewards for pollinators and visitors (Mather 1943; Stebbins 1957; Jain 1976; Lloyd 1979, 1980, etc.).

Wind pollinated species are also outcrossed, with no losses to inefficient pollinators but with the requirement of producing adequate volumes of pollen to reach the stigma. An evident cost of wind pollination is that individuals must be common or at least patchy in their distribution in order for pollen to be transferred. Furthermore, it does not function well on rainy days (Whitehead 1969).

Selfers are not constrained to patches: as far as I can see, their distribution suffers no constraints at all from limits of the pollination system. It is the lack of success of the apparently highly versatile selfers that lends credence to the idea that 'outcrossing' is a significant benefit of zoophilous and anemophilous pollination systems. Certainly, the vast majority of angiosperm species outcross, at least potentially (Lloyd 1980).

The disadvantage to selfing is not necessarily due to inbreeding. Inbreeding depression from selfing is a characteristic of obligate outcrossers rather than selfers (Mather 1943; Grant and Grant 1965; Jain 1976; Clegg 1980). If the relative fitness of outcrossed offspring is greater than that of selfers without considering detrimental recessives revealed by inbreeders, than variation is of itself beneficial. However, variation can be insured, even under complete selfing by high mutation rates, polyploidy (Fisher 1949), or complex breeding systems (such as the balanced lethals of *Oenothera* (Emerson 1935)). Perhaps introduction of new alleles or opportunity to recombine large groups of successful alleles is the advantage (Levin 1975); I leave this argument to others (William 1975; Maynard Smith 1978).

Zoophily and Selfing. If the benefits are outcrossed offspring, the costs of mutualistic pollination are shortage of suitable pollinators in some areas, and shortage of suitable days for pollinator activity, etc., leading to many unpollinated flowers and plants (Mather 1943; Stebbins 1957, 1974; Jain 1976; Levin 1979; Lloyd 1979, 1980).

For mutualism to persist, we must have

$$w_s > w_z$$

where w_z is the fitness of zoophilous individuals (mutualistically pollinated) and w_s is the fitness of selfers, and

$$w_z = pw_f + qw_u \qquad (4.1)$$

where w_f is the fitness of those that are pollinated (fertilized) and w_u the fitness of zoophilous individuals who fail to achieve pollination, and p and q are their respective frequencies (p + q = 1).

Let w = benefit minus cost. Ideally these should be measured in contribution to the next generation, for example, seeds or gametes produced, or number of surviving grandchildren. The critical step is evaluating costs and benefits in the same terms. For this I suggest calories. These can, and rigorously should be converted to gamete-equivalents or propagule-equivalents (that is, for every 200 calories of floral display another seed could be produced). It seems possible to bypass some of these problems with relative values: Both costs and benefits measured relative to the extreme values of the system.

If the benefit is $N_z v_z$, then N is the proportion of seeds set and v the relative fitness of the seeds. The maximal value of v might be that all seeds produced survive to reproduce under average conditions. The cost is I_z, the investment in floral display and rewards for pollinators, measured relative to the highest investment in the system. The variables p and N are clearly related. I choose to separate the two to emphasize the difference between plants receiving no pollen and plants with unfilled seeds in the fruits. For some kinds of zoophily, pollinator visits are few but the pollen load is sufficiently large that those flowers that are visited are fully pollinated. Treat p as per cent plants pollinated, and N per cent seeds set per flower.

Both N and v apply to both successfully (f) and unsuccessfully (u) pollinated zoophils, so that equation 4.1 becomes:

$$w_z = p(N_f v_f - I_z) + q(N_u v_u - I_z) \qquad (4.2)$$

The population will probably include both types of plants or flowers and the investment, whether successful or unsuccessful, should be the same. However, if the species is an obligate outcrosser, $N_u v_u = 0$, because the pollen does not reach the stigma without a vector and/or it is self-incompatible and does not grow.

Substituting and simplifying for obligate outcrossers:

$w_z = pN_f v_f - I_z$ and, for simplicity replacing f with z,

$$w_z = pN_z v_z - I_z \qquad (4.3)$$

Writing a similar equation for selfers

$$w_s = N_s v_s - I_s \qquad (4.4)$$

For mutualism, $w_z > w_z$ or

$$pN_z v_z - I_z > N_s v_s - I_s \qquad (4.5)$$

Evaluating this inequality, I_z should be greater than I_s: zoophilous species make a greater investment in floral displays than do selfers (Ornduff 1969; Jain 1976). Other things being equal, then, they should produce fewer seeds. There are practical problems with the comparison since, partly by definition, few species are polymorphic for obligate outcrossing and selfing, but estimates are certainly possible. N_z should be equal to or less than N_s since while zoophilous pollination might fail to occur or only a portion of the possible load of pollen be transferred, selfing should provide complete pollination. The result is that v_z must exceed v_s if there is to be mutualism: that is, the fitness of outcrossed seeds must be greater than that of inbred seeds.

In reality, there are few obligate selfers and many outcrossers are capable of selfing. There is no problem representing these intermediate strategies: let the frequency q unpollinated by animals consist of r plants that are not pollinated (n) plus s selfed plants (c).

$$w_z = pw_f + rw_n + sw_c \qquad (4.6)$$

substituting $Nv - I$ for w, and setting $N_n v_n = 0$;

$$w_z = pN_z v_z + sN_c v_c - I_z \qquad (4.7)$$

The fitness of the population is raised by the per cent selfing. If all the unsuccessfully outcrossed flowers and plants set seed by selfing, s approaches q and there are no losses except the difference between v_z and v_c in the same genetic background. However, comparing equations 4.7 and 4.4 the cost (I) will still be higher for outcrossers than selfers, so that v_z must continue to exceed v_s or mutualism and outcrossing will be lost in plants capable of facultative selfing (see also Jain 1976; Lloyd 1979). In this case, there is evidence that v_s is likely to be less than v_z, since they are subject to inbreeding depression (Levin 1984).

Some points about these equations: N_z may equal N_s when some factor other than pollination efficiency sets the number of seeds that develop. Stephenson (1981) suggests that this is a fairly common occurrence. When $N_z = N_s$, $v_z > v_s$ must be the force maintaining mutualism. The alternatives are that zoophily is maintained by phylogenetic inertia rather than selection or by interdemic selection without being favourable to individuals.

Zoophily and Anemophily. Plants can be pollinated and out-crossed non-mutualistically via anemophily (wind pollination). Establishing the analogous equations:

$$w_a = p(N_f v_f - I_a) + q(N_u v_u - I_a) \tag{4.8}$$

For an obligate outcrosser, $N_v = 0$ so that:

$$w_a = pN_a v_a - I_a \tag{4.9}$$

(No seed set without movement by wind). This is the same equation as 4.3.

For pollination to be mutualistic;

$$w_z > w_a \; or \; p_z N_z v_z - I_z > p_a N_a v_a - I_a \tag{4.10}$$

Mutualism is favoured if:

(a) $p_z > p_a$. These are the probabilities that pollination is accomplished. Both are variable (see, for example, Lloyd 1980). For p_z, the fidelity of the animal is one of the critical factors: p_z should be lower for a species visited by a variety of generalist pollinators than by one with a specializing pollinator, although generalists tend to make up in numbers what they lose in fidelity. The value of p will partly be a function of the population structure. Wind is most effective in relatively dense stands: populations of scattered individuals will have lower p's than clumped ones. However, I see no *a priori* difference between p_z and p_a: both can be very high or quite low. A high value of p is critical. The favorability of p_a in the canopy of temperate deciduous forest or prairie may be responsible for the dominance of wind pollination there.

(b) $N_z > N_a$. Greater seed set by zoophilous species clearly favours mutualism. However, the number of seeds set need not differ between mechanisms, especially where seed set depends on plant energetics rather than pollen transfer (Stephenson 1981). Pollen limitation on seed set does occur (see Willson and Burley 1983 for a review) and circumstances producing a low N for either zoophily or anemophily would radically alter selection for the breeding system.

One difference between zoophily and anemophily as forms of out-crossing is that pollen tends to be blown through the air individually while very large numbers of pollen grains can be transported on a single animal or in a single pollinium (van der Pijl and Dodson 1966; Faegri and van der Pijl 1979).

(c) $v_z > v_a$? It seems to me that outcrossed is outcrossed. There should be no difference in fitness of the seeds whether the pollen grains came via an animal or through the air. However, if there is more geitonogamy (pollination from within the plant) in one method, it may affect v. Furthermore, if the neighbouring plants are close relatives, anemophily may differ from some forms of zoophily in the neighbourhood size produced. Any pollination system that produces effective inbreeding may produce differences between v_z and v_a (Levin 1979, 1984; Lloyd 1980).

(d) $I_a:I_z$. Investments certainly differ. I_z requires both a primary and secondary attraction to attain pollination, in addition to the functional structure. The primary attractant is the actual reward (pollen or nectar). The secondary attractant is the visible or olfactory cue that informs the visitors of the presence of the primary attractant (Faegri and van der Pijl 1979). Neither is needed for anemophily. Furthermore, flowers containing nectar are subject to nectar pilferage and robbing (Inouye 1980); structures to prevent this are probably not required in anemophilous nectarless blossoms. Anemophily requires only a blossom suitable for both putting pollen into the air and, later, retrieving it from the air. While these structures can be quite elaborate, they are not *per se* more complex than those required to brush pollen onto the hairs of a bee or the face of a hummingbird and then to receive it, without selfing.

The quantity of pollen produced by anemophilous plants is clearly greater than that of zoophilous plants. However, in the light of the more complex floral structures and the energetics of primary attractant production, it would seem that I_z is not always less than I_a.

In summary, p is critical, in general N and f probably do not differ, and I is greater in zoophils. Therefore, the key seems to be $p_z:p_n$; the relative efficiency of pollination determines mutualisms.

Myrmecochory

Myrmecochory is the mutualism in which ants disperse — carry away and frequently bury — seeds. The system is mutualistic in providing mobility to the seeds, and food, presented as a food body on the seed, to the ants (Buckley 1982, and included references). The benefit of myrmecochory to the plant seems to be transport. The value of transport has variously been suggested to be (a) dispersal away from the influence of the parent plant, (b) planting of the seeds in a site rich in nutrients (especially, within the ant-nest), (c) escape from predation (because of removal from the immediate vicinity of the parent plant) (Buckley 1982; Westoby *et al.* 1982). Any of these benefits might provide additional benefit to the seedling by reducing competition with established conspecifics and increasing out-crossing through greater mixing of the population. The cost is of course the food body itself and the loss of fitness to those seeds for whom transportation makes the situation worse. I shall represent myrmecochory as follows:

$$w_m = pw_t + qw_n \qquad (4.11)$$

where m = myrmecochore, t = transported and n = non-transported, and w, p and q are as before. The population will consist of individual seeds that are successfully removed and some which are not removed by the ants, despite the food body. For myrmecochory to persist, these seeds must jointly be more fit than nonmyrmecochores of the same species, which have no mutualistic transportation away from the parent and do not invest in a food body.

The fitness of genotypes attempting mutualism will be a function of the differences in survival of seed and seedling near and far from the parent, due to reduced competition, planting, or escape from predation. These are linked in nature, since the ant nest is at once the distant location, the enriched environment and the predator-free space. Conceptually, however, they are separable. Indeed, it appears that the critical benefit of myrmecochory is different in different systems (Beattie and Culver 1981; Buckley 1982; Milewski and Bond 1982; Westoby *et al.* 1982).

It is easier first to write the equation for non-transported individuals:

$$w_n = 1_n v_n - I_m \qquad (4.12)$$

where v_n is the per cent germination, survival and reproductive success and 1_n is the per cent of the seeds not lost to predation and I_m is the investment of the mutualist in a food body, expressed, for example, as seeds sacrificed to produce it. I_m may also include reinforcement of seed coat or other modifications of the seed necessary to improve dispersal by ants, although most workers have suggested that such characteristics were produced by preadaptation (Westoby *et al.* 1982).

The fate of transported seeds is compared to that of untransported seeds. Possibly both survival rates (1) and seedling viability (v) are less favourable near the parent than at random locations, due to, among other things, seed predators being attracted to the parent plants, and pre-emption of nutrients by the parents. While fitness could be measured relative to that of random sites, I have not done that because I expect both nonmyrmecochores and unsuccessful myrmecochores to remain near the parent relative to random sites (additional dispersal mechanisms exist in both cases (Beattie and Culver 1981; Buckley 1982)).

For the myrmecochores, $w_m = pw_t + qw_n$; the specific equations being:

$$w_m = 1_t v_t - I_m \qquad (4.13)$$
$$w_n = 1_n v_n - I_m$$

Combining and simplifying:

$$w_m = p(1_t v_t) + q(1_n v_n) - I_m \qquad (4.14)$$

The nonmyrmecochore is:

$$w_n = 1_n v_n, \qquad (4.15)$$

with no investment in a food body. I consider their survival rates (1) and viability (v) to be identical with those of non-transported myrmecochorous seeds.

If $w_m > w_n$, then $p1_t v_t + q1_n v_n - I_f > 1_n v_n$, which reduces to:

$$p1_t v_t - I_f > p1_n v_n \qquad (4.16)$$

For mutualism (a) the benefit must cover the cost of the food body, (b) $v_t > v_n$, (c) $1_t > 1_n$, (d) the p's on the two sides of the equation have the same value and are relatively unimportant.

In the cases that have been investigated, the importance of the variables differs. Predation pressure appears to drive the interaction in the case of *Datura discolor* ($1_t > 1_n$) (O'Dowd and Hay 1980). For Australian sclerophyll shrubs (Westoby *et al.* 1982), removal from competition with the parent seems crucial ($v_t > v_n$), while for *Viola* spp., $v_t > v_n$ and $1_t > 1_n$ (Culver and Beattie 1978, Beattie and Culver 1981).

For the ants no coevolution has been suggested; while some species of ants are better myrmecochores than others, there is no indication of changed behaviour, and ant response can be analysed in terms of optimal foraging strategies (Beattie and Culver 1981; Buckley 1982; Westoby *et al.* 1982).

Mycorrhizal Fungi

For mycorrhizal fungi interacting as mutualists with a plant root, the benefit is a regular carbon source provided by the plant (Zak 1964; Marx 1972; Cooke 1977). The cost is provision of water, nutrients, or possibly bacteriocides for the plant (Zak 1964; Marx 1972; Barrett 1982). Structurally, mycorrhizal fungi are much modified from their free-living forms; this could be either a cost or benefit, depending upon whether it increases or decreases fungal efficiency, but presumably it is most often a benefit.

Symbolically:

$$w_{mf} = pw_m + qw_n \qquad (4.17)$$

where w's are fitnesses, p and q frequencies and mf is mutualistic fungi, the population consisting of m, mutualists and n, non-mutualistic (free-living) fungi.

Then:

$$w_m = g_o + g_m, \text{ and } w_n = g_o \qquad (4.18)$$

Here g is the growth rate. For these fungi, a realistic measure of fitness must include both vegetative and sexual reproduction. g is intended to include both; they could readily be partitioned if that is desirable. Then, g_o is the growth rate of the free-living form and g_m is the increment due to mutualism.

This describes the dynamics of the fungal population (single genotype) in the field; some will encounter suitable higher plants (with frequency p) and establish a mutualism. The remainder (q) will be free living. Since the higher plant contributes a ready carbon source $w_m > w_n$. Substituting into 4.18 and simplifying:

$$w_{mf} = g_o + pg_m \qquad (4.19)$$

The gain from mutualism over the free-living state is related to the frequency of the encounter with suitable higher plants and the size of the gain due to mutualism. The variables that determines g_m are the costs and benefits of this mutualism.

$$g_m = b_m - c_m + s_m + f_m p \qquad (4.20)$$

where b_m is the benefit received from the interaction, specifically low molecular weight carbon compounds. The cost of the mutualist service is c_m, chiefly increased uptake of nutrients and water providing to the plant, due to increased surface area of the fungal mycelia. Structually, interacting mycorrhizal fungi are quite different from free-living forms. The variable s represents relative energetic savings due to changes in structures, which are converted to fungal growth. These are probably minor effects and could conceivably represent a net cost, not benefit, in which case the sign would reverse. The final term, $f_m p$, is the feedback loop on host abundance. Since both plants and fungi can increase in size vegetatively, this feedback can take effect in a short time and on an individual basis. Population effects will probably also occur. The nutrients and water (c_m) provided to the plants raise plant growth and production, potentially increasing their abundance. The value of f_m thus represents the impact of fungal nutrients and water on higher plant numbers, expressed as p, the probability of the fungus encountering a suitable plant. Because for mutualism w_m must exceed w_n, $b_m + s_m + f_m p > c_m$ or the interaction would be quickly broken off and the population would be solely free-living.

Finally, substituting 4.20 into 4.19:

$$w_{mf} = g_o + p(b_m - c_m + s_m + f_m p) \qquad (4.21)$$

In this interaction, the options seem to be mutualism versus parasitism. For the fungus to act as a parasite:

$$w_{pf} = p'w_r + q'w_n \qquad (4.22)$$

where pf = parasitic fungi, r = parasite, and p' and q' are frequencies ($p' + q' = 1$) and n and w are as in 4.18. Then:

$w_r = g_o + g_r$, $w_n = g_o$, fitness equalling the relative growth rates, with g_r the benefit due to parasitism. Substituting into 4.22 and simplifying:

$$w_{pf} = g_o + p'g_r \qquad (4.23)$$

By the same logic used for the mutualist:

$$g_r = b_r - c_r + s_r - f_r p' \qquad (4.24)$$

For the parasite the benefit (b_r) is the carbohydrate received, and s_r is benefit due to structural changes of the interacting form, which are roughly the same as for the mutualist. However, c_r is the cost of overcoming the plant's antiparasite defenses (rather than the cost of donating nutrients and water) and $f_r p'$ is a negative value reflecting the degree to which the presence of parasitic fungi decreases plant abundance and lowers fungal fitness by making suitable hosts less available. As in any predator-prey interaction, a very strong effect of fungi on plant abundance ($f_r p'$) will eliminate the fungi as well as the plants.

Substituting 4.24 into 4.23

$$w_{pf} = g_o + p'(b_r - c_r + s_r - f_r p').$$ (4.25)

Again $b_r + s_r > c_r + f_r p'$ and $w_r > w_n$ or the interaction is selected against and discontinued.

The comparison between mutualistic and parasitic fungi envisions closely related genotypes being selected on the basis of more rapid growth and reproduction. The contrast, as represented here, is between equations 4.21 and 4.25.

(a) The values of p and p′. These represent the rate at which potentially interacting organisms encounter each other; the frequency with which fungi find roots with which to interact. I see no reason why these should differ initially between mutualists and parasites. They are at first critical to whether interaction occurs at all. However, feedback from $f_m p$ and $f_r p'$ should cause p and p′ to diverge, p increasing toward broadly interacting communities and p′ decreasing due to the negative impact of parasitism on plant abundance so that pressure on the parasite to maintain its numbers will favour non-interaction (the free-living form).

(b) The benefit received (b_m and b_r). These are the benefits, believed to be readily metabolized carbohydrate, received by the fungi from the plant. It is not clear whether they are the same or different, and, if different, which is greater. The size of the benefits, in relations to the rest of equations 4.21 and 4.23 and in relation to each other are critical to whether there is an interaction and the nature of that interaction.

(c) The costs, c_m and c_r. These ought to be very different. The cost to a mutualist is the service it provides, in this case enhanced uptake of water and nutrients. The value of these to the fungus is a critical point. Mutualism is clearly favoured if the value to the donor of the service performed is lower (and the value to the recipient is greater). If the fungi can easily assimilate water and nutrient far beyond their needs, then in terms of fungal growth traded for lack of the assimilates passed to the plant, the cost is very low.

The cost to the parasite is a totally different situation. The host plant can be expected to resist parasitism actively, for example by evolving non-permeable roots and releasing fungicides. The more serious and common the parasitism, the more specific should be the response. The distribution of parasitic fungi could be limited simply by lack of appropriate mutations to counter plant defences. Alternately, other things being equal, if nutrients were cheaper to provide than defences against plant defences, mutualism would be favoured.

(d) The change in structure (s_m and s_r) need not differ between mutualists and parasites. It may be affected differently if the plant releases fungicides and so confers a different benefit or cost to mutualists and parasites, but in the absence of such information I will assume it to be unaffected by the type of interaction.

(e) The effects of $+ f_mp$ and $- f_rp'$. Mutualism aids the host, parasitism injures it; these should affect host abundance and consequently feed back onto the fungi. The impact of f_mp and f_rp' on p and p' is discussed above in a). However, the size of f affects the net benefit of interaction. The higher f_m, the more favourable is mutualism because plants are more available for mutualism. As f_r rises, the benefits of parasitism have to be more dramatic to maintain the interaction over selection for the free-living form. The mutualistic situation would seem to begin a positive feedback loop with more benefit aiding plants more, in an accelerating cycle. This escalation would seem to be terminated chiefly by action of other limits on the pair (for example, Heithaus *et al.* 1980), generally through mutualistic loss models, or if b_r is very high, by interdemic selection. It also seems likely that the positive feedback loop is a driving force for coevolution.

The ability of vesicular-arbuscular mycorrhizal fungi to survive as free-living species and their general lack of specificity suggest p is relatively low, although higher than q (equation 4.18). Low p is also a significant block to coevolution and the spread of mutualism (equation 4.21). It does seem, however, that the positive feedback of mutualism (equation 4.21) is in place and that conditions facilitating parasitism (high b_r, low c_r and/or high c_m and low f_rp') are less common than those facilitating mutualism (high b_m, low c_m and, preferably, high f_mp).

Without modelling the host plant in detail, let me suggest that when the interaction was initiated, the fungi first parasitized the plants. From this starting point, the plant's evolutionary 'choices' are to launch a strong enough response that the fungus shifts to a free-living state, or to take advantage of the situation by establishing mutualism with the fungus. If nutrients or water are readily removed from the fungus, this might be a simple facultative response to conditions created by the mere presence of the fungus; the rewards are obvious, the feedback loop is positive and the alternative of anti-fungal defence perhaps more expensive, even if the necessary mutations are present and a really effective defence could evolve. Furthermore, for plant as well as fungus, there is a positive feedback loop, with aid to the fungus enhancing that growth, so that once initiated, it would be expected to persist and spread; competition between plant-fungal parasite and mutualist pairs would seem to favour the latter, based on the analysis here. It would seem that (1) where other factors keep suitable plant sparse, the fungi will be polymorphic for mutualists and free-living forms, (2) in competition with parasitic forms, mutualistic fungi win, opening up the conditions for (2) reinvasion by parasitic fungi with a fitness equation:

$$w_{pf} = g_o + p'(b_r + s_r - f_r p') \tag{4.26}$$

which persist in the system as parasitic mimics of the fungus-plant mutualism. This is because the presence of mutualistic fungi precludes anti-fungal defences by plant roots, so that c_r is reduced to virtually zero.

Fungus-gardening Ants

Fungus-gardening ants of the genus *Atta* collect leaves and other plant parts, which they provide to fungi within the colony for substrate, and then feed off the fungal mycelia. To model this interaction brings up the problem that this may well be a mutualism which was established only once but has subsequently undergone considerable coevolution (Wilson 1971). The interaction is both symbiotic and obligate.

There are a number of possible scenarios for the beginning of the interaction. Fungus-gardening may have evolved gradually from consumption of fungi that grew on the faecal deposits of the ants or other insects or the interaction may have begun with fungi colonizing a seed cache (Wilson 1971). Garling (1979) suggested that the fungi began as mycorrhizae using the root hairs that descended into the chambers of the ant colony.

In the beginning, the fungi can be represented by:

$$w_{fu} = pw_i + qw_n = w_n + pw_m \qquad (4.27)$$

The same tension as between free-living and interacting mycorrhizae exists here, the notation being w for fitness, p and q frequencies, for fungi, i for interacting, and n for noninteracting (free living) forms. The question lies in the values for pw_m; since all ants eat fungi, why should fungi be found near predators at all? At the beginning they were living in an ant-produced habitat, possibly using ant by-products to their advantage and certainly being preyed upon by the ants. The use of potential ant food by fungi would make the two competitors, in addition to predation on fungi by ants. If fungi consumed faecal matter or were mycorrhizal, the initial interaction was predator-prey only.

At this point, fungi are being preyed upon by ants: their 'choices' are to resist, become mutualists, or avoid encountering ants altogether. For there to be interaction, fungal growth must exceed consumption by ants, whether the interaction is mutualism or predation.

At the beginning of the cycle, for the fungi interacting with ants (i):

$$w_i = g - h \qquad (4.28)$$

where g is growth and h is harvesting by ants. These can be measured as the relative or absolute number of calories, biomass or fruiting bodies produced. For $w_m > w_n$ (m = mutualist, n = non-mutualist):

$$g_m - h_m > g_n - h_n \qquad (4.29)$$

so the net growth of mutualistic fungi must exceed that of non-mutualists despite harvesting by ants, or mutualism will not evolve. Culturing by ants raises fungal growth rates ($g_m > g_n$), and probably does not greatly raise harvesting rates ($h_m \sim h_n$), so that mutualism is favoured.

The question then becomes why obligate symbiotic ants exist, culturing and eating fungi and only fungi. Non-mutualists would be expected to eat fungi: why should they (a) become mutualists and (b) give up generalist feeding? In the beginning:

$$w_g = pw_f + qw_e \qquad (4.30)$$

where w_g is the fitness of generalist ants, w_f the fitness of that subset of the population eating fungi (or the contribution of fungi to ant fitness), w_e is the fitness from consumption of all other foods, and p and q are frequencies. What conditions caused generalists to become obligate fungivores?

Because this interaction is symbiotic, the fungi growing within the ant nests and being eaten by ants from the beginning, there is no class of noninteracting ants, and:

$$w_f = c_f s_f - r_f \qquad (4.31)$$

where f represents fungivorous ants, c is the mean number of calories brought in per food item, s is the number of food items that come in per unit time, and r is the number of the calories expended foraging for food.

For other foods, the equation is:

$$w_e = c_e s_e - r_e \qquad (4.32)$$

with all the terms as defined above. For generalist ants, then:

$$w_g = p(c_f s_f - r_f) + q(c_e s_e - r_e) \qquad (4.33)$$

When will this evolve so that $p \to 1.0$, $q \to 0$?

Depending upon the discernment of the foragers, c_e, the value of food collected is likely to be very variable. The value of fungal hyphae (c_f), on the other hand, is fairly constant. How the means compare is difficult to guess. The benefit of fungivory lies in the rate at which hyphae can be harvested (s_f), which is very high compared to the time to forage randomly in the environment (s_e) and the low value of r_f compared to r_e. Foraging for fungi within the nest is clearly faster than random foraging externally. Included in r is the cost of foragers that do not return. Thus if:

$$c_f s_f - r_f > c_e s_e - r_e, \qquad (4.34)$$

fungivory is favoured over generalized foraging.

To establish obligate mutualistic fungivory by ants from the situation in equation 4.33 requires three conditions: (1) a diet of only fungi must be nutritionally complete, (2) fungal growth rates must not limit ant growth rate, and (3) mutualism must provide greater ant fitness (growth or stability) than simple fungivory. Apparently, dietary completeness is possible (or, on the average, the diet is at least as good as the diet from general foraging) because Attine ants are obligate fungivores. Symbolically, this is included in equation 4.35 as d_{ma}, which either takes on an absolute value of 1 or 0 (complete diet or inadequate diet), or is measured relative to d_{na}, the dietary completeness of generalists (equation 4.36). There may be other costs such as increased risk of disease transmission by eating fungi raised within the nest, but those could easily be included.

Complete fungivory is only viable if the rate of fungal growth does not become a serious limit to ant growth. If it did, ants would be pressured to return to general foraging. The shift from predatory fungivory to mutualistic fungivory provides a release of this constraint, at least potentially, since the mutualists, in culturing the fungi, supply food and raise fungal growth rate. Other culturing activity, such as moving hyphae or removing competing species, will also increase fungal growth.

Thus:

$$w_{ma} = d_{ma} [c_{ma}s_{ma} - (r_{ma} + t)] \tag{4.35}$$

and

$$w_{na} = d_{na} (c_{na}s_{na} - r_{na}) \tag{4.36}$$

where w is fitness, ma = mutualistic ants and na = non-mutualistic ants (generalists; equation 4.36 can be factored into equation 4.33), in which d is dietary completeness, c is calories per food item, s is the rate food items are gathered and r is the energy expended in gathering food, as before, and t is energy expended to tend fungi.

If:

$$d_{ma} [c_{ma}s_{ma} - (r_{ma} + t)] > d_{na} (c_{na}s_{na} - r_{na}) \tag{4.37}$$

then the ants will move toward reliance on fungivory, and mutualism. As above, c_{ma} is predictable and c_{na} highly variable. Which has a higher mean is difficult to predict. However, s_{ma} should exceed s_{na}, since distances are less, so that $c_{ma}s_{ma} > c_{na}s_{na}$. More realistically, perhaps, after an initial positive response to fungi as a food source, the ant colony consumed so much of the fungus that ant growth was limited by fungal availability. If the ants were forced to revert to general foraging, obligacy and mutualism were lost. However, r_{ma} should be much less than r_{na}, it being safer to forage underground and the distances being less. It seems likely that the net benefit of $c_{ma}s_{ma} - r_{ma}$ will be sufficient to overcome the tendency of ant growth to be limited by the rate of fungal growth, especially since fungal vegetative growth is simpler and more rapid than comparable ant growth. At the same time, the single-origin nature of Attine ants indicates this was a unique event.

The value of t is critical for two reasons: (1), if t is small, greater fungal growth can be produced cheaply, and (2), t suggests the existence of a positive feedback loop in which more ant activities lead to more fungi who produce more food for ants.

A relationship in which the intrinsic rate of growth of the fungi is less than that of the ants, or in which there is a long time lag in response of ant predator to fungal prey, will make obligacy and mutualism unlikely: ant dependency on a fungal food source requires that it be reliable.

Any preconditions favouring positive feedback greatly increase the chances of mutualism. If factors limiting fungi are decreased as the ant colony increases, mutualism is facilitiated in a simple manner. This expectation can be used to evaluate critically the theories that fungus gardening ants began as (a) consuming seed caches, (b) consuming faecal deposits or middens or (c) growing off plant roots entering the colony. Only in the second case is there a precondition favouring a positive feedback. Increase in ant colonies increases the size of the midden. The number of fungi growing on plant roots is independent of ant population growth, so exhaustion of fungi as a food source is likely. In the case of consumers of seed caches, cache size may not show positive feedback with increased ant numbers, especially if ants are converting to fungivory. A prediction of this model is, then, that the fungi began as commensals consuming ant faeces.

The benefits of mutualism to fungi are clearly increased food, reduction of competing organisms and movement of the fungi by the ants. The initial costs of mutualism seem inconsequential compared to the benefits, given that they are already being preyed upon. The exploited species have little they can do to avoid exploitation, unless they can evolve a defence faster than the other species respond. This makes selection for mutualism likely; such selection might be opposed by phylogenetic inertia and lack of available alleles in the exploited species, and by lack of benefit for mutualism to the exploiters. I have suggested that in this case the ability of fungal growth rate to match or outstrip ant growth rate, and fungal food provided inadvertently by the ants, produce positive feedback with ant growth, which is necessary to keep ants from going to other food sources. Once this positive feedback is in place, further coevolution will follow. In a general way, one would expect the exploiters only to opt for or allow mutualism if they get additional benefit from the interaction. For fungus-gardening ants that seems to be the case. It is difficult to say if it is a common situation.

Mixed Feeding Flocks

In the final example, the participants are all equivalent. Mixed feed-ing flocks emphasize the idea that mutual aid can be intraspecific as well as interspecific. Mixed feeding flocks occur among many animals (for example, Wilson 1971), but have received special atten-tion among birds (for example, Wing 1946; Morse 1967, 1970; Cody 1979, etc.). The participants benefit from more efficient feeding, especially the harvesting of scattered resources (Morse 1967; Cody 1979), and less individual time and energy spent watching for predators for the same level of defence, due to the increased number of watchers (Baker 1978; Krebs 1980) and possibly lower predation (Morse 1970; Hamilton 1971; Vine 1971; Pulliam 1973; Caraco 1979, 1981). Costs are in competition between individuals; the less dominant ones may feed less well than they would in foraging alone (Baker 1978; Krebs 1980). There are probably other costs due to the intra- or interspecific aggression that proximity causes (Wing 1946; Morse 1970).

Let:

$$w_m = f_m (e_m g_m - a_m) \tag{4.38}$$

and:

$$w_s = f_s (e_s g_s - a_s) \tag{4.39}$$

where the w's are fitnesses, m and s represent mutualists and solitary individuals respectively, and f is the probability of survival for the whole period, especially with regard to predation. e is the feeding efficiency (for example, calories per hour or calories eaten per calo-ries spent searching), g is an estimate for the percentage of time spent feeding, especially relative to stopping to watch for predators. The value of g should increase with increasing N: the larger the group, the less each individual needs to watch (Pulliam 1973). In equation 4.38, a_m and a_s estimate the costs of interanimal behaviour, a complex function of the lesser food received by subdominant animals relative to solitary foraging (Baker 1978; Krebs 1980) and the costs of aggressive interactions (Morse 1970; Pulliam 1973). If:

$$f_m(e_m g_m - a_m) > f_s (e_s g_s - a_s) \tag{4.40}$$

then the flock will stay together.

Both e and a are related to the distribution of food in space and time. Cody (1979) has suggested that flocks of birds are favoured where the resources are distributed once a year and gradually removed thereafter; harvesting is more efficient if each area is stripped bare rather than slowly depleted such that late in the season harvesting is very inefficient. For $e_m > e_s$, some basic compatibility must exist between the feeding habits of the various individuals (Morse 1970). They must use resources that occur together, but do so with divergent preferences or exploit resources that are sufficiently rich that competition is not a problem. Should resources become limiting, e_s should quickly come to exceed e_m. The value of g should increase as group size increases (Hamilton 1971; Vine 1971; Pulliam 1973). Similarly, the frequency of survival, f, measured as risk to a particular individual, should decrease with increasing group size but may not be uniform throughout the flock (Hamilton 1971; Vine 1971; Pulliam 1973). For mutualism (the flock) to persist, mortality should be lower on the average for social foragers than for solitary ones. It is not clear that this is the case in feeding flocks (Morse 1970) but it is of great potential importance.

Feeding mutualisms are among the most casual that exist; individuals can participate or not very easily, indeed, within a few moments. Costs and benefits can be regularly re-evaluated, which could cause formation and dissolution of the flocks. It does seem that this is the case seasonally (Wing 1946; Morse 1967, 1970).

Discussion

I have derived equations here for five different mutualistic interactions. Roughgarden's model for the anemone-damselfish mutualism is:

$$PL_m W_{am} + (1 - PL_m) W_{sg} > PL_p W_{ap} + (1 - PL_p) W_{ss}$$

where L_m is the probability of survival of the host (sea anemone) associated with a mutualistic guest (damselfish), L_p the probability of survival of a host associated with a non-mutualistic (parasitic) guest, W_{am} is the fitness of the mutualistic guest in the associated state, W_{ap} is the fitness of the parasitic guest in the associated state, W_{sg} the fitness of the guest who has failed to find a host or whose host dies, W_{ss} the fitness of the solitary strategist and P the probability that an individual of symbiont phenotype successfully finds a host (Roughgarden 1975). My previous model for ant-extrafloral nectary mutualism is:

$$p[A(1 - D)H] > I_A$$

where p is the probability that ants will find the plant, A is the effectiveness of the ant-defence, D the effectiveness of other defences, H the intensity of herbivory and I_A the investment in nectar and nectaries (Keeler 1981). From this paper:

$$p_z N_z v_z + s_z N_c v_c - I_z > p_s N_s v_s - I_s$$

for animal pollination with regard to selfing (equations 4.5, 4.7) and

$$p_z N_z v_z - I_z > p_a N_a v_a - I_a$$

for animal pollination relative to wind pollination (4.10).

$$pl_t v_t - I_f > pl_n v_n \text{ for myrmecochory (4.16)},$$

$$(p(b_m - c_m + s_m + f_m p) > p'(b_r - c_r + s_r - f_r p') \text{ (4.21 and 4.25)}$$
for mycorrhizal fungi,

$$g_m - h_m > g_n - h_n \text{ for fungi (equation 4.29), and}$$

$$c_f s_f - r_f > c_e s_e - r_e \text{ for ants (equation 4.34) leading to:}$$

$$d_{ma}[c_{ma} s_{ma} - (r_{ma} + t)] > d_{na} (c_{na} s_{na} - r_{na})$$

for fungus-gardening ants (equation 4.36) and, finally:

$$f_m(e_m g_m - a_m) > f_s (e_s g_s - a_s) \text{ for mixed feeding flocks (4.40).}$$

A summary of the types of mutualism for which cost: benefit models exist is given in Table 4.1. The result of such varied mutualisms is a diversity of cost: benefit equations to represent them. The mutualisms were chosen to represent a broad spectrum and the models were then tailored to each situation. For symbiotic mutualists, it was necessary to understand why the interaction was a mutualism; parasitism was a serious alternative. Nonsymbiotic mutualisms, on the other hand, readily cease interacting, so parasitism was unlikely. That is, the alternatives to mutualism for symbionts are noninteraction and parasitism, or rarely, commensalism, while for nonsymbiotic mutualism, the chief alternative is noninteraction. These differences account for some of the differences in the equations. In all cases, if cost exceeds benefit, the mutualism devolves to some other interaction, or interaction ceases.

The exchanges modelled are reciprocal (mixed feeding flocks) or nonreciprocal (all others). The exchanges are of protection and feeding efficiency (mixed feeding flocks), protection for care (damselfish-sea anemone), food for protection (fungus-

Table 4.1: Mutualisms Analysed

Facultative:
    Nonsymbiotic:
        Partners Virtually Identical: mixed feeding flocks
        Partners Very Different: ant-plant mutualism at extrafloral nectaries,
            myrmecochory
    Symbiotic: mycorrhizal fungi

Obligate:
    Nonsymbiotic: pollination
    Symbiotic:
        Little coevolved: damselfish-anemone
        Coevolved: fungus-gardening ants

Partners were:
    Vertebrates: mixed feeding flocks
    Vertebrates and invertebrates: damselfish-anemone
    Invertebrate (or vertebrate) and plant: pollination, myrmecochory,
        extrafloral nectaries
    Vertebrate and fungus: fungus-gardening ants,
    Fungus and Plant: mycorrhizal fungi

---

gardening ants, extrafloral nectary-ant), food for transportation (pollination, myrmecochory) and food for food (= water and nutrient uptake for carbohydrate) (mycorrhizal fungus-plant). These are a subset of those listed by Boucher *et al.* (1982).

The cost in all cases include the investment in the mutualistic trait or service. However, for facultative mutualism, a second cost lay in the possible failure of the interaction to provide the essential service (gamete transfer, antiherbivory defence). This did not appear to be a cost of mixed species flocks, perhaps because they are so casual. However, they experienced a third type of costs, in the increase in competition and inter- and intraspecific aggression. Cost of investment (I) and chance of failure (q) should be inversely related; mutualism can occur if investment is low or failure rare. It becomes unlikely as investment and chance of failure rise; the interesting cases lie in the areas of intermediate cost and failure rates.

The greater the benefit, the more investment or failure can be tolerated. Obligate mutualisms, in which failures have zero fitness and investments are high, suggest that high investments are more tolerable than high failure rates. Mutualism is particularly favourable where the service is the least value to the donor and of the greatest value to the recipient (cf. Westoby *et al.* 1982). It would be interesting to predict such situations. One future use of cost: benefit models of mutualism would be to analyse the relative cost of the same service to donor and recipient.

Theory of selection for mutualism is not completely general but suggests the following: for individuals, there must be fitness gain from establishing the interaction. At the extremes are minor fitness gains produced with virtually no cost, versus substantial fitness gain, beneficial at even a very high cost. Situations of greater benefit for the same cost will be selected for more strongly.

There is an implicit requirement for a positive feedback loop and/or the ability to perceive the benefit. Where fitness will be lowered if the other species does not co-operate, stringent conditions are required to select for co-operation (for example, the Prisoner's Dilemma, Axelrod and Hamilton 1981). In this case, mutualism (co-operation) only results if the organism can distinguish among the individuals with which it interacts or if the interaction is long-lasting and one-to-one. Under these conditions, co-operation may develop from an existing interaction. If the interaction began as favourable to one organism and disadvantageous to the other (for example, predator-prey or parasite-host as with the fungus-gardening ants above), the exploited species would seem most likely to avoid the interaction, if possible; they are already contributing to the fitness of the exploiter. It is difficult to see what conditions favour mutualism by the exploiter. A tight positive feedback in which benefit to the 'prey' quickly results in fitness gain to themselves might work. When the initial condition is neutral or commensal, mutualism proceeds much more readily: the transition from '00' or '0 + ' to ' + + ' is much more favourable than that from ' + − '.

Mutualism is more readily established if it requires no novel behaviour or traits and/or these behaviours or traits are readily produced from existing genetic variation. Also important, but more difficult to perceive, is the absence of fitness-reducing side-effects. An example of a deleterious side-effect to mutualism is the use of extrafloral nectar by herbivores as well as by defenders (Lukefahr and Rhyne 1960; Lukefahr *et al.* 1966) or the potential invasion of mycorrhizal systems by parasitic fungi (above) side effects should be included in costs: benefit models, but they are not always evident. If the side effect is very strong, its costs will override the benefit.

Mutualism is favoured by a high probability of establishing the interaction (so that few individuals lack partners); and/or long duration of the interaction once it is established. This can compensate for low establishment rates. In Prisoner's Dilemma models, these two conditions are mirrored by the requirement for (a) a long period of 'play' and (b) that the 'play' be against a single opponent (or a series of individually recognized opponents). These rules are probably not important to symbiotic mutualisms because (a) the relative and absolute fitness of a mutualist without a partner is zero, and (b) symbiosis generally provides for the transfer of both symbionts as a unit (Boucher *et al.* 1982). In all the facultative mutualisms modelled except mixed feeding flocks, these conditions apply. Mixed feeding flocks probably differ because the exchange is reciprocal, and the interaction readily terminated.

Explaining conditions for co-operation in the game Prisoner's Dilemma, Axelrod and Hamilton (1981) conclude that the ability to respond by punishing the partner for a non-co-operative reaction is essential (in addition to factors mentioned above). I do not see this as a general rule for mutualism: I believe it follows from the particular payoff matrix of Prisoner's Dilemma where the alternatives are only ' + + ', ' + – ' ' – + ' or ' – – '. Nonreciprocity or interactions with 0 terms (suppose the matrix is ' + + ', ' + – ', ' – + ' and '00', or ' + + ', ' + 0 ', '0 – ' and '00') favour mutualism much more simply.

When the rates of formation of mutualism is the critical process and not the rate of loss (dissolution) of mutualism, these selective conditions will not be trivial and there will indeed be times when costs exceed benefits and mutualism is demonstrably selected against. Such situations can be sought (a) between environments for identical individuals, (b) in the same environment between individuals which vary for presence of the mutualistic trait, (c) between individuals making greater or less investment into providing for the partner, (d) between individuals with different life expectancies or of different ages, or (e) by creating artificial mutualists (see Bentley 1976) and looking for fitness gain. These ideas are suitable for experimental analysis. They may also explain temporal change in mutualistic interactions, such as species that are mutualistic early in the season or early in their life cycle when R_o is high, but not later after R_o declines.

The chief problem I find with cost: benefit models is the problem of analysing the mutualisms: each situation seems individual, and generalization is difficult. This may be inherent in the nature of mutualism: organisms interact mutualistically when they have something to offer each other. At the most general, for facultative mutualists, the equations are:

$p(b_m) > c_m$ for nonsymbionts, and

$p(b_m) - c_m > p'(b_r) - c_r$ for symbionts, where p and p' are the probabilities of finding the other organisms, b is the benefit, c the cost and m and r refer to mutualism and parasitism respectively. In obligate mutualism, p becomes irrelevant and as the interaction coevolves, it becomes impossible to separate the organisms, the benefit to (fitness of) organism 1 being a positive function of the benefit to (fitness of) organism 2 and vice versa.

Predicting the conditions for an exchange is desirable. Cost: benefit models attempt this, but they begin with real mutualisms in which the organisms already exchange services. Real tests will only be possible in the areas where the interaction is dissolving or where the same partners, for some reason, do not interact mutualistically.

Acknowledgements

I thank M. Kaspari, E. Routman and M. Bolick for helpful comments.

References

Axelrod, R. and D.W. Hamilton (1981) 'The Evolution of Cooperation', *Science, 211*, 1390–6

Baker, M.C. (1978) 'Flocking and Feeding in the Great Tit *Parus major* — an Important Consideration', *American Naturalist, 112*, 779–81

Barrett, J.A. (1982) 'Plant-Fungus Symbioses' in Futuyma, D.J. and Slatkin (eds), *Coevolution*, Sinauer Associates Inc., Sunderland, Mass., pp. 137–60

Beattie, A.J. and D.C. Culver (1981) 'The Guild of Myrmecochore in the Herbaceous Flora of West Virginia Forests', *Ecology, 62*, 107–15

Bentley, B.L. (1976) 'Plants Bearing Extrafloral Nectaries and the Associated Ant Community: Interhabitat Differences in the Reduction of Herbivore Damage', *Ecology, 57*, 815–20

Boucher, D.H., S. James and K.H. Keeler (1982) 'The Ecology of Mutualism', *Annual Review of Ecology and Systematics, 13*, 315–47

Buckley, R.C. (1982) 'Ant-Plant Interactions: A World Review' in Buckley, R.C. (ed.), *Ant-Plant Interactions in Australia*, Dr W. Junk Publishers, The Hague, pp. 111–82

Caraco, T. (1979) 'Time Budgeting and Group Size: a Test of Theory', *Ecology, 60*, 618–27

Caraco, T. (1981) 'Risk-Sensitivity and Foraging Groups', *Ecology, 62*, 527–31

Carlquist, S. (1965) *Island Life*, The Natural History Press, New York

Clegg, M.T. (1980) 'Measuring Plant Mating Systems', *Bioscience, 30*, 814–18

Cody, M.L. (1979) 'Finch Flocks in the Mohave Desert', *Theoretical Population Biology, 2*, 142–58

Cooke, R.C. (1977) *The Biology of Symbiotic Fungi* John Wiley and Sons Inc., London

Covich, A.P. (1976) 'Analyzing Shapes of Foraging Areas: Some Ecological and Economic Theories', *Annual Review of Ecology and Systematics*, 7, 235–57

Culver, D.C. and A.J. Beattie (1978) 'Myrmecochory in *Viola*: Dynamics of Seed-Ant Interactions in some West Virginia Species', *Journal of Ecology*, 66, 53–72

Dean, A.M. (1983) 'A Simple Model of Mutualism', *American Naturalist*, 121, 409–17

Emerson, S.H. (1935) 'The Genetic Nature of deVries' Mutations in *Oenothera lamarkiana*', *American Naturalist*, 69, 545–59

Faegri, K. and L. van der Pijl (1979) *The Principles of Population Ecology*, 3rd edn, Pergamon Press, Oxford

Feinsinger, P. (1983) 'Coevolution and Pollination' in Futuyma, D.J. and M. Slatkin (eds), *Coevolution*, Sinauer Associates Inc., Sinderland, Mass., pp. 282–310

Fisher, R.A. (1949) *The Theory of Inbreeding*, Oliver and Boyd, Edinburgh

Garling, L. (1979) 'Origin of Ant-Fungi Mutualism: a New Hypothesis', *Biotropica*, 11, 284–91

Grant, V. and A. Grant (1965) Flower Pollination in the Phlox Family, Columbia University Press, New York

Hamilton, W.D. (1971) 'Geometry of the Selfish Herd', *Journal of Theoretical Biology*, 31, 295–311

Heithaus, E.R., D.C. Culver and A.J. Beattie (1980) 'Models of some Ant-Plant Mutualisms', *American Naturalist*, 116, 347–61

Holldobler, B. and C.J. Lumsden (1980) 'Territorial Strategies in Ants', *Science*, 210, 732–9

Huey, R.B. and M. Slatkin (1976) 'Costs and Benefits and Lizard Thermoregulation', *Quarterly Review of Biology*, 51, 363–84

Inouye, D.W. (1980) 'The Terminology of Floral Larceny', *Ecology*, 61, 1251–3

Jain, S.K. (1976) 'Evolution of Inbreeding Plants', *Annual Review of Ecology and Systematics*, 7, 469–95

Keeler, K.H. (1981) 'A Model of Selection for Facultative, Non-Symbiotic Mutualism', *American Naturalist*, 118, 488–98

Kevan, P.G. and H.G. Baker (1983) 'Insects as Flower Visitors and Pollinators', *Annual Review of Entomology*, 28, 407–55

Kodric-Brown, A. and J.H. Brown (1978) 'Influence of Economics, Interspecific Competition, and Sexual Dimorphism on Territoriality of Migrant Rufous Hummingbirds', *Ecology*, 59, 285–96

Krebs, J.R. (1980) 'Flocking and Feeding in the Great Tit: A Reply to Baker', *American Naturalist*, 115, 147–9

Levin, D.A. (1975) 'Pest Pressure and Recombination Systems in Plants' *American Naturalist*, 109, 437–52

Levin, D.A. (1979) 'Pollinator Foraging Behavior: Genetic Implications for Plants' in Solbrig, O.T., G.B. Johnson and P.H. Raven (eds.), *Topics in Plant Population Biology*, Columbia University Press, New York, pp. 131–56

Levin, D.A. (1984) 'Inbreeding Depression and Proximity-Dependent Crossing Success in *Phlox drummondii*', *Evolution*, 38, 116–27

Lloyd, D.G. (1979) 'Some Reproductive Factors Affecting the Selection of Self-Fertilization in Plants', *American Naturalist*, 113, 67–79

Lloyd, D.G. (1980) 'Demographic Factors and Mating Patterns in Angiosperms' in Solbrig, O.T. (ed.), *Demography and Evolution in Plant Populations*, Blackwell Scientific Publications, Oxford, pp. 67–88

Lukefahr, M.J. (1960) 'Effects of Nectariless Cottons on Population of three Lepidopterous Insects'. *Journal of Economic Entomology*, 53, 242–4

Lukefahr, M.J., C.B. Cowan, T.R. Frimmer and L.W. Noble (1966) 'Resistance of Experimental Cotton Strain 1514 to the Bollworm and Cotton Fleahopper'. *Journal of Economic Entomology*, 59, 393–5

Magnusen, J.J., J.B. Crowder and P.A. Medrick (1979) 'Temperature as an Ecological Resource', *American Zoologist*, *19*, 331–44

Marx, D.H. (1972) 'Ectomycorrhizae as Biological Deterrents to Pathogenic Root Infections', *Annual Review of Phytopathology*, *10*, 429–54

Mather, K. (1943) 'Polygenic Inheritance and Natural Selection', *Biological Review*, *18*, 32–64

Maynard Smith, J. (1978) *The Evolution of Sex*, Cambridge University Press, Cambridge

Maynard Smith, J. (1982) *Evolution and the Theory of Games*, Cambridge University Press, Cambridge

Milewski, A.F. and W.J. Bond (1982) 'Convergence of Myrmecochory in Mediterranean Australia and South Africa' in Buckley, R.C. (ed.), *Ant-Plant Interactions in Australia*, Dr W. Junk Publisher, The Hague

Morse, D.H. (1967) 'Foraging Relationships of Brown-Headed Nuthatches and Pine Warblers', *Ecology*, *48*, 94–103

Morse, D.H. (1970) 'Ecological Aspects of some Mixed-Species Foraging Flocks of Birds', *Ecological Monographs*, *40*, 119–68

O'Dowd, J.D. and M.E. Hay (1980) 'Mutualism Between Harvester Ants and a Desert Emphemeral: Seed Escape from Rodents', *Ecology*, *61*, 531–40

Ornduff, R. (1969) 'Reproductive Biology in Relation to Systematics', *Taxon*, *18*, 121–33

Pijl, L. van der (1972) *Principles of Dispersal in Higher Plants*, 2nd edn, Springer-Verlag, New York

Pijl, L. van der and C. Dodson (1966) *Orchid Flowers. Their Pollination and Evolution*, University of Miami Press, Coral Gables

Pulliam, H.R. (1973) 'On the Advantages of Flocking', *Journal of Theoretical Biology*, *38*, 419–22

Pyke, G.H., H.R. Pulliam and E.L. Charnov (1977) 'Optimal Foraging: a Selective Review of Theory and Tests', *Quarterly Review of Biology*, *52*, 137–54

Reichert, S.E. and P. Hammerstein (1983) 'Game Theory in the Ecological Context', *Annual Review of Ecology and Systematics*, *14*, 377–410

Roughgarden, J. (1975) 'Evolution of Marine Symbiosis — a Simple Cost-Benefit Model', *Ecology*, *56*, 1201–8

Rubeinstein, D.I. (1981) 'Population Density, Resource Patterning and Territoriality in the Everglades Pygmy Sunfish', *Animal Behavior*, *29*, 155–72

Schoener, T.W. (1971) 'Theory of Feeding Strategies', *Annual Review of Ecology and Systematics*, *2*, 369–464

Slatkin, M. and J. Maynard Smith (1979) 'Models of Coevolution', *Quarterly Review of Biology*, *54*, 233–63

Stapanian, M.A. and C.C. Smith (1978) 'A Model for Seed Scatterhoarding: Coevolution of Fox, Squirrels and Black Walnuts', *Ecology*, *59*, 884–96

Stebbins, G.L. (1957) 'Self-Fertilization and Population Variability in Higher Plants', *American Naturalist*, *91*, 337–54

Stebbins, G.L. (1974) *Flowering Plants: Evolution Above the Species Level*, Belknap Press, Cambridge, Mass.

Stephenson, A.G. (1981) 'Flower and Fruit Abortion: Proximate Causes and Ultimate Functions', *Annual Review of Ecology and Systematics*, *12*, 253–80

Stiles, F.G. (1975) 'Ecology, Flowering Phenology and Hummingbird Pollination of some Costa Rican *Heliconia* Species', *Ecology*, *56*, 285–301

Templeton, A.R. and L.E. Gilbert (1975) 'Population Genetics and the Coevolution of Mutualism', this volume, Chapter 5

Vine, I. (1971) 'Risk of Visual Detection and Pursuit by a Predator and the Selective Advantage of Flocking Behavior', *Journal of Theoretical Biology*, *30*, 405–22

Waddington, C.H. (1957) *The Strategy of the Genes*, Allen & Unwin, London

Westoby, M., B. Rice, J.M. Shelley, D. Haig and J.L. Kohen (1982) 'Plants Use of

Ants for Dispersal at West Head, New South Wales' in Buckley, R.C. (ed.), *Ant-Plant Interactions in Australia*, Dr W. Junk Publisher, The Hague

Whitehead, D.R. (1969) 'Wind Pollination in the Angiosperms: Evolutionary and Environmental Considerations', *Evolution, 23*, 28–35

William, G.C. (1975) *Sex and Evolution*, Princeton University Press, Princeton

Willson, M.F. and N. Burley (1983) *Mate Choice in Plants: Tactics, Mechanisms and Consequences*, Princeton University Press, Princeton

Wilson, D.S. (1983) 'The Effect of Population Structure on the Evolution of Mutualism: a Field Test Involving Burying Beetles and their Phoretic Mites', *American Naturalist, 121*, 851–70

Wilson, E.O. (1971) *The Insect Societies*, Belknap Press of Harvard University, Cambridge, Mass.

Wing, L. (1946) 'Species Association in Winter Flocks', *Auk, 63*, 507–10

Zak, B. (1964) 'Role of Mycorrhizae in Root Disease', *Annual Review of Phytopathology, 2*, 377–92

5 POPULATION GENETICS AND THE COEVOLUTION OF MUTUALISM

Alan R. Templeton and Lawrence E. Gilbert

Introduction

Boucher *et al*. (1982) define mutualism as 'an interaction between species that is beneficial to both', and they point out that two definitions of 'benefit' have coexisted in the ecological literature for decades. The first definition defines benefit on the individual level, in terms of the relative fitness *within* a species of individuals participating to various degrees in the mutualistic interaction. The second definition considers the dynamics of whole population densities or sizes. Both definitions of benefit are useful depending upon what biological problem is being investigated, but it is critical to note that these definitions can be incompatible with one another in defining the nature of the interspecific interaction. For example, Gilbert (1977) pointed out that in some butterfly-plant mutualisms the butterfly may allow more constant individual seed set, but does *not* necessarily produce more adult plants. Likewise, the pollen or nectar provided by the plant helps to increase and make constant the individual butterfly's egg production, but does *not* necessarily lead to more adult butterflies. Many potential mortality factors intervene between propagule and reproductive adult both for the plants and the butterflies, and these other factors may control the actual population dynamics. As another potential example, consider the bacterial endosymbionts that exist in many *Drosophila* species (Ehrman 1983). When flies are 'cured' of their endosymbionts, they die. Obviously, an individual fly has an absolute requirement for the endosymbiont, but there is not the slightest indication that such an endosymbiont affects population dynamics in any natural population: rather, factors such as weather and larval substrate availability appear critical in determining *Drosophila* population dynamics (Templeton and Johnston 1982).

Even though mutualisms like those discussed above exert no direct effect on population dynamics, the mutualism may be essential for the continued existence of both species in the community (as is certainly true for the *Drosophila*/endosymbiont example) and may exert strong selective forces upon individuals within each species (as is certainly the case for *Drosophila* in the endosymbiont example as shown by the genetic studies of Ehrman 1983). As a result, it is very important for any investigator of mutualism to state clearly what definition of benefit is appropriate for the biological phenomenon being investigated.

As Boucher *et al.* (1982) point out, the individual-level definition is the one appropriate for questions of evolution. In this chapter, we are concerned with the coevolution of mutualistic traits, and accordingly, we consider only the individual-level definition of benefit. We do not mean to imply by this that a mutualistic coevolutionary process will never influence population dynamics. Rather, any influence on population size must be regarded as a *secondary* result of the coevolutionary process (which could feed back onto the individual-level fitness differences), but the course of the coevolutionary process is determined only through the individual benefits within each species.

Given that individual-level fitnesses differences are the appropriate measure for any coevolutionary model, it must still be recognized that the significance of individual fitness lies in the fact that they influence a population-level process: i.e. evolution. Intraspecific evolution is manifest only at the level of a reproducing population, and natural selection is the mechanism by which individual-level fitnesses are translated into population-level effects. As a result, models that phrase 'Evolution' solely in terms of individual fitnesses are just as incompatible with neo-Darwinian theory as are the models based only upon population dynamic criteria. Both of these modelling approaches effectively ignore the role of natural selection as an agent mediating between two levels of biological hierarchy: the individual and population levels.

Consequently, coevolutionary models must measure interspecific interactions by intraspecific individual-level costs and benefits, but then place these individual costs and benefits in the context of a reproducing population. As soon as this is done, several constraints enter the model that would otherwise be ignored. One important constraint is that only individual-level fitness differences that have a genetic basis can be translated into population-level effects via the mediating agent of natural selection. The genetic constraint alone allows many factors other than individual fitness to play a direct role in the adaptive process mediated by natural selection. First, in a reproducing population, the genetic architectures underlying the individual-level fitness phenotypes are broken down and recombined during meiosis. Hence, the number of genes, their linkage relationships, etc. directly enter into the transformation from individual level fitness differences to selective effects in a reproducing population. Next, the gametes produced during meiosis are paired during fertilization. The pattern and probability of various potential fertilizations is determined by the genetic population structure of the species: its system of mating, the number of potential mates, and the pattern and extent of gene flow. Finally, the new genotypes appearing after fertilization interact with the environments to which the developing individual is exposed in producing the individual's fitness. At this stage, environmental interactions, dominance, epistasis, pleiotropy, etc. enter into the evolutionary process mediated by natural selection. Hence, the role of natural selection in evolution should never be phrased in terms of fitness alone.

The genetic constraint also introduces the problem of the unit of selection; namely, what is the appropriate level of genetic organization to which we should ascribe fitness measures? Mayr (1970) has long championed the idea that the individual is the 'target of selection'. This is certainly true in the sense that fitness, as well as any other phenotype, only takes on a biological reality in terms of actual individuals. Moreover, the fitness of an individual is the result of all of its genes and their co-ordinated response to the environment in which that individual developed. Unfortunately, all this is rather irrelevant to the problem of the unit of selection. In most species, an individual's genotype is unique. Meiosis and sexual reproduction generally insure an individual's total genotype will never appear again in the history of the species. Consequently, the genotype is not an appropriate unit of selection since it has no continuity over time, and evolution is a process that can only occur over time. Rather, the unit of selection must be a level of genetic organization to which can be ascribed fitness measures that allow an accurate description of the effects of natural selection over time. Because of the many connotations the phrase 'the unit of selection' has acquired, it is perhaps best to describe such levels of genetic organization as units of selective response.

One advantage of phrasing this problem in terms of selective response is that such a response can be measured, and hence the appropriate genetic unit can be experimentally manipulated. A series of experiments using *Drosophila mercatorum* (Templeton *et al.* 1976; Templeton 1979; Annest and Templeton 1978) have revealed that the unit of selective response represents the dynamic balance between forces building up gene complexes (epistasis, tight linkage, restricted outcrossing, intense selection) versus those breaking down gene complexes (recombination, outcrossing, gene flow, large number of potential mates, etc.). One implication of these results is that if the genetic architectures underlying two phenotypes in an outcrossing population are not characterized by tight linkage, strong epistasis, or direct pleiotropy, then the two phenotypes will effectively represent independent units of selective response.

Phenotypes represent gene-by-environment interactions at the individual level, and in the models considered here the environment refers to the interactions an individual has with individuals of another species. We will examine the situation in which two interacting species simultaneously define more than one unit of selective response.

A Potential Example of Different, but Simultaneous, Coevolutionary Interactions

To illustrate the fact that different units of selective response can simultaneously exist with respect to traits involved in interspecific interactions, consider the case of sympatric species of butterflies in the genus *Heliconius* (Gilbert 1975, 1977, 1979). We emphasize at the onset that we are not trying to produce a detailed or even realistic model of these butterfly communities. Rather, we are using knowledge and insights gained from studies on these butterflies as a conceptual springboard for the development of our coevolutionary model. In some cases we are forced to make assumptions about the system that are of unknown validity, but nevertheless we find this system to be a useful conceptual vehicle for illustrating general properties about coevolution.

The larvae of these butterflies use various species of *Passiflora* as host plants, and, generally, sympatric species use non-overlapping sets of host species. Hence, from the point of view of larval traits, these is no interspecific interaction. Moreover, as with many Lepidoptera, the population sizes are probably determined by density-dependent factors acting on the larval stages (Gilbert 1983).

Hence, the most likely *population dynamic* interaction between these species is neutralism, although it is possible for the larvae to interact in other ways such as through shared predators or parasitoids. We will initially assume a population dynamic neutralism even though its validity in natural communities has not yet been established.

The adult *Heliconius* butterflies use the cucurbit vines from the genera *Gurania* and *Anguria* as a source of nectar and pollen, and usually all sympatric species use the same plants as pollen and nectar sources. Moreover, the nectar and pollen are critical for maintaining adult survival and in producing eggs (Dunlap-Pianka *et al.* 1977). Hence, any trait which increases adult foraging efficiency or competitive ability should be strongly selected, and the nature of the coevolutionary interaction for such traits would be one of competition. Traits in *Heliconius* that can be explained by such competition selection include the highly developed visual system, related head size, learning ability, early morning flight and trap-lining behaviour (Gilbert 1975), and the fact that different species have significantly differing abilities to utilize small grained or large grained pollens (Boggs *et al.* 1981).

By using amino acids derived from pollen for cyanogenic-glycoside synthesis (Nahrstedt and Davis 1983) and possibly by incorporating larval stored *Passiflora* allelochemics, the adult butterflies are distasteful to predators. This distastefulness presumably explains the evolution of warning coloration so characteristic of the various species of *Heliconius*. Moreover, selective pressures in a coevolutionary context would favour the sympatric species forming Müllerian complexes, which has in fact occurred (Turner 1977; Gilbert 1983). Consequently, with respect to the traits of wing coloration and pattern, the coevolution of the sympatric *Heliconius* species has apparently been governed by mutualism.

Adult sympatric individuals of various *Heliconius* species are therefore simultaneously competitors and mutualists with respect to different suites of traits. As long as the genetic systems underlying these different suites of traits define independent units of selective response, it is erroneous and misleading to speak of a single interspecific interaction as characterising the net relationship between the sympatric species with respect to coevolution.

In its most basic sense, coevolution depends upon the facts that (1) fitness represents a gene-by-environment interaction and (2) the types of responses an individual has to members of other species are an important aspect of the environment. Quite a large number of traits can contribute to fitness, and there is no reason to assume that the coevolutionary pressures influencing the genes underlying different traits must operate in the same manner or direction. This idea may be regarded as a genetic extension of the work of Levins (1975) that recognizes that the important elements of an ecological community are not necessarily species, but processes, life history stages, plant parts, etc.

A Coevolutionary Model of the *Heliconius* Example

Additional insight into the implications of having more than one simultaneous unit of coevolutionary response can be achieved by modelling the *Heliconius* example described in the previous section. For simplicity, consider only two traits: one involved with wing coloration/pattern and the other involved with adult foraging efficiency or pollen size utilization.

Much work has been done on the genetic basis of wing colour and pattern in *Heliconius* butterflies (Turner 1979; Sheppard *et al.* 1984; Gilbert unpublished). These studies have revealed that the convergence to a common pattern among sympatric species is primarily accomplished through a very few major gene loci, and usually through the replacement of recessive alleles by dominant alleles. In light of these results, consider the following model of coevolution of Müllerian mimicry between two species, say 1 and 2. Let N_1 represent the density of species 1 and N_2 the density of species 2. In species 1 let there be a recessive phenotype associated with the genotype aa, and likewise assume there is a recessive phenotype associated with the genotype bb in species 2. These two recessive phenotypes are assumed to be so different that predators would never confuse them. However, within each species, assume there is a dominant allele (A in species 1 and B in 2) such that the dominant phenotypes resemble each other sufficiently so as to cause some predator confusion. The dominant can also resemble, but to a lesser degree, the recessive phenotype in their own species and, to even a smaller degree, the recessive phenotype in the other species. Hence, the dominant phenotypes represent an intermediate trait state between the two recessive phenotypes. We will now assume that the fitness contribution of a given wing phenotype is proportional to the number of individuals resembling that phenotype times some coefficient measuring the degree of resemblance. Thus, the fitness of aa individuals with respect to the wing trait unit of selective response is:

$$ag_{aa}N_1 + b(1 - g_{aa})N_1 + c(1 - g_{bb})N_2 \qquad (5.1)$$

where g_{aa} is the genotype frequency of aa in species 1, g_{bb} is the genotype frequency of bb in species 2, and a, b and c are constants that measure the proportional fitness contributions to aa individuals from single individuals of the same phenotype, the other intraspecific phenotype, and the dominant phenotype of species 2 respectively. Given the previously stated resemblance assumptions, a > b > c. Similarly, the fitness of the dominant phenotype, A – (A – = AA and Aa), in species 1 is:

$$a(1 - g_{aa})N_1 + bg_{aa}N_1 + s(1 - g_{bb})N_2 + dg_{bb}N_2 \qquad (5.2)$$

where s and d measure the proportional fitness contributions to A – individuals from a resemblance to the dominant and recessive phenotype in the other species respectively. Given the previously stated resemblance assumptions, a > s > b or d. Similar fitness equations can be written for the species 2 phenotypes.

Much less information exists concerning the genetic basis of foraging efficiency or pollen size usage. Hence, we will use a black box quantitative genetic model. The fitness measure for this unit of selective response will be a standard, Lotka-Volterra function; namely, the fitness associated with a competitive trait in species i is:

$$1 + k_i(K_i - N_i - \Theta_{ij}N_j) \qquad (5.3)$$

where k_i and K_i are constants, and Θ_{ij} measures the competitive ability of individuals of species i relative to those of species j. We assume that the Θ's represent the trait undergoing coevolutionary change with additive genetic variance V_a in species i.

Although we have assumed that the mutualistic and competitive traits are independent units of selective response *given* the appropriate fitness measures (that is, the traits are not linked directly through strong epistasis, pleiotropy or genetic linkage and the population structure of both species is sufficiently outcrossing), the fitness measures can be correlated in this case because they depend upon common environmental parameters, namely the number of individuals in one's own and the other species. We emphasize that this linkage of the evolutionary fates of these traits is accomplished at the level of the gene-by-environment interaction and not at the level of the genetic architecture *per se*. Consequently, given the densities of the two species, we can still model the evolution of each trait as an independent unit.

The simplest assumption concerning these densities, and the most realistic for the *Heliconius* example, is to assume that the N's are completely unaffected by either of the traits undergoing coevolution. To simplify the situation even further, we now assume the N's are constants. We now investigate the evolution of the mutualistic wing trait and the competitive resource-use trait under these assumptions.

Letting p be the frequency of the A allele (the allele conferring a mutualistic resemblance to species 2), the change in allele frequency in a large population over one generation can always be expressed as (Templeton 1982):

$$\Delta P = p a_A / w \tag{5.4}$$

where w is the average fitness of the population and a_A is the average fitness excess of the A allele. The average fitness excess of allele A is the conditional mean fitness of individuals given they bear an A allele, minus w. Since p and w are always positive, the direction of evolution is determined solely by the sign of a_A. From equations 5.1 and 5.2, the average excess of the A allele is:

$$g_{aa}[N_1(1 - 2g_{aa})(a - b) + dg_{bb}N_2 + (1 - g_{bb})N_2(s - c)] \tag{5.5}$$

Equation 5.5 will always be positive (and therefore A will always increase in frequency) when $g_{aa} < 1/2$. However, under these conditions, the increase in the frequency of the A allele occurs because it is so common within species 1 that it is the dominant warning pattern. Thus, even in the absence of any interspecific interaction (that is, set $N_2 = 0$ or $d = s = c = 0$ in 5.5), p will increase when $g_{aa} < 1/2$. When $g_{aa} \geq 1/2$, p can only increase if the interspecific interactions are large enough. Hence, the interspecific interactions are only important in determining the evolution of wing coloration or pattern when p is initially small. It is reasonable to assume that both species were initially near fixation for the recessive phenotype, so the interspecific interaction is critical in determining the fate of the A allele. To see this more clearly, let g_{aa} and g_{bb} be close to 1 in equation 5.5. Then, the A allele will increase in frequency whenever:

$$N_2 d > N_1 (a - b) \tag{5.6}$$

Obviously, the increase of the A allele in species 1 depends critically upon the numbers of species 2 and the strength of the interspecific mutualism as measured by d. Similar conditions can be derived for the increase of the B allele in species 2. For coevolution to occur (the increase of both the A and B alleles), it is best for both N's to be roughly the same size, otherwise the necessary inequality will be difficult to satisfy in both species.

Turning now to the evolution of competitive ability, it follows from equation 5.3 and the standard quantitative genetic model (Falconer 1960), that the change in Θ_{ij} over one generation due to selection in species i is given by:

$$\Delta\Theta_{ij} = -kN_j V_a(i)/[1 + k_i(K_i - N_i - \Theta_{ij}N_j)] \tag{5.7}$$

It is obvious from the above that Θ_{ij} is always reduced by the action of natural selection as long as there is some additive variance and as long as the other species is present. Moreover, the more individuals of the other species, the more intense the selection for reducing Θ_{ij}. Such a reduction could be accomplished either through niche partitioning and/or greater competitive ability in excluding the other species from the resource.

Equations 5.6 and 5.7 describe the coevolutionary relationship between these two species. As long as inequality 5.6 is satisfied and there is additive variance for Θ, a species can simultaneously enhance a mutualistic benefit and reduce a competitive detriment. In this case with fixed N's, there is absolutely no interference between these qualitatively different coevolutionary processes.

This situation can change if one of the traits undergoing coevolution is also a determinant of population size. For example, suppose the mutualism is density limiting. Then, we would expect the N's to be increasing functions of p and u. As can be seen from equation 5.7, an increase in the N's will increase the intensity of selection on Θ. Hence, we have the interesting situation that the coevolution of the mutualism *intensifies* the competition between the species with respect to their foraging traits.

On the other hand, if the competitive traits were density limiting, a variety of outcomes are possible depending upon the biological meaning of a reduction in Θ. If this reduction is accomplished by increasing foraging competitiveness such that the other species tends to be excluded, the size of one species population would increase but that of the other species would decrease. As is evident from inequality 5.6, this would decrease the chance for an enhancement of the interspecific mutualism through coevolution. Instead, any evolution of similarity would be through the rarer species converging to the pattern of the more common. If the reduction in Θ is accomplished through niche partitioning, the N's may remain either unchanged or perhaps increase. However, unless the ratio between N_1 and N_2 is altered by this coevolution of the competitive trait will have little effect on the coevolution of the mutualistic trait. Hence, when the competitive trait is density-limiting, there is generally no enhancement of the coevolution for the mutualistic trait, and indeed, the coevolution of the competitive trait can prevent the coevolution of the mutualistic trait. Consequently, the competitive and mutualistic traits influence the coevolution of each other in a highly asymmetric fashion when one of them is density limiting.

Some Generalizations

Although the above model is phrased specifically in terms of the *Heliconius* example, many other potential examples exist of species having units of selective response that are qualitatively different with respect to an interspecific interaction. For example, different species of trees that live in the same area often display interspecific synchrony of seed set (Bock and Lepthien 1976), a phenomenon that can be interpreted as a coevolved mutualistic adaptation to reduce seed predation. However, the same tree species that mast together can also be competing for space, nutrients, sunlight, etc. Indeed, it is easy to envision a situation in which the masting mutualism increases the number of seeds surviving to germination, which in turn would accentuate competitive interactions for seedling traits. Another potential example is provided by species of plants (O'Dowd and Williamson 1979) and animals (Clark and Robertson 1979) that are aggregated to form mutualistic defences against predators. Despite the mutualism with respect to predator defence, these same plants and animals could be competing for pollinators, space, nutrients, etc., and once again, the very adaptation of aggregation could accentuate the coevolution of competitive traits.

It is also important to realize that what is often regarded as a single trait or fitness component can have more than one independent unit of selective response. For example, consider the fitness component of larval viability in flies of the genus *Drosophila*. Many experiments have shown that larvae of closely related species, such as *D. melanogaster* and *D. simulans*, compete for larval food resources, and that the outcome of such nutritional competition can influence the chances of surviving to the pupal stage (for example, Miller 1964). This represents exploitation competition for shared resources. On the other hand, Budnik and Brncic (1975) have shown that the waste products produced by the larvae often have no effect on the viability of conspecifics but lower the viability of other *Drosophila* species. This is interference competition. Finally, *Drosophila* larvae need to obtain sterols from the yeasts they ingest in order to develop. Bos *et al.* (1976) demonstrated that different species of *Drosophila* differ in which yeast sterols they are able to metabolise into usable *Drosophila* sterols. Bos *et al.* (1977) then showed that larval monocultures of *D. simulans* could barely survive on certain strains of yeast, and monocultures of *Drosophila melanogaster* larvae could not survive at all. Nevertheless, when the two species were raised together, the larvae of both species could survive with high viabilities. This survivorship was interpreted as a mutualistic crossfeeding of diffusable metabolic products involved in sterol biosynthesis. Hence, with respect to the genes controlling energy intake and processing, *Drosophila* larvae experience the coevolutionary pressures of exploitative competition; with respect to the genes controlling biotic waste products, they experience interference competitive coevolution; and with respect to the genes controlling sterol biosynthesis, they can experience mutualistic coevolution. As long as the genes governing these various metabolic systems are not strongly linked and the population structure is sufficiently outcrossing, all of these gene systems can define independent units of selective response even though they are all contributors to the single fitness component of larval viability.

There is nothing new, of course, to the idea that two species cannot be characterized by a single qualitative type of interspecific interaction. It has long been recognized that different stages or age categories can experience qualitatively different types of interactions with other species and that the nature of the interaction can change through ecological time or space as density or community composition fluctuate. Nevertheless, the dominant idea in the literature is that at a given point in space and time, there is a single, net interaction between any species pair. The primary motivation for this widespread belief is that there is a single measure of benefit or cost; that is, one at the individual level or one at the population level. As long as the cost or benefit is one-dimensional, it automatically follows that there is such a thing as a *net* cost or benefit; and hence a *net* type of interaction.

The strength of the concept of net interaction is most clearly seen in papers in which the authors recognized that different traits were coevolving under qualitatively different types of interactions, yet still clung to the idea of a net interaction. For example, Brown and Kodric-Brown (1979) discussed the evolution of temperate, humming-bird-pollinated plants. They argued that the population sizes of such plants are generally limited by edaphic requirements and not pollination, but that pollination could influence individual fitness. These plants generally share the same pollinators, and there has been a convergence in flower colours and size among the plant species. Brown and Kodric-Brown explained this convergence in terms of mutualism. However, they also argued that these plants have interference competition through the male function (pollen dispersal) when a plant species' own pollen is mixed in with other species through sharing a pollinator. To reduce this interspecific interference in pollen transport, they argued that the species evolved character displacement in the orientation of the anthers and stigmas so that pollen is transported on different parts of the hummingbird. To summarize their views, Brown and Kodric-Brown (1979) viewed the coevolution of flower colour, shape and size in terms of interspecific mutualism, and viewed the coevolution of anther and stigma orientation in terms of interference competition. Nevertheless, they still explain the overall coevolution in terms of 'net' effects. Thus, the early coevolution is a competitive one, and 'once interference competition among these flower species has been reduced to some tolerable level, the advantages of converging to use the same pollinators outweigh the disadvantages of diverging to coevolve species-specific associations'. Although this coevolutionary scheme is certainly possible, other evolutionary routes exist if the coevolving traits represent different units of selective response. First, both traits could be coevolving simultaneously and the interaction between the species was both mutualistic and competitive from the onset. Second, the mutualistic sharing of pollinators could easily have *intensified* the coevolutionary response to interference competition — just the opposite of the idea that the mutualism must 'outweigh' the competition in order to coevolve. The genetic concept of different units of selective response makes the principle of a 'net' advantage or disadvantage misleading when discussing coevolution, or any other evolutionary responses to another species.

Genetics introduces many other complications into understanding the factors governing the coevolution of a mutualistic trait. For example, in the previous section we developed a model for the evolution of mutualistic wing patterns assuming an underlying dominant allele. Note that in constructing the equations describing the evolution of such an allele (Equations 5.1, 5.2, 5.4 and 5.5), we did not make any assumptions about the system of mating such as Hardy-Weinberg genotype frequencies. Moreover, the inequality governing the initial evolution, inequality 5.6, depends only upon the resemblance patterns and the densities of the two species and is invariant to the exact nature of the genotype frequencies. However, this invariance depends very much upon the dominance of the mutualistic allele. Suppose instead that the mutualistic alleles, A and B, were recessive. With Hardy-Weinberg genotype frequencies or an excess of heterozygotes above Hardy-Weinberg expectations, the average excess of the A allele can *never* be positive under the initial conditions that the mutualistic alleles are rare. Hence, the mutualism can never coevolve. On the other hand, let the genotype frequencies in species 1 now have the form: $g_{AA} = p^2 + pqf$, $g_{Aa} = 2pq(1 - f)$ and $g_{aa} = q^2 + pqf$ where p is the frequency of the recessive mutualistic allele, $q = 1 - p$ and f measures the deviation from Hardy-Weinberg genotype frequencies such that $f > 0$. Once again, assume the mutualitic alleles are initially very rare in both species. Then, the average excess of the A allele under these limiting conditions is:

$$a_A = f[N_2d - N_1(a - b)] \qquad (5.8)$$

Hence, the same condition as described in inequality 5.6 must now be satisfied for the evolution of the mutualistic trait; i.e. $N_2d > N_1(a - b)$. Consequently, a recessive allele can form the basis of a coevolved mutualism, but it requires the additional constraint that $f > 0$. This constraint can be satisfied if the population is inbreeding, if there is positive assortative mating for the wing pattern trait coded for by this locus, and/or if there is considerable population subdivision resulting in a significant amount of genetic differentiation between local populations. As this simple constrast reveals, whether a mutualistic trait can evolve depends critically upon the genetic architecture of the trait (in this case dominance vs. recessiveness) and upon the population structure of the coevolving species (system of mating and population subdivision). The important conclusion is that the coevolution of the trait is *not* completely determined by the nature of the interspecific interaction. Obviously, an interspecific interaction is *necessary* for coevolution to occur, but it is not *sufficient*. This also implies that factors other than the interspecific interaction can initiate or impede the coevolution of a trait. For example, suppose two species have Hardy-Weinberg population structures and recessive, mutualistic alleles such that $N_2d > N_1(a - b)$, with a similar inequality for species 2. The mutualism in this case will not coevolve. Now suppose a climatic change occurred that subdivided both species to create a Wahlund effect but without changing the ratio of their densities nor the nature of the interspecific interaction. Now the mutualism would coevolve. Obviously, genetic architecture and intraspecific population structure are important determinants of the course of coevolution.

Another important factor determining the nature of interspecific coevolution is the intraspecific fitness effects of the coevolving trait. For example, the evolution of the competitive trait in the *Heliconius* could result in the extinction or exclusion of one species from the community, and the mutualistic trait would not interfere with this process. Indeed the mutualism would progressively decrease in importance. The reason for this result is that the trait of wing pattern is mutualistic both inter- and intraspecifically. Hence, as the other species is eliminated from the community, other individuals from one's own species can provide the mutualistic benefit, making the coevolutionary option less and less likely. On the other hand, many interspecific mutualisms are at least potentially characterized by intraspecific competition for the mutualistic benefit, such as plant-pollinator mutualisms.

To see the impact of intraspecific competition for a interspecific mutualistic trait, consider the following modification of the *Heliconius* fitness models (equations 5.1 and 5.2). Let the fitness of the aa genotype be given by:

$$c(1 - g_{bb}N_2 - k(1 - g_{aa})N_1N_2[dg_{bb} + s(1 - g_{bb})] \qquad (5.9)$$

and the fitness for the dominant phenotype (Aa and AA) be:

$$s(1 - g_{bb})N_2 + dg_{bb}N_2 - kg_{aa}N_1N_2c(1 - g_{bb}) \qquad (5.10)$$

In these fitness models, only interactions with the other species can confer a fitness benefit, and the extent of the benefit is reduced through intraspecific competition as measured by k, the number of conspecifics of the alternative phenotype, and the extent of the fitness benefit the alternative phenotype receives from interactions with the other species. With these fitnesses, the average excess of the

A allele is given by $g_{aa}N_2W_A$ where W_A equals

$$(1 - g_{bb}) [s(1 + k(1 - g_{aa})N_1) - c(1 + kg_{aa}N_1)] + dg_{bb}(1 + k(1 - g_{aa})N_1) \qquad (5.11)$$

Hence, the intensity of the potential coevolutionary response depends upon the density of the other species, N_2, but the sign of the evolutionary response depends only upon W_A. Consequently, in great contrast to the case of a mutualism which is both inter- and intraspecific, the density of the other species does not enter as a determinant of whether coevolution will occur or not. In particular, retaining the assumption that $s > c$ and that intraspecific competition is occurring ($k > 0$), a mutualistic coevolution will always be initiated. Hence, coevolution of an interspecific mutualism is more probable when the trait is associated with intraspecific competition rather than intraspecific mutualism.

Conclusions

A mutualistic benefit cannot be properly ascribed either to individuals or to populations when dealing with evolutionary phenomena. Rather, the benefit must be associated with a genetic unit of selective response. Several different units of selective response can be simultaneously evolving in response to the interactions occurring between a single pair of species. Hence, the coevolutionary relationships between two species cannot be described in terms of some 'net' cost or benefit, and it is often meaningless or even erroneous to argue that one fitness component can 'outweigh' another. Such statements are incompatible with the fact that the individual genotype is not a unit of selective response in most sexually reproducing species, and are therefore incompatible with the process of evolution.

The concept of the unit of selective response forces one to acknowledge that the evolutionary interactions between two species are not one-dimensional but potentially multi-dimensional. The entries of this multi-dimensional response can often differ qualitatively in the nature of their fitness interactions with the other species. Hence, mutualism and competition are not coevolutionary alternatives but can exist simultaneously. Nevertheless, because all the entries of this multi-dimensional response depend upon a common environmental factor (the other species), their joint evolution must be considered. In particular, the coevolution of mutualistic traits can often accentuate the coevolution of competitive traits either by increasing the density of the other species or by increasing the intensity of the association with the other species. On the other hand, the coevolution of the competitive trait can either diminish, accentuate or have no effect on the coevolution of a mutualistic response.

Finally, the coevolution of a mutualistic trait is governed only in part by the existence of the interspecific interaction. The nature of the genetic architecture of the unit of selective response as well as the population structure of the coevolving species are important determinants of the course of coevolution. Moreover, the nature of the intraspecific fitness effects is critical, with the most likely interspecific mutualisms to coevolve being those associated with intraspecific competition rather than intraspecific mutualism. Hence, any theory of coevolution that is based only on the nature of the interspecific interaction is *a priori* inadequate.

Acknowledgements

This work was supported in part by NIH grant RO1 AG02246 to ART. We also wish to think Drs James Bull and Michael Singer for their valuable comments on an earlier draft of this manuscript.

References

Annest, J.L. and A.R. Templeton (1978) 'Genetic Recombination and Clonal Selection in *Drosophila mercatorum*', *Genetics*, *89*, 193–210

Bock, C.E. and L.W. Lepthien (1976) 'Synchronous Eruptions of Boreal Seed-Eating Birds', *American Naturalist*, *110*, 559–71

Boggs, C.L., J.T. Smiley and L.E. Gilbert (1981) 'Patterns of Pollen Exploitation by *Heliconius* Butterflies', *Oecologia*, *48*, 284–9

Bos, M., B. Burnet, R. Farrow and R.A. Woods (1976) 'Development of *Drosophila* on Sterol Mutants of the Yeast *Saccharomyces cerevisiae*', *Genetic Research*, *28*, 163–76

Bos, M., B. Burnet, R. Farrow and R.A. Woods (1977) 'Mutual Facilitation Between Larvae of the Sibling Species *Drosophila melanogaster* and *D. simulans*', *Evolution*, *31*, 824–8

Boucher, D.H., S. James and K.H. Keeler (1982) 'The Ecology of Mutualism', *Annual Review of Ecology and Systematics*, *13*, 315–47

Brown, J.H. and A. Kodric-Brown (1979) 'Convergence, Competition, and Mimicry in a Temperate Community of Hummingbird-Pollinated Flowers', *Ecology*, *60*, 1022–35

Budnik, M.Y. and D. Brncic (1975) 'Effects of Larval Biotic Residues on Viability in Four Species of *Drosophila*', *Evolution*, *29*, 777–80

Clark, K.L. and R.J. Robertson (1979) 'Spatial and Temporal Multi-Species Nesting

Aggregations in Birds as Anti-Parasite and Anti-Predator Defenses', *Behavioural Ecology and Sociobiology*, 5, 359–71

Dulap-Pianka, H., C.L. Boggs and L.E. Gilbert (1977) 'Ovarian Dynamics in Heliconiine Butterflies: Programmed Senescence Versus Eternal Youth', *Science*, *197*, 487–90

Ehrman, L. (1983) 'Endosymbiosis' in Futuyma, D. and M. Slatkin (eds.), *Coevolution*, Sinauer, Sunderland, Mass., pp. 128–36

Falconer, D.S. (1960) *Introduction to Quantitative Genetics*, Ronald Press, New York

Gilbert, L.E. (1975) 'Ecological Consequences of a Coevolved Mutualism Between Butterflies and Plants' in Gilbert, L.E. and P. Raven (eds.), *Coevolution of Animals and Plants*, University of Texas Press, Austin, pp. 210–40

Gilbert, L.E. (1977) 'The Role of Insect-Plant Coevolution in the Organization of Ecosystems' in Labyrie, V. (ed.), *Comportement des Insectes et Milieu Trophique*, CNRS, Paris, pp. 399–413

Gilbert, L.E. (1979) 'Development of Theory in the Analysis of Insect-Plant Interactions' in Horn, D.J., R.D. Mitchell and G.R. Stairs (eds.), *Analysis of Ecological Systems*, Ohio State University Press, Columbus, pp. 177–54

Gilbert, L.E. (1983) 'Coevolution and Mimicry' in Futuyma, D. and M. Slatkin (eds.), *Coevolution*, Sinauer, Sunderland, Mass., pp. 263–81

Levins, R. (1975) 'Evolution in Communities Near Equilibrium', in Cody, M. and J. Diamond (eds.), *Ecology and Evolution of Communities*, Belnap Press, Cambridge, Mass., pp. 16–50

Mayr, E. (1970) *Populations, Species, and Evolution*, Belknap Press, Cambridge, Mass.

Miller, R.S. (1964) 'Larval Competition in *Drosophila melanogaster* and *Drosophila simulans*', *Ecology*, 45, 132–48

Nahrstedt, A. and R.H. Davis (1983) 'Occurrence, Variation and Biosynthesis of the Cyanogenic Glucosides Linamarin and Lotaustralin in Species of *Heliconiini* (Insecta: Lepidoptera)', *Comparative Biochemistry and Physiology*, *75B*, 65–73

O'Dowd, D.J. and G.B. Williamson (1979) 'Stability Conditions in Plant Defense Guilds', *American Naturalist*, *114*, 379–83

Sheppard, P.M., J.R.G. Turner, K.S. Brown, W.W. Benson and M.C. Singer (1984) 'Genetics and The Evolution Of Mullerian Mimicry in *Heliconius* Butterflies', *Philosophical Transactions of the Royal Society of London*, *B*, in press

Templeton, A.R. (1979) 'The Unit of Selection in *Drosophila mercatorum*. II. Genetic Revolutions and The Origin of Coadapted Genomes in Parthenogenetic Strains', *Genetics*, *92*, 1265–82

Templeton, A.R. (1982) 'Adaptation and The Integration of Evolutionary Forces', in Milkman, R. (ed.), *Perspectives on Evolution*, Sinauer, Sunderland, Mass., pp. 15–31

Templeton, A.R., C.F. Sing and B. Brokaw (1976) 'The Unit of Selection and Meiosis in Parthenogenetic Strains', *Genetics*, *82*, 349–76

Templeton, A.R. and J.S. Johnston (1982) 'Life History Evolution Under Pleiotropy and K-Selection in a Natural Population of *Drosophila mercatorum*', in Barker, J.S.F. and W.T. Starmer (eds.) *Ecological Genetics and Evolution: The Cactus-Yeast-Drosophila Model System*, Academic Press, New York, pp. 225–39

Turner, J.R.G. (1977) 'Butterfly Mimicry: The Genetical Evolution of An Adaptation' in Hecht, M.K., W.C. Steere and B. Wallace (eds.), *Evolutionary Biology*, *10*, Plenum, New York, pp. 163–206

Turner, J.R.G. (1979) 'Contrasted Modes of Evolution in the Same Genome: Alloenzymes and Adaptive Changes in *Heliconius*', *Proceedings of the National Academy of Science of the USA*, 76, 1924–8

6 EVOLUTION IN A MUTUALISTIC ENVIRONMENT

Richard Law

Introduction

The evolution of interactions between a population of living organisms and its environment raises some interesting issues for population biology (Thompson 1982; Lewontin 1983). The simplest view is that there is no interaction — that the population is thrown passively in directions determined by forces of natural selection quite outside its sphere of influence. Although such a view may give a good approximation to the evolutionary behaviour of a population when its environment is predominantly abiotic, it has shortcomings when the population is embedded in a community of other living organisms with which it interacts. The reason for this is that, as the population undergoes evolutionary change, it alters the environments of these other species because it is part of their environments, just as they are part of its environment. If this alteration leads to genetic change in the other species, the result will be an evolutionary feedback, in which the population is both altered by and brings about alterations to its environment. The biotic environment, therefore, unlike the abiotic one, has the crucial property to evolve through its interaction with the population.

Evolution of the biotic environment of a population is a well-known phenomenon when the environment is composed of species with which the population has antagonistic interactions, such as predator/prey, host/pathogens and competitors. For the reasons given later, we expect a population to bring about an evolutionary deterioration of such an environment. However, evolution of a mutualistic environment has been the subject of rather little investigation, and it is this issue on which I focus here. I begin with a simple conjecture that a population will tend to bring about an evolutionary improvement in this mutualistic environment, in contrast to the deterioration more familiar in antagonistic ones. Several predictions then follow from this concerning the rate of evolution, the evolution of sex and the tendency for specificity to evolve between a population and components of its mutualistic environment. The material in this chapter is largely taken from Law and Lewis (1983), to which the reader should refer for further details.

Evolution in Antagonistic and Mutualistic Environments

To clarify the reasoning behind the expectation that antagonistic environments should tend to deteriorate, consider the following sequence of events. Suppose that a population of a certain species (say species A) at some point in time contains a set of genetically distinct phenotypes, some of which are more frequent than others. Now, suppose that important components of the environment of individuals in this population are species with which it has antagonistic interactions such as predators/prey, host/pathogens and competitors. Since species A is itself part of the environment of these antagonists, it can, in principle, act as an agent of natural selection on them. Furthermore, it is the phenotypes of species A at high frequency that exert the strongest pressures, since it is these with which the antagonists are most likely to interact. Thus, for example, the predators are selected for their capacity to attack and consume the high frequency phenotypes of species A, and so on. If the selected antagonists are genetically different from their less successful counterparts, the frequent phenotypes of species A will therefore experience a deterioration in their environment. This tendency for deterioration was noted by Fisher (1958: 44–5) and has been discussed by Van Valen (1973) and Maynard Smith (1976). It leads to a view of evolution as a existential game (Slobodkin 1968), in which the winners are those organisms which, through reducing the extent of deterioration in their environment by continually changing, remain to run in the race in the future, just as the Red Queen of *Alice through the Looking Glass* had to run as fast as she could to remain standing where she was (Van Valen 1973).

By way of contrast, consider species A in an environment in which the only other species (B) is one with which it interacts in a mutualistic manner. As before species A, being a component of the mutualist's environment, will act as a selective force on B, with phenotypes at high frequency in A exerting the strongest pressures. Thus, in species B, one expects to find selection for phenotypes which are particularly successful in partnership with the frequent phenotypes of species A. The crucial difference from before is that, if these selected mutualists are genetically different from their less

Table 6.1: Mutualistic Symbioses for Testing Predictions

Symbioses		Period of origin	
Rhizobium and plants	Sprent 1979; Allen and Allen 1981	? Upper Cretaceous	Norris 1956, 1958; see also Graham 1964
Frankia and plants	Bond 1963; Torrey 1978; Akkermans and Roelofsen 1980	? Lower Cretaceous	Bond 1963
Cyanobacteria and plants	Whitton 1973; Stewart *et al.* 1980		
Cyanobacteria and fungi	Ahmadjian 1967; Hale 1974		
Green algae and fungi	Àhmadjian 1967; Hale 1974	Mid Tertiary	Smith 1921; see also Richardson and Green 1965; Hawkworth and Hill 1984
Chlorella and invertebrates	Karakashian and Karakashian 1965	? Recent	Law and Lewis 1983
Dinoflagellates and invertebrates	Taylor 1973a; Trench 1981	Mid Jurassic	Rosen 1977
Ericoid mycorrhizas	Harley and Smith 1983; Read 1983	? Lower Tertiary	Raven and Axelrod 1974
Vesicular-arbuscular mycorrhizas	Gerdemann 1968; Mosse 1973; Harley and Smith 1983	Mid Devonian	Nicolson 1975

(? = no direct fossil evidence)

successful counterparts, the environment will tend to improve for the high frequency phenotypes of species A. We therefore expect the deterioration arising from antagonistic interactions to be replaced by a tendency towards improvement in the environment. Extending Van Valen's regal metaphor, this could be likened to King Midas, since, whenever species A encounters an environment of this kind, it tends to transform it into a state from which greater rewards are reaped. However, as we shall see, the Midas touch of mutualism does not have the destabilizing effects that it had for King Midas himself.

It should be understood at the outset that this argument focuses on just one facet of the general issue of evolution in a mutualistic environment. Clearly, the mutualistic environment must already be in existence, so the forces giving rise to it in the first place are not considered here. The argument does not address the complexities which arise when there are both mutualistic and antagonistic elements in the environment (Templeton and Gilbert 1985). Neither does it consider the potential for invasion of non-mutualistic mutants into populations in mutualistic environments (Soberon and Martinez del Rio 1985). Nor does it have anything to say about the general quantitative evolutionary dynamics of mutualistic species; progress here requires solution of systems of differential (or difference) equations of at least four dimensions (for example, Roughgarden 1979: Chapter 23). The argument is intended for heuristic purposes only.

A Data Set

Before turning to the predictions that stem from the above conjecture about evolution in mutualistic environment, it is desirable to define a data set against which the predictions can be compared. On the face of it, this appears to be a hopeless task because the mutualists with which a population interacts are usually but one component of an environment containing many antagonists as well. Attempts to determine the effect of the mutualists are therefore likely to be frustrated by the presence of antagonists. However, there are certain mutualistic relationships which by their very nature restrict the intensity of antagonistic forces for one of the partners. These are mutualistic symbioses in which for a substantial part of its life one partner lives partly or wholly inside the other. The internal partner, which I will call the inhabitant (after Starr 1975 and Lewis 1985), thus lives in an environment in which the external partner gives some measure of protection from the antagonistic forces that would otherwise be experienced. The external partner (exhabitant — after Starr 1975), on the other hand, remains unbuffered from antagonistic interactions with predators/prey, host/pathogens and competitors and, therefore, does not live in a mutualistic environment. The resulting asymmetry between the inhabitants and exhabitants is an interesting feature in itself, leading to the expectation that their patterns of evolution should differ.

My choice of mutualistic symbioses is guided by four criteria. First, the association must be demonstrably mutualistic, in the sense that the fitness of individuals is greater when they are together than when they are apart (see Roughgarden 1975; Margulis 1981: 162; Lewin 1982, Law and Lewis 1983, Lewis 1985). Second, the association should involve one partner living partly or wholly inside the other for most of its lifespan, so that the inhabitant's environment is dominated as far as possible by mutualistic forces. Third, there should be evidence of a substantial evolutionary history of the symbiosis, so that any general patterns of coevolution of the partners should have emerged. Fourth, there should be an accessible body of information about the association on which one can draw to test the predictions.

Using these criteria, the set of mutualistic symbioses listed in Table 6.1 has been chosen. In certain cases research workers who have specialized on particular symbioses would question the mutualistic categorization given to them (for example, Ahmadjian and Jacobs 1981); a justification for their inclusion is given in Law and Lewis (1983). The chosen set is heavily influenced by criterion four but, as far as is known, this does not lead to a bias in favour of the predictions. It should also be understood that none of the inhabitants live in the 'ideal' mutualistic environment; they must all interact with antagonists to some extent. Even if they lived entirely within the tissues of their exhabitants, they indirectly experience their hosts' antagonists and if, as is more common, they have a free-living phase, they interact with their own antagonists. Where they differ from other organisms is in the possession of a mutualistic refuge from their own antagonists for a substantial part of their lives. These symbioses are therefore not ideal material on which to test theories of evolution in a mutualistic environment; they are simply closer to the ideal than any others of which I know.

Clearly, the symbioses that have been chosen represent no more than a small subset of the mutualistic associations known in the living world. None the less, most of them play crucial roles in the biosphere, so their evolution is far from a trivial issue. In addition, they have the useful feature of being diverse with respect to function and taxonomic group. This means that they represent independent lines of evolution and can therefore be envisaged as independent replicates of a natural evolutionary experiment.

Rate of Evolution

Predictions

The conjecture that the tendency for deterioration of antagonistic environments is transformed into a tendency for improvement of mutualistic ones leads to the prediction that the rate of genetic change should be lower when mutualistic forces predominate. To see why this should be so, let us return to species A in its antagonistic environment (p. 146). One would expect that those progeny of high frequency phenotypes which are different from their parents should have an advantage over those that are the same because, simply due to this difference, they should be able to escape from the full force of the antagonists already selected by their parents. By contrast, in a mutualistic environment, it is those progeny of high frequency phenotypes which are the *same* as their parents that have the advantage, because they are most compatible with the mutualists that have already been selected. Therefore, we expect that the rate of genetic change in mutualistic environment should be lower than in an antagonistic one. In the context of our set of mutualistic symbioses with their asymmetrical environments, the prediction is that the inhabitants should evolve more slowly than their exhabitants.

To test this prediction in a rigorous way, one needs information on the rate of genetic change of inhabitants and exhabitants — information which, as far as I know, does not exist. The only kind of data which is generally at our disposal is the number of extent taxa of inhabitants and exhabitants. This is a poor substitute, but it is probably related, albeit distantly, to the rate of genetic change, for the following reason. During the geographical spread of a successful mutualistic partnership, the different selection pressures experienced by the exhabitant in different populations should lead to radiation and speciation. In contrast, the pressures for constancy of the inhabitant should prevent this from happening to the same degree, so that, over the course of time, it radiates and speciates to a lesser extent. As a result, we would expect to find fewer taxa of inhabitants than of their exhabitants.

There are some obvious and serious problems about using this kind of data to test the prediction that inhabitants evolve more slowly than their exhabitants (see Discussion below), and it is important to bear in mind the limitations of the data set. I would emphasize that I use it simply because I know of no better data generally available. However, in view of the magnitude of the differences between inhabitants and exhabitants (see Results below), it is unlikely that serious errors are introduced through its use.

Results

Table 6.2 lists the number of genera and species of inhabitants and exhabitants belonging to each kind of mutualistic symbiosis. It is clear that there is much less taxonomic variation in inhabitants than exhabitants. The taxonomy of some groups is a matter of debate by specialists, so the figures should be taken as no more than approximate (see Law and Lewis (1983) for details of their calculation.) However, the difference between inhabitants and exhabitants in number of taxa is so large that alterations to taxonomic groupings are most unlikely to change the message from the table. The data are therefore strongly supportive of the prediction, in so far as the differences in number of taxa can be taken as the outcome of differences in the rate of genetic change.

Discussion

The disparity in numbers of taxa of inhabitants and exhabitants in Table 6.2 is open to interpretations other than the one above. First, it could be simply a taxonomic artefact. Clearly, taxonomic criteria must vary considerably over groups of organisms as diverse as these and it is just conceivable that this, in itself, could account for the results. However, I would argue that this is improbable, because it is unlikely to lead to such a consistent difference between inhabitants and exhabitants. Perhaps the most likely possibility is that convergent evolution of inhabitants with different origins, dictated by the similar requirements of a particular symbiotic way of life, could be the cause. Yet, for this to be so, exhabitants would also have to be bad taxonomists because, as we will see later, inhabitants within each kind of symbiosis generally lack strong specificity to particular

Table 6.2: Number of Genera and Species of Inhabitants and Exhabitants of Each Mutualistic Symbiosis

Symbiosis	Approximate Number of Taxa of Inhabitants			Approximate Number of Taxa of Exhabitants		
	Genera	Species		Genera	Species	
Rhizobium and plants	1	4	Graham 1964; Elkan 1981	600	17,500	Allen and Allen 1981
Frankia and plants	1	?	Lechevalier and Lechevalier 1979	17	150	Torrey 1978; Akkermans and Roelofsen 1980
Cyanobacteria and plants	2	?	Stewart *et al.* 1980	20	150	Moore 1969; Silvester and McNamara 1976; Stewart *et al.* 1980
Cyanobacteria and fungi	9	?	Ahmadjian 1967; Wetmore 1970; see also Stewart *et al.* 1980	70	1,500	Poelt 1973; James and Henssen 1976
Green algae and fungi	21	?	Ahmadjian 1967; Archibald 1975 Tschermak-Woes 1970, 1978, 1980a, b	300	13,500	Poelt 1973; James and Henssen 1976
Chlorella and invertebrates	1	1	D.C. Smith pers. comm.	14	14	Droop 1963; D.C. Smith pers. comm.
Dinoflagellates and invertebrates	2	4	Freudenthal 1962; Taylor 1974	200	650	Droop 1963; Taylor 1974
Ericoid mycorrhizas	2	2	Burgeff 1961; Read 1974; Couture *et al.* 1983	25	2,000	Largent *et al.* 1980; Malajczuk and Lamont 1981
VA mycorrhizas	4	78	Gerdemann and Trappe 1975; Hall and Fish 1979; Walker pers. comm.; Trappe 1982	11,000	225,000	Gerdemann 1968; Stebbins 1974

species of exhabitant. In view of this, I would argue that the result is more than just a taxonomic artefact.

A second alternative is that the result is due to the different genetic systems of inhabitants and exhabitants. This is worth mentioning because the next section shows that inhabitants tend to be asexual and exhabitants to be sexual. Stanley (1975) argued that most evolutionary change occurs during speciation and that speciation in its usual sense is only possible in sexual species, so that sexual lineages should be more diverse and species-rich. This prediction is in accordance with the observed result. However, it is debatable whether appropriate variation to drive such evolution exists in appreciable amounts (Lande 1980). Furthermore even if the difference in genetic systems is a significant force, the result is ultimately the outcome of the premium on genetic constancy and change in mutualistic and antagonistic environments respectively, in so far as this generates the advantage for asexual or sexual reproduction.

A third alternative is that there exist species with a predisposition for becoming the mutualistic inhabitants of different exhabitants. In an extreme case, each association observed with an exhabitant species could have arisen independently, in the absence of any subsequent coevolution of the partners. This argument has some force because of the disjunct taxonomic distribution of the exhabitants in certain kinds of mutualistic symbioses. For example, consider the nitrogen-fixing associations between cyanobacteria and plants, found in certain bryophytes, ferns, cycads and angiosperms; the most reasonable interpretation of this is that each exhabitant group corresponds to at least one independent origin of the symbiosis. Moreover, algae and fungi form associations rather easily, so lichen-like symbioses are probably strongly polyphyletic (Smith 1921: 272; but see also Ahmadjian 1970).

On the other hand, one should bear in mind that this argument places no constraints on the rate of evolution of the inhabitant, once an association has become established. Thus, unless there are additional forces generated by the association, one would expect appreciable radiation and speciation of the inhabitant over the course of time, as in the exhabitant. Most of the symbioses in the present survey have ancient origins (Table 6.1) and a high degree of structural and physiological co-ordination, so they are most likely to represent the outcome of extended periods of coevolution between the partners. That there should now be so few inhabitant taxa relative to exhabitant ones is more likely to be a result of the interactions between partners, notwithstanding polyphyletic origins of the symbioses.

Therefore, despite alternative possibilities, I would suggest that, at present, it is the pressures for different rates of genetic change that provide the most plausible explanation for the different number of taxa of inhabitants and exhabitants.

Evolution of Sex

Predictions

The tendency of antagonistic environments to deteriorate has consequences for the evolution of sex. This is worth considering because the selective forces that maintain sex in natural populations represent one of evolutionary biology's more profound mysteries (Ghiselin 1974; Williams 1975; Maynard Smith 1978; Bell 1982). To appreciate why a deteriorating environment should be important, let us return to the population of species A, with its set of genetically distinct phenotypes at different frequencies, in an antagonistic environment (see p. 146). We have already seen that the feedback, brought about by species A through its selective effect on the environment, has the result that progeny are born into an environment already selected for deleterious properties in combination with the common parental phenotypes. If progeny of the common parental phenotypes are produced asexually, they will be the same as their parents and experience the full deleterious effects of the environment. However, if they are produced by sexual reproduction, they escape from the full deleterious effects simply because they are different from their parents. Thus, there should be selection for sexual reproduction in environments in which antagonistic forces predominate (Levin 1975; Solbrig 1976; Glesener and Tilman 1978; Jaenike 1978; Hamilton 1980; Hamilton *et al.* 1981; Hudson and Law 1981; Bell 1982; Rice 1983).

If, on the other hand, species A lives in a biotic environment consisting solely of another mutualistic species (B), selection on sex operates in the reverse direction. The selective forces of species A on its mutualistic environment tend to make the environment improve for the phenotypes at high frequency, so that their progeny are born into an environment already selected for its compatibility with the parents. Thus, the more similar these progeny are to their parents, the better suited they will be to the pre-selected mutualists of species

B. Therefore, there should be selection for *asexual* reproduction in environments where mutualistic forces predominate. In the context of our set of mutualistic symbioses, this mean that sexual reproduction should be utilized less in inhabitants than in their exhabitants.

However, there are certain objections that can be raised against the argument. First, there is no advantage to reproduction without sex for individuals with phenotypes which occur at low frequency in species A. Indeed, unless they have sex, they cannot produce progeny with phenotypes matching those selected in species B. Clearly, the relative merits of sexual and asexual reproduction depend very much on the frequency of the phenotype. But therein lies the key to this dilemma; selection against sex in high-frequency phenotypes should be a stronger force than selection for sex in low-frequency ones, simply because the former represent a greater part of the population.

A second objection is that, within the array of progeny genotypes produced by sexual parents, it is still possible that, by chance, phenotypes could arise better matched to species B than the asexually-produced progeny of high-frequency phenotypes. The strength of this argument depends very much on whether the chance of producing such a phenotype, together with its gain in fitness, outweighs the average loss in fitness resulting from sex (as in the 'Sisiphean' genotypes of Williams (1975)). It could well be that sex is advantageous in the very early stages of establishing a mutualistic association but, since species B is being selected for its compatibility with the phenotypes of A currently at high frequency, it would seem that the probability of producing such a phenotype would soon become low.

Third, one might well ask what happens if species B is sexual, since this is to be expected in exhabitants of our mutualistic symbioses. At any point in time, species A is selecting those phenotypes in B that are particularly successful with its own high frequency phenotypes. But if B is sexual, the effect of this selection is undone when it reproduces. In such circumstances, it can be argued that the advantage of uniformity in the progeny of A will be lost and that progeny matching the new array of phenotypes in B are more likely to be produced if A is sexual. I see no completely satisfactory resolution of this problem (but see Discussion in Law and Lewis 1983). However, it is worth pointing out that the minimal requirement for selection against sex is simply that the progeny of high frequency phenotypes should find themselves in an environment which their parents have already selected. One can envisage conditions under which this would be so, even if species B is sexual. For example, suppose that individuals of species A have shorter lifespan than those of B (typical in our set of mutualistic symbioses). For most of the time, progeny of A will still find themselves in environments previously selected by their parents, giving an advantage to asexual reproduction that could counteract its disadvantage when B reproduces sexually.

Despite these objections, the prediction that sexual reproduction should be less utilized in mutualistic inhabitants than in their exhabitants does provide a reasonable point from which to initiate the discussion. The sections that follow are therefore developed in this spirit.

Results

Table 6.3 summarizes information about the prevalence of sex in inhabitants and exhabitants of the set of mutualistic symbioses given in Law and Lewis (1983). The table shows that there are marked differences between inhabitants and exhabitants. Sex occurs widely in all exhabitants but, in those inhabitants for which there is information, it is very restricted, if it occurs at all. Only the green algal inhabitants of lichens may run counter to this pattern; although sex has never been observed in the intact thallus, sexual fusions have been seen in cultures of two of the most common genera, *Trebouxia* (Ahmadjian 1959) and *Trentepohlia* (McCoy 1978). In addition, one should bear in mind the possibility of hyphal fusion leading to parasexual recombination in the fungal inhabitants of ericoid and VA mycorrhizas. Apart from these putative exceptions, the evidence is strongly supportive of the prediction.

Discussion

However, it must be pointed out that these are other reasons why, in principle, different systems of reproduction could have arisen in inhabitants and exhabitants. The first is that the inhabitants usually come from so-called 'lower' taxonomic groups than their exhabitants and it could be that their lack of sex simply reflects a tendency for sexual mechanisms to be less well developed in such taxa. To evaluate this assertion one needs taxa encompassing both inhabitants and exhabitants. There is only one such taxon in our set of mutualistic symbioses — the phylum Ascomycota, members of which are found as inhabitants of ericoid mycorrhizas and as

Table 6.3: Prevalence of Sex in Exhabitants, Inhabitants and Free Living Taxa Related to Inhabitants

Symbiosis	Occurrence of sex in:			Free living taxa, related to inhabitant	
	Exhabitant	Inhabitant			
Rhizobium and plants	widespread	present but limited	Hirsch 1979; Beringer et al. 1980; Kondorosi and Johnston 1981	present but extent unknown	Waksman 1967
Frankia and plants	widespread	absent	Lalonde, pers. comm.	present but extent unknown	Bazin 1968; Stewart and Singh 1975; see also Delaney et al. 1976
Cyanobacteria and plants	widespread	not known		present but extent unknown	
Cyanobacteria and fungi	widespread	not known		''	''
Green algae and fungi	widespread	present in some taxa, absent in others	Ahmadjian 1967; McCoy 1978	present in some taxa, absent in others	Lewin 1976
Chlorella and invertebrates	widespread	absent	Lewin 1976	absent	
Dinoflagellates and invertebrates	widespread	absent or very rare	Taylor 1973b Schoenberg and Trench 1980a	present but extent unknown	Beam and Himes 1980
Ericoid mycorrhizas	widespread	absent or very rare	Read, pers. comm.	widespread	Dennis 1968
VA mycorrhizas	widespread	absent or very rare		widespread	

exhabitants of lichens. Fruiting bodies have never been observed under natural conditions when they are inhabitants but they occur regularly when they are exhabitants. An alternative test, for which there is more data, compares inhabitants with their free-living relatives, since the transition to their mutualistic environment should lead to a reduction in sex. Table 6.3 shows that there are four kinds of mutualistic symbiosis in which sex is probably absent in inhabitants and present in their free-living relatives; there are two kinds in which sex, while present in inhabitants, is probably reduced relative to their free-living relatives; there is one kind in which there is no reduction because both the inhabitants and their free-living relatives are asexual; finally there are two kinds in which there is no information on sex in the inhabitant. In view of all this evidence, I would argue that we need to look beyond the difference in taxonomic groups of inhabitants and their exhabitants, in order to account for the difference in genetic systems.

If one accepts that the different reproductive systems of inhabitants and exhabitants result from selective forces associated with their partnership, the asymmetry between their antagonistic and mutualistic environments is only one of a number of possibilities. Another alternative is that internalization of the inhabitant tends to isolate it from potential partners for sexual reproduction, generating a strong selective pressure for reproduction without sex. Such a pressure should exist whether the inhabitant lives in a mutualistic or antagonistic environment. It is therefore of interest that many fungal-like and true fungal pathogens with varying degrees of internalization retain sexual reproduction. Although in certain cases sex is reduced in internal antagonists (Price 1980:76), the reduction is much less than in the mutualistic inhabitants considered here. There are evidently important differences in the selective forces experienced by inhabitants if they are mutualistic rather than antagonistic.

Another possibility is that internalization provides the inhabitant with more constant conditions by protecting it from fluctuations of the abiotic environment. Since it has been suggested that abiotic fluctuations provide a selective force for maintenance of sex (Maynard Smith 1971, 1980; Charlesworth 1976), the capacity of an exhabitant to buffer its inhabitant could account for the pattern. However, the conditions that need to be met for the environment to generate the required force seem unlikely to be satisfied in reality (Maynard Smith 1978). Moreover, it is arguable how much protection the exhabitant gives its inhabitant from alterations in the *abiotic* environment. For example, Schoenberg and Trench (1980a) note that, in the coral symbiosis, abiotic fluctuations are experienced by both inhabitant and exhabitant. The same probably holds true for lichens. Thus, inhabitants may receive little protection from abiotic fluctuations and, even if they do, it is not at all clear from theory whether this would alter selection in favour of sex to selection against it.

I therefore suggest that the key to the difference in sex between mutualistic inhabitants and their exhabitants lies in their contrasting biotic environments. Mutualistic environments, with their tendency to improve, generate a selective force against sex, in contrast to the selective force for sex in deteriorating antagonistic environments.

Evolution of Host Specificity

Predictions

The conjecture that antagonistic environments tend to deteriorate and mutualistic ones to improve leads to a prediction that there should be less specialization between the partners of coevolving mutualistic symbioses than between those of antagonistic ones (Vanderplank 1978: 151; Harley and Smith 1983: 357 *et seq*.). The reason for this becomes clear if we return again to species A in its contrasting environments of antagonistic and mutualistic species (p. 146). (Notice that the arguments in this section are limited to a comparison between interactions where A is a pathogen or predator and interactions where A is a mutualist.)

When the environment of species A is composed of a set of host or prey species (B, C, D, etc.), there is selection for phenotypes in B, C, D, etc. that 'resist' formation of associations with phenotypes of A currently at high frequency. Thus, as long as the selected phenotypes of B, C, D are genetically different from their less successful counterparts, these species evolve resistance to the common phenotypes of A. *A priori*, there is no reason to expect that the same mechanism for resistance should evolve in B,C,D, etc; this will depend solely on the sets of phenotypes that happen to be available within their populations-sets, which are most unlikely to be the same in different species. Species A, faced with the evolution of different forms of resistance, is unlikely to contain a phenotype that can overcome them all. There are two possible consequences of this in species A. Either there is selection for a single phenotype which overcomes resistance in say, B rather than C or D, etc, or there is disruptive selection for more than one phenotype of A, each of which overcomes the resistance of a subset of B, C, D, etc. Both of these consequences have the effect that species A evolve strains with increased specificity towards a particular subset of B, C, D, etc. (as long as the selected phenotypes in A are genetically different from the others).

By contrast, when the environment of A is composed of mutualistic species, those phenotypes 'resisting' formation of associations with phenotypes of A at high frequency are clearly at a disadvantage. In all the mutualists there is selection for phenotypes which are compatible with the same set of phenotypes in A — that is, those currently at high frequency in A. In this way, the divisive evolutionary forces of antagonism are replaced by the cohesive forces of mutualism, and specialized associations should be less likely to evolve.

This prediction can be tested using the set of mutualistic symbioses in Table 6.1. Before doing so, it should be noted that we cannot simply count the number of exhabitant species in which a single inhabitant species lives, because the lower rate of evolution of inhabitants in itself leads one to expect a single inhabitant in different exhabitants. However, in spite of their low rate of evolution, it is still possible that, within an inhabitant species, strains specific to particular exhabitants could arise. Thus, the method I have adopted is to determine the number of exhabitant species with which a single inhabitant strain associates, information which is readily available for some of the symbioses.

Results

The number of exhabitant taxa that can be successfully inoculated with a single strain of inhabitants for various symbioses is listed in Table 6.4. The list is by no means exhaustive but it is representative of the results usually obtained and it illustrates the low level of specificity of the inhabitant strains. If there is any bias, it is towards an underestimate of the number of exhabitant species, since only a small proportion of potential exhabitants has been tested. The results stand in marked contrast to those of many antagonistic symbioses (see, for example, Day 1974: 96–7) for evidence of high levels of specificity between crop plant species and their parasites), and lend strong support to the prediction of a lower degree of spe-

Table 6.4:   Numbers of Exhabitants Successfully Reinfected by Single Strains of Inhabitants

Symbiosis	Exhabitant from which inhabitant strain isolated	Exhabitant taxa reinfected[1] number	proportion of those tested	Reference
Rhizobium/legume	*Cassia chamaecrista*	41 species	?	Wilson 1944[2]
	Dalea alopecuroides	40 species	?	''
	Desmodium canadense	40 species	?	''
	Glycine max	24 species	?	''
	Lespedeza striata	45 species	?	''
Frankia/plants	*Alnus*	4 genera	50%	Akkermans and Roelofsen 1980[3]
	Casuarina	1 genus	12.5%	''
	Coriaria	1 genus	20%	''
	Elaeagnus	4 genera	100%	''
	Hippophae	5 genus	71%	''
Ericoid mycorrhizas	*Calluna vulgaris*	6 species	100%	Pearson and Read 1973
	Vaccinium myrtillus	6 species	100%	''
	V. oxycoccus	6 species	100%	''
	V. macrocarpon	6 species	100%	''
	Erica cinerea	6 species	100%	''
	Rhododendron ponticum	6 species	100%	''
VA mycorrhizas	?[4]	19 species	?%	Mosse 1973

Notes:
1.   This includes the exhabitant from which the inhabitant strain was first isolated (with the exception of VA mycorrhizas — see 4).
2.   A sample of Wilson's (1944) results in his Table 1.
3.   A sample of the results of Akkermans and Roelofsen (1980); the full set is in their Table II. They do not give the number of species of exhabitant, but since most species within an exhabitant genus usually form the symbiosis, it is likely that the inhabitant strain will infect most of the species.
4.   Mosse (1973) refers to the strain as 'yellow spore vacuolate', but does not state the host plant species from which it was isolated.

cialization between partners of coevolving mutualistic symbioses than between those of antagonistic ones.

There is insufficient information on the other five kinds of symbiosis to include them in the table. However, dinoflagellate strains in association with marine invertebrates are also likely to exhibit low levels of specificity, in view of the fact that Schoenberg and Trench (1980b) were able to infect a single exhabitant species with nine out of twelve inhabitant strains taken from a range of sites and exhabitants. Bonnett and Silvester (1981) showed that cyanobacterial associations with *Gunnera manicata* could be formed by six out of eight strains of *Nostoc* tested. However, these strains came from exhabitants in different phyla and, in one case, from a free-living source in the soil, so the low level of specificity here is most unlikely to be the direct outcome of coevolution. Notice, though, that there is evidence of greater specificity in associations between *Chlorella* and certain invertebrates (McCauley and Smith 1982). In lichens, the difficulties of growing the fungal partner in isolation have prevented tests of inhabitant strain specificity.

A note of caution should be added to these results. The demonstration of low specificity in culture does not mean that a strain is equally successful with all exhabitants in nature. Population densities of inhabitants in alien hosts may be lower than in the hosts from which they were originally isolated (Schoenberg and Trench 1980b) and they may not always develop functional mutualistic partnerships (D.J. Read, pers. comm.). The results demonstrate only that exhabitants of mutualistic symbioses have a relatively low capacity to discriminate between inhabitant strains.

Discussion

As in previous sections, the results are open to alternative interpretations. The first is that, conceivably, the low levels of specificity could be a product of the inhabitants' asexual genetic systems. However, if there is any effect of asexual reproduction, it is arguable that it would lead to increased rather than reduced specificity in the inhabitants because, in the absence of sex, there is less tendency for blocks of genes associated with particular hosts to be broken down (Mitter and Futuyma 1979; Mitter *et al.* 1979). It is worth noting that ectomycorrhizal fungi (not considered in the survey in view of their limited penetration into root tissue, see p. 164), in which sexual reproduction is well developed, also show low levels of specificity (Harley and Smith 1983: 360 et seq.). Thus the lack of specificity cannot simply be the outcome of asexual reproduction; indeed, there must be other forces sufficiently strong to outweigh any propensity for specialization in asexual species.

Certain purely ecological considerations also need to be borne in mind when considering the evolution of host specificity. Although specialization carries with it the possible advantage of closer integration with and more efficient utilization of the host, there are also costs which, if sufficiently strong, could prevent host specificity from evolving. The first of these becomes important if an inhabitant's capacity for dispersal is limited. In a population containing genotypes with different degrees of host specificity, the chance of a specialized one making contact with a compatible partner through passive dispersal is less than that of one with lower host specificity. As a result, genotypes with greater host specificity could be selected against. Clearly, the capacity for dispersal is critical and, since all the associations in Table 6.4 occur below ground level from where dispersal is difficult, this could be a powerful force in the evolution of specificity (Harley and Smith 1983: 376). However, ectomycorrhizal fungi again provide an interesting comparison, because many have efficient above-ground mechanisms for dispersal of their spores; yet most still retain low levels of host specificity (Harley and Smith 1983: 360 *et seq.*), suggesting that there are other forces more important than dispersal. Lichens which produce propagules containing both partners (Bowler and Rundel 1975) would provide a critical test of the role of dispersal but unfortunately the problem of culturing the mycobiont in isolation at present precludes this.

Another ecological consideration is the diversity of potentially mutualistic species in the environment. As diversity increases, the chances of a specialist genotype contacting a compatible partner through passive dispersal decreases and an advantage for genotypes of lower host specificity is generated. Mycorrhizal fungi again provide a test of this theory, because vesicular-arbuscular (VA) mycorrhizas are found predominantly in plant communities of high diversity whereas ectomycorrhizas are more characteristic of low diversity plant communities (Malloch *et al.* 1980). Since the specificity of VA mycorrhizal fungi is not noticeably lower than that of the others (Harley and Smith 1983: 369), it seems unlikely that this force is critical for the evolution of specificity.

The evidence available at the moment therefore points towards the tendency of mutualistic environments to undergo evolutionary improvement as the most important force in preventing the development of host specificity. Such environments generate a cohesive force that discourages the evolution of specialized interactions with individual mutualistic species.

Exceptions?

Ectomycorrhizas

Ectomycorrhizas, although important mutualistic symbioses, have been excluded from the data set. The reason for this is that the fungal associates develop largely outside the plant root forming a pseudoparenchymatous sheath around it, with rather limited progression into the root epidermis and cortex and virtually no intracellular penetration (Marks and Kozlowski 1973). Our criterion of internalization is therefore not satisfied for the fungal components of ectomycorrhizas. (It should be borne in mind that ericoid and VA mycorrhizas also have extensive hyphal growth outside the plant root, but they lack the sheath around the root and show much greater penetration into the cells of the plant (Harley and Smith 1983).)

However, the comparison between ectomycorrhizas and the other mutualistic symbioses is instructive. In contrast with the low numbers of taxa of mutualistic inhabitants in Table 6.2, there are at least 100 genera and 3,500 species of ectomycorrhizal fungi, mostly from the Basidiomycota (Trappe 1962; Hacskaylo 1971; Singer 1975; R. Watling pers. comm.). Moreover, sex is as widespread in these mycorrhizal Basidiomycotes as it is in non-mycorrhizal ones (Singer 1975). These observations fit well with the theory that it is only the mutualistic environment within the exhabitant that generates pressures against genetic change. On the other hand, ectomycorrhizal fungi, like the other symbioses, usually have a low level of host specificity, the selective forces, in this case, not being dependent on internalization.

Agaonid Wasp Pollinators of Ficus

Associations between plants and their pollinators usually lack internalization and are much too transitory to warrant inclusion in the data set. However, the mutualism of Agaonid wasps and *Ficus*, in which the wasp life cycle occurs largely within the fig inflorescence (Faegri and van der Pijl 1979; 176 *et seq.*), is an interesting exception

(Boucher pers. comm.). In all respects the results from the mutualism run counter to our predictions; the wasps (inhabitants) are taxonomically diverse, sexual, and have a high degree of specificity to particular species of *Ficus* (Wiebes 1963). I would suggest that there are two contributory factors leading to this anomalous result. First, the degree of internalization of the wasp is lower and more transitory than that of the mutualistic symbioses in the data set. Second, and more important, the association is exceptional in its direct control over reproductive isolation in the plant (see discussion of pollination mutualisms in Thompson 1982: 121). Any lack of specificity in pollination by a species of fig results in loss of fitness to its normal host. There may also be a cost to the wasp if normal development of inflorescences is prevented through lack of fertilization. The parallel speciation which has evidently occurred in this association is discussed by Kiester *et al.* (1984).

Conclusion

The three major results of this study are that mutualistic inhabitants are taxonomically depauperate relative to their exhabitants, that they rarely, if ever, undergo sexual reproduction and that they lack strong specificity to particular host species. All these results are predicted from the conjecture that mutualistic environments have a tendency to undergo evolutionary improvement. There are clearly alternative interpretations for the results (some of which I have tried to evaluate in the chapter) but, at present, they appear to be less consistent with the observations than the one advocated. My conclusion is therefore that the conjecture of evolutionary improvement of mutualistic environments, although not itself tested here, is compatible with important features of mutualisms in nature. This may help to shed light in our studies on interactions between certain mutualistic species, just as the notion of environmental deterioration has illuminated studies of interactions between antagonistic species.

Acknowledgements

This chapter would not have been written without the help and stimulus of D.H. Lewis. I would like to think him together with A.H. Fitter and J.H. Lawton for critical comments on the manuscript and E. Gibson for preparing the typescript.

References

Ahmadjian, V. (1959) 'Experimental Observations on the Algal Genus *Trebouxia* de Puymaly', *Svensk Botanisk Tidskrift*, *53*, 71–80

Ahmadjian, V. (1967) *The Lichen Symbiosis*, Blaisdell, Waltham, MA.

Ahmadjian, V. (1970) 'The Lichen Symbiosis: Its Origin and Evolution', *Evolutionary Biology*, *4*, 163–84

Admadjian, V. and J.B. Jacobs (1981) 'Relationship Between Fungus and Alga in the Lichen *Cladonia cristatella* Tuck', *Nature*, *289*, 169–72

Akkermans, A.D.L. and W. Roelofsen (1980) 'Symbiotic Nitrogen Fixation by Actinomycetes in *Alnus*-type Root Nodules' in Stewart, W.D.P. and J.R. Gallon (eds.), *Nitrogen Fixation*, Academic Press, London, pp. 279–99

Allen, O.N. and E.K. Allen (1981) *The Leguminosae. A Source Book of Characteristics, Uses, and Nodulation*, University of Wisconsin Press, Madison

Archibald, P.A. (1975) '*Trebouxia* de Puymaly (Chlorophyceae, Chlorococcales) and *Pseudotrebouxia* gen. nov. (Chlorococcales, Chlorosarcinales)', *Phycologia*, *14*, 125–37

Bazin, M.J. (1968) 'Sexuality in a Blue-Green Alga: Genetic Recombination in *Ancystis nidulans*', *Nature*, *218*, 282–3

Beam, C.A. and M. Himes (1980) 'Sexuality and Meiosis in Dinoflagellates' in Levandowski, M. and S.H. Hunter (eds.), *Biochemistry and Physiology of Protozoa*, 2nd edn, vol. 3, Academic Press, New York, pp. 171–206

Bell, G. (1982) *The Masterpiece of Nature: the Evolution and Genetics of Sexuality*, Croom Helm, London; University of California Press, Berkeley & Los Angeles

Beringer, J.E., N.J. Brewin and A.W.B. Johnston (1980) 'The Genetic Analysis of *Rhizobium* in Relation to Symbiotic Nitrogen Fixation', *Heredity*, *45*, 161–86

Bond, G. (1963) 'The Root Nodules of Non-Leguminous Angiosperms', *Symposia of the Society for General Microbiology*, *13*, 72–91

Bonnett, H.T. and W.B. Silvester (1981) 'Specificity in the *Gunnera-Nostoc* Endosymbiosis', *New Phytologist*, *89*, 121–8

Bowler, P.A. and P.W. Rundel (1975) 'Reproductive Strategies in Lichens', *Botanical Journal of the Linnean Society*, *70*, 325–40

Burgeff, H. (1961) *Mikrobiologie des Hochmores*, Fisher Verlag, Stuttgart

Charlesworth, B. (1976) 'Recombination Modification in a Fluctuating Environment', *Genetics*, *83*, 181–95

Couture, M., J. Fortin and Y. Dalpe (1983) '*Oidiodendron griseum* Robak: an Endophyte of Ericoid mycorrhiza in *Vaccinium* spp.', *New Phytologist*, *95*, 375–80

Day, P.R. (1974) *Genetics of Host-Parasite Interaction*, Freeman, San Francisco

Delany, S.F., M. Herdman and N.G. Carr (1976) 'Genetics of Blue-Green Algae' in Lewin, R.A. (ed.), *The Genetics of Algae*, Botanical Monographs, vol. 12, Blackwell, Oxford, pp. 7–28

Dennis, R.W.G. (1968) *British Ascomycetes*, Cramer, Lehre

Droop, M.R. (1963) 'Algae and Invertebrates in Symbiosis', *Symposia of the Society for General Microbiology*, *13*, 171–99

Elkan, G.H. (1981) 'The Taxonomy of the Rhizobiaceae', *International Review of Cytology*, Supplement 13, 1–14

Faegri, K. and van der Pijl, L. (1979) *The Principles of Pollination Ecology*, 3rd edn, Pergamon Press, Oxford

Fisher, R.A. (1958) *The Genetical Theory of Natural Selection*, 2nd edn, Dover, New York

Freudenthal, H.D. (1962) '*Symbiodinium* gen. nov. and *Symbiodinium microadriaticum*, sp. nov., a Zooxanthella: Taxonomy, Life Cycle and Morphology', *Journal of Protozoology*, *9*, 45–52

Gerdemann, J.W. (1968) 'Vesicular-Arbuscular Mycorrhiza and Plant Growth', *Annual Review of Phytopathology*, 6, 397–418

Gerdemann, J.W. and J.M. Trappe (1975) 'Taxonomy of the Endogonaceae' in Sanders, F.E., B. Mosse and P.B. Tinker (eds), *Endomycorrhizas*, Academic Press, London, pp. 35–51

Ghiselin, M.T. (1974) *The Economy of Nature and the Evolution of Sex*, University of California Press, Berkeley

Glesener, R.R. and D. Tilman (1978) 'Sexuality and the Components of Environmental Uncertainity: Clues from Geographic Parthenogenesis in Terrestrial Animals', *American Naturalist*, 112, 659–73

Graham, P.H. (1964) 'The Application of Computer Techniques to the Taxonomy of the Root-Nodule Bacteria of Legumes', *Journal of General Microbiology*, 35, 511–17

Hacskaylo, E. (1971) 'The Role of Mycorrhizal Associations in the Evolution of the Higher Basidiomycetes' in Petersen, R.H. (ed.), *Evolution in the Higher Basidiomycetes*, University of Tennessee Press, Knoxville, pp. 217–37

Hale, M.E. (1974) *The Biology of Lichens*, 2nd edn, Arnold, London

Hall, I.R. and B.J. Fish (1979) 'A Key to the Endogonaceae', *Transactions of the British Mycological Society*, 73, 261–70

Hamilton, W.D. (1980) 'Sex Versus Non-Sex Versus Parasite', *Oikos*, 35, 282–90

Hamilton, W.D., P.A. Henderson and N.A. Moran (1981) 'Fluctuation of Environment and Coevolved Antagonist Polymorphism as Factors in the Maintenance of Sex' in Alexander, R.D. and D.W. Tinkle (eds), *Natural Selection and Social Behaviour: Recent Research and Theory*, Chiron Press, New York, pp. 363–81

Harley, J.L. and S.E. Smith (1983) *Mycorrhizal Symbiosis*, Academic Press, London

Hawksworth, D.L. and D.J. Hill (1984) *The Lichens*, Blackie (in press)

Hirsch, P.R. (1979) 'Plasmid-determined Bacteriocin Production by *Rhizobium leguminosarum*', *Journal of General Microbiology*, 113, 219–28

Hutson. V. and R. Law (1981) 'Evolution of Recombination in Populations Experiencing Frequency-Dependent Selection with Time Delay', *Proceedings of the Royal Society Series B*, 213, 345–59

Jaenike, J. (1978) 'An Hypothesis to Account for the Maintenance of Sex Within Populations', *Evolutionary Theory*, 3, 191–4

James, P.W. and A. Henssen (1976) 'The Morphological and Taxonomic Significance of Cephalodia' in Brown, D.H., D.L. Hawksworth and R.H. Bailey (eds), *Lichenology: Progress and Problems*, Academic Press, London, pp. 27–77

Karakashian, S.J. and M.W. Karakashian (1965) 'Evolution and Symbiosis in the Genus *Chlorella* and Related Algae', *Evolution*, 19, 368–77

Kiester, A.R., R. Lande and D.W. Schemske (1984) 'Models of Coevolution and Speciation in Plants and Their Pollinators', *American Naturalist*, 124, 220–43

Koudorosi, A. and A.W.B. Johnston (1981) 'The Genetics of Rhizobium', *International Review of Cytology, Supplement 13*, 191–224

Lande, R. (1980) 'Microevolution in Relation to Macroevolution', *Paleobiology*, 6, 235–8

Largent, D.L., N. Sugihara and C. Wishner (1980) 'Occurence of Mycorrhizae on Ericaceous and Pyrolaceous Plants in Northern California', *Canadian Journal of Botany*, 58, 2274–9

Law, R. and D.H. Lewis (1983) 'Biotic Environments and the Maintenance of Sex — Some Evidence from Mutualistic Symbioses', *Biological Journal of the Linnean Society*, 20, 249–76

Lechevalier, M.P. and H.A. Lechevalier (1979) 'The Taxonomic Position of the Actinomycetic Endophytes' in Gordon, J.C., C.T. Wheeler and D.A. Perry (eds), *Symbiotic Nitrogen Fixation in the Management of Temperate Forests*, Forestry Research Laboratory, Oregon State University, Corvallis, pp. 111–22

Levin, D.A. (1975) 'Pest Pressure and Recombination Systems in Plants', *American Naturalist, 109*, 437–51

Lewin, R.A. (1976) 'Introduction' in Lewin, R.A. (ed.), *The Genetics of Algae*, Blackwell, Oxford, pp. 1–6

Lewin, R.A. (1982) 'Symbiosis and Parasitism — Definitions and Evaluations', *Bioscience,32*, 254–9

Lewis, D.H. (1985) 'Symbiosis and Mutualism: Crisp Concepts and Soggy Semantics', this volume, Chapter 2

Lewontin, R.C. (1983) 'Gene, Organism and Environment' in Bendall, D.S. (ed.), *Evolution from Molecules to Men*, Cambridge University Press, pp. 273–85

McCauley, P.J. and D.C. Smith (1982) 'The Green Hydra Symbiosis. V. Stages in the Intracellular Recognition of Algal Symbionts by Digestive Cells', *Proceedings of the Royal Society, London, Series B, 216*, 7–23

McCoy, G.A. (1978) *'Nutritional, Morphological, and Physiological Characteristics of* Trentepohlia, *(I.U. 1227) in Axenic Culture on Defined Media*, PhD thesis, Oregon State University

Malajczuk, N. and B.B. Lamont (1981) 'Specialized Roots of Symbiotic Origin in Heathlands' in Specht, R.L. (ed.), *Ecosystems of the World, Volume 9B. Heathlands and Related Shrublands. Analytical Studies*, Elsevier, Amsterdam, pp. 165–82

Malloch, D.W., K.A. Pirozynski and P.H. Raven (1980) 'Ecological and Evolutionary Significance of Mycorrhizal Symbioses in Vascular Plants (a Review)', *Proceedings of the National Academy of Sciences, USA, 77*, 2113–8

Margulis, L. (1981) *Symbiosis in Cell Evolution*, Freeman, San Francisco

Marks, G.C. and T.T. Kozlowski (eds.) (1973) *Ectomycorrhizae: Their Ecology and Physiology*, Academic Press, New York

Maynard Smith, J. (1971) 'What Use is Sex?', *Journal of Theoretical Biology, 30*, 319–35

Maynard Smith, J. (1976) 'A Comment on the Red Queen', *American Naturalist, 110*, 325–30

Maynard Smith, J. (1978) *The Evolution of Sex*, Cambridge University Press, Cambridge

Maynard Smith, J. (1980) 'Selection for Recombination in a Polygenic Model', *Genetical Research (Cambridge), 35*, 269–77

Mitter, C. and D.J. Futuyma (1979) 'Population Genetic Consequences of Feeding Habits in Some Forest Lepidoptera', *Genetics, 92*, 1005–21

Mitter, C., D.J. Futuyma, J.C. Schneider and J.D. Hare (1979) 'Genetic Variation and Host Plant Relations in a Parthenogenetic Moth', *Evolution, 33*, 777–90

Moore, A.W. (1969) *'Azolla*: Biology and Agronomic Significance', *Botanical Review , 35*, 17–34

Mosse, B. (1973) 'Advances in the Study of Vesicular-Arbuscular Mycorrhiza', *Annual Review of Phytopathology, 11*, 171–96

Nicolson, T.H. (1975) 'Evolution of Vesicular-Arbuscular Mycorrhizas', in Sanders, F.E., B. Mosse and P.B. Tinker (eds.), *Endomycorrhizas*, Academic Press, London, pp. 25–34

Norris, D.O. (1956) 'Legumes and the *Rhizobium* Symbiosis', *Empire Journal of Experimental Agriculture, 24*, 247–70

Norris, D.O. (1958) 'Lime in Relation to the Nodulation of Tropical Legumes' in Hallsworth, E.D. (ed.)', *Nutrition of the Legumes*, Academic Press, New York, pp. 164–82

Pearson, V. and D.J. Read (1973) 'The Biology of Mycorrhiza in the Ericaceae. I. The Isolation of the Endophyte and Synthesis of Mycorrhizas in Aseptic Culture', *New Phytologist, 72*, 371–9

Poelt, J. (1973) 'Classification' in Ahmadjian, V. and M.E. Hale (eds), *The Lichens*, Academic Press, New York, pp. 599–632

Price, P.W. (1980) *Evolutionary Biology of Parasites*, Princeton University Press, Princeton

Raven, P.H. and D.I. Axelrod (1974) 'Angiosperm Biogeography and Past Continental Movements', *Annals of the Missouri Botanical Garden*, 61, 539–673

Read, D.J. (1974) '*Pezizella ericae* sp. nov., the Perfect State of a Typical Endophyte of Ericaceae', *Transactions of the British Mycological Society*, 63, 381–419

Read, D.J. (1983) 'The Biology of Mycorrhiza in the Ericales', *Canadian Journal of Botany*, 61, 985–1004

Rice, W.R. (1983) 'Parent-Offspring Pathogen Transmission: a Selective Agent Promoting Sexual Reproduction', *American Naturalist*, 121, 187–203

Richardson, D.H.S. and B.H. Green (1965) 'A Subfossil Lichen', *Lichenologist*, 3, 89–90

Rosen, B.R. (1977) 'The Depth Distribution of Recent Hermatypic Corals and its Palaeontological Significance', *Mémoires du Bureau de Recherches Géologiques et Minières*, 89, 507–17

Roughgarden, J. (1975) 'Evolution of Marine Symbiosis — a Simple Cost-Benefit Model', *Ecology*, 56, 1201–8

Roughgarden, J. (1979) *Theory of Population Genetics and Evolutionary Ecology: an Introduction*, MacMillan, New York

Schoenberg, D.A. and R.K. Trench (1980a) 'Genetic Variation in *Symbiodinium* (= *Gymnodinium*) *microadriaticum* Freudenthal, and Specificity in its Symbiosis with Marine Invertebrates. I. Isozyme and Soluble Patterns of Axenic Cultures of *Symbiodinium microadriaticum*', *Proceedings of the Royal Society, London, Series B*, 207, 405–27

Schoenberg, D.A. and R.K: Trench (1980b) 'Genetic Variation in *Symbiodinium* (= *Gymnodinium*) *microadriaticum* Freudenthal, and Specificity in its Symbiosis with Marine Invertebrates. III. Specificity and Infectivity of *Symbiodinium microadriaticum*', *Proceedings of the Royal Society, London, Series B*, 207, 445–60

Silvester, W.B. and P.J. McNamara (1976) 'The Infection Process and Ultrastructure of the *Gunnera-Nostoc* Symbiosis', *New Phytologist*, 77, 135–41

Singer, R. (1975) *The Agaricales in Modern Taxonomy*, Cramer, Vaduz

Slobodkin, L.B. (1968) 'Toward a Predictive Theory of Evolution' in Lewontin, R.C. (ed.), *Population Biology and Evolution*, Syracuse University Press, New York, pp. 187–205

Smith, A.L. (1921) *Lichens*, Cambridge University Press, London

Soberon, J.M. and C. Martinez del Rio (1985) 'Cheating and Taking Advantage in Mutualistic Associations', this volume, Chapter 8

Solbrig, O.T. (1976) 'On the Relative Advantages of Cross- and Self-Fertilization', *Annals of the Missouri Botanical Garden*, 63, 262–76

Sprent, J.I. (1979) *The Biology of Nitrogen-Fixing Organisms*, McGraw Hill, London

Stanley, S.M. (1975) 'Clades Versus Clones in Evolution: Why we Have Sex', *Science*, 190, 382–3

Starr, M.P. (1975) 'A Generalized Scheme for Classifying Organismic Associations', *Symposia of the Society for Experimental Biology*, 29, 1–20

Stebbins, G.L. (1974) *Flowering Plants: Evolution Above the Species Level*, Arnold, London

Stewart, W.D.P. and H.N. Singh (1975) 'Transfer of Nitrogen-Fixing (*nif*) Genes in the Blue-Green Alga *Nostoc muscorum*', *Biochemical and Biophysical Research Communications*, 62, 62–9

Stewart, W.D.P., P. Rowell and A.N. Rai (1980) 'Symbiotic Nitrogen-Fixing Cyanobacteria' in Stewart, W.D.P. and J.R. Gallon (eds), *Nitrogen Fixation*, Academic Press, London, pp. 239–77

Taylor, D.L. (1973a) 'Algal Symbionts of Invertebrates', *Annual Review of Microbiology*, 27, 171–87

Taylor, D.L. (1973b) 'The Cellular Interactions of Algal-Invertebrate Symbiosis', *Advances in Marine Biology, 11*, 1–56

Taylor, D.L. (1974) 'Symbiotic Marine Algae: Taxonomy and Biological Fitness' in Vernberg, W.B. (ed.), *Symbiosis in the Sea*, University of South Carolina, Columbia, pp. 245–62

Templeton, A.R. and L.E. Gilbert (1985) 'Population Genetics and Coevolution of Mutualism', this volume, Chapter 5

Thompson, J.N. (1982) *Interaction and Coevolution*, Wiley, New York

Torrey, J.G. (1978) 'Nitrogen Fixation by Actinomycete-Nodulated Angiosperms', *BioScience, 28*, 586–92

Trappe, J. (1962) 'Fungus Association of Ectotrophic Mycorrhizae', *Botanical Review, 28*, 538–606

Trappe, J. (1982) 'Synoptic Keys to the Genera and Species of Zygomycetous Mycorrhizal Fungi', *Phytopathology, 72*, 1102–8

Trench, R.K. (1981) 'Cellular and Molecular Interactions in Symbioses Between Dinoflagellates and Marine Invertebrates', *Pure and Applied Chemistry, 53*, 819–35

Tschermak-Woess, E. (1970) 'Über wenig bekannte und neue Flechtengonidien V. Der Phycobiont von *Verrucaria aquatilis*, and die Fortpflanzung von *Pseudopleurococcus arthopyreniae*', *Österreichische Botanische Zeitschrift, 118*, 443–55

Tschermak-Woess, E. (1978) 'The Phycobionts in the Section *Cystophora* of *Chaenotheca*, especially *Dictyochloropsis splendida* and *Trebouxia simplex*, spec. nova.', *Plant Systematics and Evolution, 129*, 185–208

Tschermak-Woess, E. (1980a) *Asterochloris phycobiontica*, nov. gen., nova spec., the Phycobiont of the Lichen *Varicellaria carneonivae* (Anzi) Erichs', *Plant Systematics and Evolution, 135*, 279–94

Tschermak-Woess (1980b) '*Elliptochloris bilobata*, gen. et spec. nov., the Phycobiont of *Catolechia wahlenbergii*', *Plant Systematics and Evolution, 136*, 63–72

Vanderplank, J.E. (1978) *Genetic and Molecular Basis of Plant Pathogenesis*, Springer-Verlag, Berlin

Van Valen, L. (1973) 'A New Evolutionary Law', *Evolutionary Theory, 1*, 1–30

Waksman, S.A. (1967) *The Actinomycetes. A Summary of Current Knowledge*, Ronald Press, New York

Wetmore, C.M. (1970) 'The Lichen Family Heppiaceae in North America', *Annals of the Missouri Botanical Garden, 57*, 158–209

Whitton, B.A. (1973) 'Interactions with Other Organisms' in Carr, N.G. and B.A. Whitton (eds), *The Biology of the Blue-Green Algae*, Blackwell, Oxford, pp. 415–33

Wiebes, J.T. (1963) 'Taxonomy and Host Preferences of Indo-Australian Fig Wasps of the Genus *Ceratosolen*, (Agaonidae). *Tijdschrift voor Entomologie, 106*, 1–112

Williams, G.C. (1975) *Sex and Evolution*, Princeton University Press, Princeton

Wilson, J.K. (1944) 'Over Five Hundred Reasons for Abandoning the Cross-Inoculation Groups of the Legumes', *Soil Science, 58*, 61–9

7 EQUILIBRIUM POPULATIONS AND LONG-TERM STABILITY OF MUTUALISTIC ALGAE AND INVERTEBRATE HOSTS

Clayton B. Cook

Introduction

Associations between invertebrate cells and unicellular algae are classical examples of mutualism, but they have received little attention from ecologists or theoreticians interested in coevolutionary processes. They are products of long periods of evolutionary history; reef-building corals have probably possessed endosymbiotic algae since the Triassic (Wells 1956; Goreau 1963; see Law, this volume). They are ecologically important: the dinoflagellates (zooxanthellae) symbiotic with corals and other reef invertebrates are among the major producers on coral reefs (Muscatine 1980a), and play important roles in nutrient cycling and conservation in oligotrophic waters (Muscatine and Porter 1977).

The intracellular habitat of the algae doubtless provides a refuge from predation, and the aerobic metabolism of the host cell reduces the problem of photorespiration (Phipps and Pardy 1982). As the algae are isolated from the external environment by at least two layers of host cell membranes, the host cell is a source of potential nutrients, and it is significant that endosymbiotic algae are often associated with cells which have a nutritive function. Yonge (1957: 437) referred to algal symbionts as 'imprisoned phytoplankton'; his analogy might be more aptly rephrased as 'refugee phytoplankton'. In turn the symbionts are energy sources for the host, providing lipids and other small molecular weight photosynthetic products (Smith *et al.* 1969; Patton and Burris 1983) as well as products of nitrogen and sulphur metabolism (Burris 1983; Cook 1983).

There has been much experimental work on cellular, physiological and metabolic aspects of these associations (reviewed by Smith *et al.* 1969; Taylor 1974, Trench 1979, 1981; Cook 1983) which is not part of the general ecological literature. Price's (1980) review of the evolutionary ecology of parasites provides a useful framework for considering this work in the light of the evolutionary ecology of mutualism. His conclusions, based on several basic ecological and evolutionary concepts, stress the importance of short-term, fragmented populations in the evolution of parasites. In this chapter I apply his concepts to the available information on algae-invertebrate mutualisms and find that long-term ecological stability of populations has produced coevolved specific associations.

The Ecological Concepts

Endosymbiotic Algae Exist in Continuous Environments

Parasites exist in small, discontinuous environments. Given the balance which must be struck between exploiting and over-exploiting a resource, they are rarely associated with a host for the duration of the host's life cycle, and hardly ever through more than one generation without reinfecting a second host. Hence, parasites have evolved a number of life history characteristics which allow them to deal with discontinuous environments (dispersal mechanisms, long-lived resting stages, alternate host strategies).

In contrast, symbiotic algae are usually associated with a single host for most of the host's life, if not through generations. Those symbiotic algae which are similar to *Chlorella* lack a motile dispersive stage. Transmission of symbionts via asexual reproduction assures an environmental continuum for the algae (Smith 1980): most protozoan hosts of these algae lack sexual reproduction, and green hydra primarily reproduce asexually. Sexual reproduction occurs sporadically in green hydra: in most strains, the algae are acquired by the eggs during gamete maturation (Muscatine and McAuley 1982). The nutritional selection against symbiont-free green hydra (Muscatine and Lenhoff 1965; Kelty and Cook 1976) makes it unlikely that these hydra occur naturally. It is not known if symbiotic chlorellae exist naturally outside of a host.

Most non-*Chlorella*-like symbiotic algae possess flagella (hence motile dispersive stages) at some point in their life histories. In most, the flagella are absent during the symbiotic phase (Parke and Manton 1967; McLaughlin and Zahl 1966). The most familiar and widespread of these are the dinoflagellate zooxanthellae (*Symbiodinium = (Gymnodinium) microadriaticum*) which occur ubiquitously with invertebrate hosts in warm, shallow-water marine habitats.

Many hosts (including most scleractinian corals and sea

anemones) contain zooxanthellae continuously throughout the life history, from egg to egg. Individual hosts may be extremely long-lived: coral colonies can be hundreds of years old, and asexually reproducing cnidarians such as sea anemones (*Aiptasia* sp., the clone-forming *Anthopleura elegantissima*) and fragmenting corals (Porter *et al.* 1981; Neigel and Avise 1983) provide a long-term environment for symbiont populations. These species may periodically release motile symbionts (Steele 1975, 1976), although the fate of these algae is unclear.

Other hosts for zooxanthellae produce symbiont-free gametes, and acquire symbionts shortly after larval metamorphosis. Two notable example are gorgonian corals (Kinzie 1974) and tridacnid bivalves (LaBarbera 1975; Fitt and Trench 1982). Tridacnids obtain zooxanthellae from the plankton while filter feeding. Motile zooxanthellae stimulate ingestive behaviour in gorgonians although gorgonians are not normally herbivorous (Kinzie 1974). Tridacnids and gorgonians retain symbionts throughout the rest of their lives, as is normally the case for metazoan hosts. Periodic, catastrophic losses of zooxanthellae occur from corals when heavy freshwater runoff lowers the salinity of inshore waters (Goreau 1964; Egana and DiSalvo 1982). Presumably the symbionts are replaced either by proliferation of residual algae or by dispersive forms.

In culture, motile zooxanthellae are produced by isolates from virtually every host. In the lab this process is circadian, with motile forms generally during the light phases of photoperiods (Fitt *et al.* 1981, Lerch and Cook 1984). They are probably released naturally by most, if not all, hosts (Trench 1983), but how long they persist is not known.

The prasinophyte symbionts associated with acoelous flatworms in the genus *Convoluta* also have a flagellated, free-living stage. Worm gametes and embryos in egg cases do not contain symbionts, but symbionts are sometimes found associated with the surface of the egg cases (Holligan and Gooday 1975). Reports that the algae were attracted to egg cases have not been confirmed in recent experimental studies (Douglas and Gooday 1982). The flagellated algae are ingested by the young worms at some point after hatching, and the worms cease to feed after this time (Keeble 1910).

While the vast majority of algae-invertebrate mutualists form long-lasting associations, more transitory associations can occur. Some nudibranchs obtain zooxanthellae from cnidarian prey (Rudman 1981; Kempf 1984). The symbionts proliferate in slug cells and probably contribute to the energy of the host (Hoegh-Guldberg 1983; Kempf 1984). The extent to which this association is obligatory or specific, and the nutritional benefits of the association to the nudibranchs, are not known.

Environmental Grain of Endosymbiotic Algae and Hosts

Price (1980) concluded that parasites inhabit 'coarse-grained' environments, as they are small relative to hosts, have generally limited powers of dispersal, and must find hosts which are relatively uncommon (q.v. MacArthur and Wilson 1967). Species-specific host preferences which evolve as a consequence of variability of host species in both space and time restrict parasites to specific prey resources; thus among predators parasites represent the 'extreme in specialized resource exploitation' (Price 1980: 19)

The relevance of this to algae-invertebrate mutualisms depends on the perspective of the mutualist. For the invertebrate 'host', potential symbionts in the environment are a fine-grained resource: suspension feeders will encounter a diversity of algae, of which few are likely to be suitable. This has been shown for young *Convoluta* (Douglas and Gooday 1982), and filter feeding larval tridacnids (Fitt and Trench 1982). Free-living symbionts probably have a coarser-grained environment, which includes less suitable hosts and potential predators. Indigestibility is one way to circumvent herbivory; symbiotic *Chlorella* will pass through the guts of predators intact and photosynthetically competent (data of Hohman cited in Muscatine *et al.* 1975). Both symbiotic *Chlorella* (Hohman *et al.* 1982) and zooxanthellae (Fitt and Trench 1983a) interfere with host cell digestive functions. The situation is analogous to gelatinous green algae which utilize herbivore guts as a nutrient resource (Porter 1976b). Perhaps algae which are resident symbionts in animal digestive cells represent the extreme in this sort of nutrient exploitation.

Endosymbiotic Algae are Equilibrium Populations

If one considers hosts as island patches and parasites as colonizers, parasite populations in general are non-equilibrium systems and host-parasite systems are non-equilibrium communities (Price 1980, *sensu* MacArthur and Wilson 1967). Individual populations have a low probability of being successfully established and a high probability of extinction, given the patchiness of host availability, development of host defences, and other factors. With high extinction rates, changes in species composition within hosts would be expected.

The algae-invertebrate mutualisms which have been examined appear to be in equilibrium at both population and community levels. As sexual reproduction is rare among symbiotic algae (p. 172; see also Law, this volume), the dispersive forms released from a given host species should be genetically very similar (cf. Schoenberg and Trench 1980a). Thus colonization rates by a particular symbiont genotype should be relatively high, as compared with parasites. The success of colonization success is demonstrated by those species which must obtain symbionts at some time during the life cycle: these virtually always contain the 'right' symbiont in natural populations (cf. Schoenberg 1980a). This situation would also occur if the association were obligate for the host, and uninfected hosts did not survive.

Once the association between host cells and appropriate symbionts has been established, the symbiont population becomes stable — both on a host cell basis (probably the most relevant colonizable unit) and considering the total population within the host. The digestive cells of a given strain of green hydra have constant algal populations under defined conditions (Figures 7.1a, 7.2; Pardy 1974, McAuley 1981c; Neckelmann and Muscatine 1983). Their level is a balance between cell division rates and the expulsion of excess or moribund algae, and may be regulated by nutrient availability within the host cell (McAuley 1981a, b, c; Muscatine and Pool 1979; Neckelmann and Muscatine 1983). If the intracellular population is experimentally raised, it is subsequently restored, primarily by the action of the host cell in attacking or expelling excess algae (McAuley and Smith 1982b; Neckelmann and Muscatine 1983; Figure 7.2). Hydras of defined developmental and nutritional states have constant cell numbers (Otto and Campbell 1977) — the net result of cell division and losses via bud production and tissue sloughing. The total symbiont population of developmentally similar green hydra is thus constant (Pardy 1974; McAuley 1981a).

There are fewer data for zooxanthellae, but measurements of symbiont numbers per host biomass in sea anemones (Taylor 1969; Steele 1976) indicate that the populations are likely to be constant under given conditions. The intracellular dynamics of zooxanthellae populations have not been investigated.

Experimental studies of the 'infection' of symbiont-free hosts provide insight into colonization of host cells and the maintenance

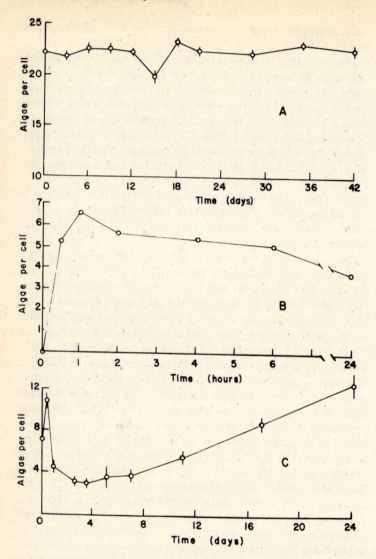

Figure 7.1: Populations of Symbiotic Algae in Digestive Cells of the European Strain of Green Hydra (Hydra viridis). 7.1A Algae per digestive cell in symbiotic green hydra maintained under standard conditions. The populations are constant over time; mean of 5 hydra, ± one standard error (Data of McAuley 1981c,

with permission of P.J. McAuley and Alan R. Liss). 7.1B and 7.1C Algae per cell in aposymbiotic (algae-free) *H. viridis* at intervals after the introduction of hydra symbionts into the digestive cavity, illustrating the colonization of host cells by symbionts; 7.1B Population levels in digestive cells during the first 24 hours; 7.1C population levels over 24 days. Note the initial 'sorting' phenomenon (Days 1–4) and the subsequent repopulation of host cells. (Data of Jolley and Smith 1980, with permission of D.C. Smith and the Royal Society.)

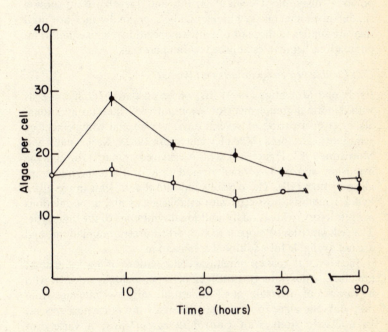

Figure 7.2: 'Invasion' of Symbiotic Green Hydra (Florida Strain) by Hydra Symbionts ● --- ●: symbiont populations in digestive cells of green hydra after the introduction of supernumerary algae; o --- o: control hydra without extra algae. Note the restoration of the normal symbiont population. Mean of 5 hydra, ± one standard error. (Data of Neckelmann and Muscatine 1983, with permission of N. Neckelmann and the Royal Society).

of symbiont populations (Smith 1980, 1981; Cook 1980; Muscatine 1982). The host cells are generally non-specific phagocytes (McNeil 1981; McNeil *et al*. 1982; Fitt and Trench 1983a) which avidly engulf potential symbionts. The digestive cells take up fewer symbionts than the normal complement, which is restored as the symbiont population divides more rapidly than host cells (green hydra: Pardy and Muscatine 1973; Pardy 1974; Muscatine *et al*. 1975; Jolley and Smith 1980; Figures 7.1B, 7.1C). Populations of zooxanthellae introduced into symbiont-free sea anemones (Trench 1971b; Kinzie and Chee 1979; Schoenberg and Trench 1980c) and scyphozoan polyps (Colley and Trench 1983; Fitt and Trench 1983a) show a similar phenomenon: symbiont numbers gradually increase until they are similar to those of of native associations, with equilibrium populations being re-established within host cells.

Can Established Associations be Invaded?

Pardy and Muscatine (1973) first observed that green hydra cells with a normal complement of symbionts will take up additional algae. Over a period of several days, the normal complement of symbionts is restored (McAuley and Smith 1982a; Neckelmann and Muscatine 1983; Figure 7.2). This is achieved primarily through the expulsion and/or lysosomal degradation of the supernumerary algae by host cells. These results indicate that at least in the green hydra symbiosis, host cells with established symbiont populations are refractory to invasion by additional symbionts of the same type. The cellular 'islands' appear to support saturated populations, and are not available for additional colonization.

Such is not the case if unnatural combinations of hosts and algae are made. Aposymbiotic hosts can be experimentally infected with a variety of heterologous (unnatural) algae — ranging from nonsymbiotic algae to symbionts isolated from other host species. In a classic study with *Convoluta roscoffensis*, a variety of heterologous combinations was produced. Secondary introductions of *Platymonas convolutae* (the natural symbiont) resulted in the displacement of other algae in worm tissues so that only the native symbiont remained (Provasoli *et al*. 1968).

I know of no similar experiments with zooxanthellae such as *S. microadriaticum*. Schoenberg (1980a) makes the point that while a variety of strains of zooxanthellae can establish experimental associations with a given host species, natural populations of hosts only contain the 'right' symbiont (zooxanthellae show few phenotypic changes in response to differing host environment, Schoenberg and Trench 1980c). He suggests that heterologous associations are formed in nature, at least when algae-free hosts first acquire algae. Once subsequent invasion by the proper symbiont occurs, these should outcompete other algae. Heterologous combinations of zooxanthellae and hosts show reduced fitness relative to the native symbiosis (Kinzie and Chee 1979; Schoenberg and Trench 1980c; Fitt and Trench 1982; Colley and Trench 1983). There appears to be only one report of natural hosts simultaneously harbouring two different symbiont populations: the sea anemones *Anthopleura xanthogrammica* and *A. elegantissima* (Muscatine 1971). The population dynamics of these algae would be most interesting to study.

Taken together, the evidence on symbiont standing crop, the regulation of algal numbers, and establishment of associations and the specificity of natural populations indicates that populations and communities of endosymbiotic algae are at equilibrium.

The Evolutionary Concepts

Are Evolutionary and Speciation Rates Lower Among Endosymbiotic Algae Than in Parasites?

The ecological characteristics of parasite populations promote isolated gene pools and the evolution of sibling species (Price 1980). As populations of symbiotic algae tend to be stable in size through time, one would expect lower speciation rates among endosymbiotic algae. Evolutionary conservatism in symbiotic algae should also result from the virtual absence of sexual reproduction. The *Chlorella*-like algae belong to the Chlorococcales, which are characterized by the absence of sexual forms. Various life history schemes for zooxanthellae have included putative sexual forms (McLaughlin and Zahl 1966; Taylor 1973; Steele 1976), but meiosis has not been documented in *S. microadriaticum*, and the ploidy of motile or non-motile forms is unknown (cf. Schoenberg and Trench 1980a).

On the other hand, selective forces of the host cell environment and coevolutionary considerations would favour genetically distinct populations. Relatively few sorts of algae have become mutualistic with invertebrates, but these few have been singularly successful in establishing symbioses with a wide variety of hosts. Speciation among these forms is a problematic taxonomic tangle: depending on how 'species' are defined, these algae have either evolved into many species, or into many genetically distinct populations of a few species. These taxonomic questions are presently unresolved.

The sorting out of these relationships has been the result of culture studies, so that phenotypic responses under controlled conditions and in differing host environments could be determined. Lists of symbiotic *Chlorella* strains in culture are found in Muscatine *et al.* (1967), Karakashian and Karakashian (1965), Pardy and Muscatine (1973) and Jolley and Smith (1980); isolates of zooxanthellae are listed in Schoenberg (1976, 1980b), Schoenberg and Trench (1980a), and Colley and Trench (1983).

The *Chlorella*-like algae have not been formally assigned to a genus. They are by convention referred to the genus *Chlorella*. Many of these strains have consistent morphological and physiological differences, but others show phenotypic plasticity which may be expressed in response to different hosts. These plastic strains tend to acquire similar characteristics in host cells of the same species (Pardy 1976; Pool 1979), so the degree of genetic difference between them is not clear. Electrophoretic studies of both cultured and native symbionts might help to clarify the situation.

Endozoic dinoflagellates have been considered to comprise three species: two amphidinoid species are associated with pelagic hosts (Taylor 1971a, b) while the gymnodinoid zooxanthellae have been assigned to a single species of pantropical distribution, *Symbiodinium* (*Gymnodinium*) *microadriaticum* (Freudenthal 1962; Taylor 1974; but see Loeblich and Sherley 1979). However, an intensive series of studies by R.K. Trench and his students has made it very clear that populations of *S. microadriaticum* from a variety of hosts have consistent differences both in the host and in culture. These differences include morphology (size, cell covering, organelle morphology; Schoenberg and Trench 1980b), the kinds of photosynthetic products which are released (Trench 1971a), isoenzyme patterns (Schoenberg 1976, 1980: Schoenberg and Trench 1980a), pigment-protein complexes (Chang and Trench 1982), sterol synthesis (Kokke *et al.* 1981; Withers *et al.* 1982), growth rates (Fitt and Trench 1983b), motility patterns (Fitt *et al.* 1981) and infectivity in other hosts (Schoenberg and Trench 1980c; Colley and Trench 1983). Some isolates from the same host species are similar in allopatric populations, while samples from other host species may show geographic variation (cf. Chang and Trench 1982).

Adaptive Radiation Among Endosymbiotic Algae

Adaptive radiation is widespread among parasites, due to (Price 1980): (1) host diversity, both taxonomically and spatially, (2) the size of the host target, both in determining founder population size and the the number of competing populations on a 'host island', (3) the amount of evolutionary time available, (4) the extent of coevolutionary modification. It is clear that genetically distinct populations of endosymbiotic algae have evolved; what is not clear is how the differences which have been found are adaptive.

How is Symbiont Diversity Related to Host Diversity?

We have seen that there is specificity in the establishment and maintenance of algae-invertebrate symbioses. The reasons for this are not clear, as different mechanisms operate in different systems, and may even operate in different strains of the same species (green hydra: McNeil and Smith 1982; McAuley and Smith 1982a). Specificity during establishment may result from differences in symbiont behaviour (that is, motility rhythms: Fitt *et al.* 1981) or ingestive behaviour of hosts (Kinzie 1974). The uptake of potential symbionts by phagocytic host cells may depend on immunological or lectin-binding receptor sites (Pool 1979; Meints and Pardy 1980), on surface charge distribution (McNeil *et al.* 1981), or on other factors (McNeil and Smith 1982; Colley and Trench 1983).

After establishment, the fate of symbiont populations may be affected by a variety of interactions (Smith 1980, 1981), but how host cell diversity is involved has not been intensively studied. In some *Chlorella* symbioses an initial 'sorting' phenomenon occurs in which some symbionts disappear soon after they are taken up by host cells (Karakashian and Karakashian 1965; McAuley and Smith 1982a; Figure 7.1B). It is not clear whether 'sorting' (which evidently involves both host lysomal and exocytotic activity) discriminates between genetically different algae, between genetically similar algae which may be developmentally different, or if it is selective at all (cf. McAuley and Smith 1982a). The phenomenon has not been reported in systems involving other kinds of algal symbionts.

Given the equilibrium nature of symbiont populations, genetic diversity is most likely to be introduced during the infective process (Schoenberg and Trench 1980c; Schoenberg 1980a), or by the accumulation of genetic differences and differential growth in long-term populations. It seems intuitive that host cell diversity would influence the sorting out and competition of genetic variants, but this has not been demonstrated. Long term selection between host and symbionts would involve subtle interactions such as the interplay of nutrient flow between partners (Smith 1981), although there are no clear indications as to how this would influence the diversity of associations. Balances must be struck between the flow of photosynthetic carbon from the algae (p. 183) and the use of host cell nutrients by symbionts (for example, Cook 1971, Carroll and Blanquet 1984a, b).

One result of such diversity would be a spectrum of host-zooxanthellae associations based on the degree of photo-autotrophy and heterotrophy. Porter (1976a) has hypothesized that such a range exist among corals, with more heterotrophic species having adaptations favouring food capture. A similar situation may exist in terms of nitrogen utilization. Nitrate uptake occurs in some corals and zoanthids (for example, D'Elia *et al.* 1983; D'Elia and Cook, unpublished), but not in some sea anemones (for example, Wilkerson and Muscatine 1984). The use of dissolved nitrate as a nitrogen source by a symbiotic unit should imply nitrogen 'autotrophy', in contrast to sea anemones which typically have large heterotrophic components in their diets. Some adaptations which appear to have evolved in host cells as part of this nutritional spectrum are protein which may control the flux of nutrients to algal symbionts (for example, alanine in the jellyfish *Cassiopea xamachana*: Carroll and Blanquet (1984b) and 'host-factors' which may control the amount of photosynthate released by the algae (for example, Trench 1971b). The distribution of such control mechanisms in the context of a nutritional spectrum is a promising avenue for further work.

Effects of Size and Amount of Evolutionary Time Available

Hosts with increased host body size support a larger number of coexisting parasite species, as would be predicted by the MacArthur-Wilson model (Price 1980). This model does not seem to hold at all for algae-invertebrate symbioses. Unialgal populations are the rule in hosts which range in size from protozoans to tridacnid clams and large coral colonies. Large hosts (for example, vertebrates and trees) offer a diversity of habitats, and represent 'large' colonizable islands for parasites. Large invertebrate hosts are colonized by algal symbionts during early phases or their development, as larvae or post-larvae, if the gametes themselves are not infected.

Thus the 'islands' are initially small, and symbiont populations rapidly spread to other host cells. As discussed above, subsequent invasions of a host by algae which are less than 'optimal' symbionts are generally unsuccessful.

Some algae-invertebrate mutualisms (scleractinian corals) must be quite ancient, with a long evolutionary history. If one grants the evolution of eukaryotic cells through endosymbiotic events (Margulis 1981), then the propensity for cells to enter into endosymbiotic relationships is old indeed. All of which is to say that these mutualisms have in all likelihood had a long history, with much time for coevolutionary modification. Given the coevolutionary features discussed in the next section, it would be most unlikely for such mutualisms to dissociate into separate partners over evolutionary time.

Selective Pressures for Coevolutionary Modification

Of all the ecological and evolutionary concepts considered by Price (1980), these are probably the most important factors in the success and continuity of algae-invertebrate mutualisms. Specific integrating mechanisms such as those controlling nutrient flux have probably evolved after the formation of the symbiosis. Some of these adaptations favour fitness of the established associations, while lowering the fitness of the isolated partners; these features would serve to make the associations both more obligate and more specific for each partner.

The outstanding feature of endosymbiotic algae (and of lichen algae as well) is the release of substantial amounts of fixed carbon which can be utilized by the heterotrophic partner (Smith *et al.* 1969). The amounts are greater than those of free-living algae, and the compounds often are unusual photosynthetic products. A corollary is lowered growth rates: Fitt and Trench (1983b) reported doubling times for zooxanthellae cultured from the anemone *Aiptasia tagetes* of 3.3 to 5.7 days, while Wilkerson *et al.* (1983) found that the *in situ* doubling times for the symbionts of *Aiptasia pulchella* to be 42 days. Some endosymbiotic algae show a structural reductions such as loss of flagella (p. 172) and cell coverings when entering into symbiosis.

Some hosts become developmentally dependent on the possession of symbionts: strobilation in some jellyfish polyps will not occur without zooxanthellae (see Colley and Trench 1983), and symbiotic algae are required for *Convoluta roscoffensis* to reach sexual maturity (Keeble 1910). A few hosts have reduced feeding or digestive ability; the best-known cases are among the xeniid soft corals (Gohar 1940; Schlichter 1982).

The uptake of dissolved inorganic nutrients is another case in point. Algae-invertebrate mutualisms take up a variety of inorganic nutrients from solution; these include compounds such as ammonium, which is normally toxic and excreted by animals (Muscatine 1980b). In the case of corals and sea anemones, it is thought that the algae are responsible, because symbiont-free hosts cannot effect uptake (Muscatine *et al*. 1979; Wilkerson and Muscatine 1984). Green hydra are exceptional: they continue to take up sulphate when algae-free, although it is not metabolized and may be toxic (Cook 1981 and in preparation). The kinetics of inorganic nitrogen uptake by zooxanthellae associations have peculiar 'diffusional components', and differ from the saturation kinetics typical of isolated symbionts and free-living phytoplankters (cf. D'Elia *et al*. 1983; Wilkerson and Muscatine 1984). These anomalies may be related to the passage of ammonium through the host cell, although ammonium in particular is toxic to animal cells. The uptake and recycling of inorganic nutrients is an important feature of algae-invertebrate mutualisms which occur in nutrient-poor environments such as coral reef waters (Muscatine and Porter 1977).

The release of fixed carbon from some intact associations may be a similar phenomenon. Measurements of fixed carbon losses from corals (primarily mucus) range from 6 per cent (Muscatine *et al*. 1983) to 50–60 per cent (Crossland *et al*. 1980) of short-term symbiont production. While some mucus is used for hygienic, defensive or other purposes, the loss of as much as 50 per cent of total fixed carbon from such productive associations seems extravagant. One intriguing explanation is that the energy-rich mucus attracts fish and other organisms, and these serve as local sources of nitrogen and phosphorus by producing excreta (Meyer *et al*. 1983). The extent of such losses in other zooxanthellar associations has not been examined.

Do the Stability and Diversity of Algae-Invertebrate Endosymbioses Indicate a General Phenomenon Among Mutualisms?

The evidence presented in this chapter indicates that algae-invertebrate symbioses are highly coevolved mutualisms. Given long-term (both in terms of individual lifespans and evolutionary time) host environments, endozoic algae appear to have undergone long periods of evolutionary refinement. In turn, there have been host coevolutionary responses which have resulted in specificity and diversity among natural populations.

How do these findings relate to other mutualisms? Lichens and mycorrhizae (*sensu latu*) have been classically considered to be 'mutualistic', and others (for example, Law, this volume) have argued that in these associations 'inhabitants' (lichen algae, or fungi in the case of vesicular-arbuscular mycorrhizae) should have little genetic diversity, as they are protected from external selective forces by the 'exhabitant'. In contrast, genetically distinct populations of zooxanthellae and some *Chlorella*-like symbionts have evolved notwithstanding the taxonomic problems associated with endozoic algae (see pp. 179–80).

If endozoic algae are more genetically diverse than other 'inhabitants', this may reflect the 'degree' of coevolved mutualism. Algae-invertebrate symbioses (particularly those associations with zooxanthellae) have typically evolved in nutrient-poor environments (Muscatine and Porter 1977). The resulting selective forces would tend to integrate and optimize mechanisms of nutrient acquisition, utilization and partitioning for both partners. There is increasing opinion that lichens, at least, are not truly 'mutualistic', given the lack of critical evidence for nutrient flow from the fungus (Smith 1975) and the occurrence of fungal parasitism (Ahmadjian and Jacobs 1981). The apparent lack of specificity among lichens (Barrett 1983) suggests that similar integrated regulatory mechanisms have not evolved.

One further point should be made regarding diversity among 'inhabitants'. The apparent lack of diversity at higher taxonomic levels among endozoic algae may reflect the difficulties in establishing an evolutionarily stable association within a host cell. Long-term acceptance of an intracellular symbiont by a host cell may mean that the original colonizing cells had to possess certain unique characteristics. It is conceivable that once these founder populations became established, they could spread to other appropriate hosts. Subsequent radiation from these quite special cells could have produced the genetic diversity which now exists.

Summary

Algae-invertebrate mutualisms appear to be highly coevolved systems. They are characterized by long-term stability of symbiont populations and probably are evolutionarily conservative. The integrating mechanisms which promote the maintenance and re-establishment of specific associations have been evidently refined during long periods of evolution. As suboptimal combinations of algae and hosts can be produced at least in the laboratory, these should be ideal systems to test models of coevolved mutualism (for example, Roughgarden 1975, 1983).

Acknowledgements

I think D.A. Schoenberg and R.K. Trench for stimulating discussion. The preparation of this manuscript was supported in part by grants from the Bermuda government and British Petroleum Ltd to Anthony H. Knap. I am grateful to Wolfgang Sterrer, Director of the Bermuda Biological Station, for use of facilities.

This is Contribution #969 of the Bermuda Biological Station for Research Inc.

References

Ahmadjian, V. and J. Jacobs (1981) 'Relationship Between Fungus and Algae in the Lichen *Cladonia cristatella* Tuck', *Nature, 289*, 169–72
Barrett, J.A (1983) 'Plant-Fungus Symbioses' in Futuyma, D.J. and M. Slatkin (eds), *Coevolution*, Sinauer Associates, Concord, Ma., pp. 137–60
Burris, R.H. (1983) 'Uptake and Assimilation of $^{15}NH_4$ by a Variety of Corals', *Marine Biology, 75*, 151–6
Carroll, S. and R.S. Blanquet (1984a) 'Alanine Uptake by Isolated Zooxanthellae of the Mangrove Jellyfish, *Cassiopea xamachana*. I. Transport Mechanisms and Utilization', *Biological Bulletin, 166*, 409–18
Carroll, S. and R.S. Blanquet (1984b) 'Alanine Uptake by Isolated Zooxanthellae of the Mangrove Jellyfish, *Cassiopea xamachana*. II. Inhibition by Host Homogenate Fraction', *Biological Bulletin, 66*, 419–26
Chang, S.S. and R.K. Trench (1982) 'Peridinin-Chlorophyll *a* Proteins From the Symbiotic Dinoflagellate *Symbiodinium (Gymnodium) microadriaticum*, Freudenthal', *Proceedings of the Royal Society of London Series B, 215*, 191–210
Colley, N.J. and R.K. Trench (1983) 'Selectivity in Phagocytosis and Persistence of Symbiotic Algae by the Scyhistoma Stage of the Jellyfish *Cassiopae xamachana*, *Proceedings of the Royal Society of London Series B, 219*, 61–82
Cook, C.B. (1971) 'Transfer of ^{35}-S Labeled Material from Food Ingested by

Aiptasia sp. to Its Endosymbiotic Zooxanthellae' in Lenhoff, H.M., L. Muscatine and L.V. Davis (eds), *Experimental Coelenterate Biology*' University of Hawaii Press, Honolulu, pp. 218–24

Cook, C.B. (1980) 'Infection of Invertebrates with Algae' in Cook, C.B., P.W. Pappas and E.D. Rudolph (eds), *Cellular Interactions in Symbiosis and Parasitism*, Ohio State University Press, Columbus, pp. 47–74

Cook, C.B. (1981) 'Adaptations to Endosymbiosis in Green Hydra', *Annals of the New York Academy of Science*, 361, 273–83

Cook, C.B. (1983) 'Metabolic Interchange in Algae-Invertebrate Symbiosis', *International Review of Cytology Supplement*, 14, 177–210

Crossland, C.J., D.J. Barnes, T. Cox and M. Devereaux (1980) 'Compartmentation and Turnover of Organic Carbon in the Staghorn Coral *Acropora formosa*', *Marine Biology*, 59, 181–7

D'Elia, C.F., S.L. Domotor and K.L. Webb (1983) 'Nutrient Uptake Kinetics of Freshly Isolated Zooxanthellae', *Marine Biology*, 75, 157–67

Douglas, A.F. and G.W. Gooday (1982) 'The Behaviour of Algal Cells Toward Egg Capsules of *Convoluta roscoffensis* and Its Role in the Persistence of the *Convoluta* — Algae Symbiosis', *British Phycological Journal*, 17, 383–8

Egana, A.C. and L.H. DiSalvo (1982) 'Mass Expulsion of Zooxanthellae by Easter Island Corals', *Pacific Science*, 36, 61–3

Fitt, W.K. and R.K. Trench (1982) 'Spawning, Development and Acquisition Zooxanthellae by *Tridacna squamosa* (Mollusca, Bivalva)', *Biological Bulletin*, 161, 213–35

Fitt, W.K. and R.K. Trench (1983a) 'Endocytosis of the Symbiotic Dinoflagellate *Symbiodinium microadriaticum* Freudenthal by Endodermal Cells of the Scyphistomae of *Cassiopae xamachana*, *Journal of Cell Science*, 64, 195–212

Fitt, W.K. and R.K. Trench (1983b) 'The Relation of Diel Patterns of Cell Division to Diel Patterns of Motility in the Symbiotic Dinoflagellate *Symbiodinium microadriaticum* Freudenthal in Culture', *New Phytologist*, 94, 421–32

Fitt, W.K., S.S. Chang and R.K. Trench (1981) 'Motility Patterns of Different Strains of the Symbiotic Dinoflagellate *Symbiodinium (Gymnodinium) microadriaticum* (Freudenthal) in Culture', *Bulletin of Marine Science*, 31, 436–43

Freudenthal, H.D. (1962) '*Symbiodinium* gen. nov. and *S. microadriaticum* sp. nov., a Zooxanthella: Taxonomy, Life Cycle and Morphology' *Journal of Protozoology*, 9, 45–52

Gohar, H.A.F. (1940) 'Studies on the Xenilivae of the Red Sea', *Pobl. Mar. Biol. Sta. Ghardaqa*, 2, 25–118

Goreau, T.F. (1963) 'Calcium Carbonate Deposition by Coralline Algae and Corals in Relation to Their Role as Reef-Builders', *Annals of the New York Academy of Science*, 109, 127–67

Goreau, T.F. (1964) 'Mass Expulsion of Zooxanthellae from Jamaican Reef Communities After Hurricane Flora', *Science*, 145, 383–6

Hoegh-Guldberg, I.O. (1983) 'Photosynthesis and Translocation of Newly Fixed Carbon by Zooxanthellae in the Nudibranch, *Pteraeolidia ianthina*, *American Zoologist*, 23, 923 (Abstract)

Hohman, T.C., P.L. McNeil and L. Muscatine (1982) 'Phagosome-Lysosome Fision Inhibited by Algal Symbionts of Hydra viridis', *Journal of Cell Biology*, 94, 56–63

Holligan, P.M. and G.W. Gooday (1975) 'Symbiosis in *Convoluta*', *Symposium of the Society for Experimental Biology29*, 205–27

Jolley, E. and D.C. Smith (1980) 'The Green Hydra Symbionts. II. The Biology of the Establishment of the Symbiosis' *Proceedings of the Royal Society of London Series B*, 207, 311–33

Karakashian, S.J. and M.W. Karakashian (1965) 'Evolution and Symbiosis in the Genus *Chlorella* and Related Algae', *Evolution*, 19, 368–77

Keeble, F. (1910) *Plant-Animals: A Study in Symbiosis*, Cambridge University Press

Kelty, M.O. and C.B. Cook (1976) 'Survival During Starvation of Symbiotic, Aposymbiotic and Nonsymbiotic Hydra' in Mackie, G.O. (ed.), *Coelenterate Ecology and Behavior*, Plenum Press, New York, pp. 409–14

Kempf, S.C. (1984) 'Symbiosis Between the Zooxanthella *Symbiodinium (Gymnodinium) microadriaticum* (Freudenthal) and Four Species of Nudibranchs', *Biological Bulletin*, *166*, 110–26

Kinzie, R.A. (1974) 'Experimental Infection of Aposymbiotic Gorgonian Polyps With Zooxanthellae', *Journal of Experimental Marine Biology and Ecology*, *15*, 335–45

Kinzie, R.A. and G.S. Chee (1979) 'The Effects of Different Zooxanthellae on the Growth of Experimentally Reinfected Hosts', *Biological Bulletin*, *156*, 315–27

Kokke, W.C., W. Fenical, L. Bohlin and C. Djerassi (1981) 'Sterol Synthesis in the Host-Symbiont Association in Caribbean Gorgonians', *Comparative Biochemistry and Physiology*, *B68*, 281–7

LaBarbera, M. (1975) 'Larval and Post-Larval Development of the Giant Clams *Tridacna maxima* and *Tridacna squamosa* (Bivalvia: Tridacnidae)', *Malacologia*, *15*, 69–79

Lerch, K.A. and C.B. Cook (1984) 'Some Effects of Photoperiod on the Motility Rhythm of Cultured Zooxanthellae', *Bulletin of Marine Science*, *34*, 477–83

Loeblich, A.R. III and J.L. Sherley (1979) 'Observations on the Theca of Free-Living and Symbiotic Isolates of *Zooxanthella microadriatica* (Freudenthal) comb. nov.', *Journal of the Marine Association of the UK*, *59*, 195–205

MacArthur, R.H. and E.O. Wilson (1967) *The Theory of Island Biogeography*, Princeton University Press

McAuley, P.J. (1981a) 'What Determines the Population Size of Intracellular Symbionts in the Digestive Cells of Green Hydra?' *Experientia*, *37*, 346–47

McAuley, P.J. (1981b) 'Control of Cell Division of the Intracellular Chlorella Symbionts in Green Hydra', *Journal of Cell Science*, *47*, 197–206

McAuley, P.J. (1981c) 'Ejection of Algae in the Green Hydra Symbiosis', *Journal of Experimental Zoology*, *217*, 23–31

McAuley, P.J. and D.C. Smith (1982a) 'The Green Hydra Symbiosis. V. Stages in the Intracellular Recognition of Algal Symbionts by Digestive Cells', *Proceedings of the Royal Society of London Series B*, *216*, 7–23

McAuley, P.J. and D.C. Smith (1982b) 'The Green Hydra Symbiosis. VII. Conservation of the Host Cell Habitat by the Symbiotic Algae', *Proceedings of the Royal Society of London Series B*, *216*, 415–26

McLaughlin, J.J.A. and P.A. Zahl (1966) 'Endozoic Algae' in Henry, S.M. (ed.), *Symbiosis*, vol. 1, Academic Press, New York, pp. 257–97

McNeil, P.L. (1981) 'Mechanisms of Nutritive Endocytosis. I. Phagocytic Versatility and Cellular Recognition in *Chlorohydra* Digestive Cells, a Scanning Electron Microscope Study', *Journal of Cell Science*, *49*, 311–29

McNeil, P.L. and D.C. Smith (1982) 'The Green Hydra Symbiosis. IV. Entry of Symbionts Into Digestive Cells', *Proceedings of the Royal Society of London Series B*, *216*, 1–6

McNeil, P.L., T.C. Hohman and L. Muscatine (1981) 'Mechanisms of Nutritive Endocytosis. II. The Effect of Charged Agents on Phagocytic Recognition by Digestive Cells, A Scanning Electron Microscope Study', *Journal of Cell Science*, *52*, 243–69

Margulis, L. (1981) *Symbiosis in Cell Evolution*, W.H. Freeman, San Francisco

Meints, R.H. and R.L. Pardy (1980) 'Quantitative Demonstration of Surface Involvement in a Plant-Animal Symbiosis: Lectin Inhibition of Reassociation', *Journal of Cell Science*, *43*, 239–51

Meyer, J.L., G.T. Schultz and G.S. Helfman (1983) 'Fish Schools: an Asset to Corals', *Science*, *220*, 1047–9

Muscatine, L. (1971) 'Experiments on Algae Coexistent in a Sea Anemone', *Pacific*

Science, 25, 13–21

Muscatine, L. (1980a) 'Productivity of Zooxanthellae', in Falhowski, P. (ed.), *Primary Productivity in the Sea*, Plenum, New York, pp. 31–402

Muscatine, L. (1980b) 'Uptake, Retention and Release of Dissolved Inorganic Nutrients by Marine Alga-Invertebrate Associations' in Cook, C.B., P.W. Pappas and E.D. Rudolph (eds), *Cellular Interactions in Symbiosis and Parasitism*, Ohio State University Press, Columbus, pp. 229–44

Muscatine, L. (1982) 'Establishment of Photosynthetic Eukaryotes as Symbionts in Animal Cells' in Schiff, J.A. (ed.), *On the Origins of Chloroplasts*, Elsevier-North-Holland, Amsterdam, pp. 77–92

Muscatine, L. and H.M. Lenhoff (1965) 'Symbiosis of Hydra and Algae. II. Effects of Limited Food and Starvation on Growth of Symbiotic and Aposymbiotic Hydra', *Biological Bulletin, 129,* 316–28

Muscatine, L. and P.J. McAuley (1982) 'Transmission of Symbiotic Algae to Eggs of Green Hydra', *Cytobios, 33,* 111–24

Muscatine, L. and R.R. Pool (1979) 'Regulation of Numbers of Intracellular Algae', *Proceedings of the Royal Society of London Series B, 204,* 131–9

Muscatine, L. and Porter, J.W. (1977) 'Reef Corals: Mutualistic Symbioses Adapted to Nutrient-Poor Environments', *Bioscience, 27,* 454–60

Muscatine, L., S.J. Karakashian and M.W. Karakashian (1967) 'Soluble Extracellular Products of Algae Symbiotic With a Ciliate, a Sponge and a Mutant Hydra', *Comparative Biochemistry and Physiology, 20,* 1–22

Muscatine, L., C.B. Cook, R.L. Pardy and R.R. Pool (1975) 'Uptake, Recognition and Maintenance of Symbiotic *Chlorella* by *Hydra viridis*', *Symposium of the Society of Experimental Biology, 29,* 175–204

Muscatine, L., P. Falkowski and Z. Dubinski (1983) 'Carbon Budgets in Symbiotic Associations' in Schwemmler, W. and H.E.A. Schenk (eds), *Endocytobiology*, vol. 2, Walter de Gruyter, Berlin (in press)

Neckelmann, N. and L. Muscatine (1983) 'Regulatory Mechanisms Maintaining the Hydra-*Chlorella* Symbiosis', *Proceedings of the Royal Society of London Series B, 219,* 193–210

Neigel, J.E. and J.C. Avise (1983) 'Clonal Diversity and Population Structure in Reef-Building Coral, *Acropora cervicornis*: Self-Recognition Analysis and Demographic Interpretation', *Evoltuion, 37,* 437–53

Otto, J.J. and R.D. Campbell (1977) 'Tissue Economics of Hydra: Regulation of the Cell Cycle, Animal Size and Development by Controlled Feeding Rates', *Journal of Cell Science, 28,* 117–32

Pardy, R.L. (1974) 'Some Factors Affecting the Growth and Distribution of the Algal Symbionts of *Hydra viridis*, *Biological Bulletin,* 147, 105–18

Pardy, R.L. (1976) 'The Morphology of Green Hydra Symbionts as Influenced by Host Strain and Environment', *Journal of Cell Science, 20,* 655–69

Pardy, R.L. and L. Muscatine (1973) 'Recognition of Symbiotic Algae by *Hydra viridis*: a Quantitative Study of the Uptake of Living Algae by Aposymbiotic *H. viridis*', *Biological Bulletin, 145,* 565–79

Parke, M. and I. Manton (1967) 'The Specific Identity of the Algal Symbiont of *Convoluta roscoffensis*', *Journal of the Marine Biological Association of the UK, 47,* 445–64

Patton, J.S. and R.E. Burris (1983) 'Lipid Synthesis and Extrusion by Freshly Isolated Zooxanthellae', *Marine Biology, 75,* 131–6

Phipps, D.W. and R.L. Pardy (1982) 'Host Enhancement of Symbiont Photosynthesis in the Hydra-Algae Symbiosis', *Biological Bulletin, 162,* 83–94

Pool, R.R. (1979) 'The Role of Algal Antigenic Determinants in the Recognition of Potential Algal Symbionts by Cells of *Chlorohydra*', *Journal of Cell Science, 35,* 367–79

Porter, J.W. (1976a) 'Autotrophy, Heterotrophy and Resource Partitioning in Caribbean Reef-Building Corals', *American Naturalist*, *110*, 731–42

Porter, J.W., J.D. Woodley, G.J. Smith, J.E. Neigel, J.F. Battey and D.G. Dallmeyer (1981) 'Population Trends Among Jamaican Reef Corals', *Nature*, *294*, 249–50

Porter, K.G. (1976b) 'Enhancement of Algal Growth and Productivity by Grazing Zooplankton', *Science*, *192*, 1332–4

Price, P.W. (1980) *Evolutionary Biology of Parasites*, Princeton University Press, Princeton, NJ

Provasoli, L., T. Yamasu and I. Manton (1968) 'Experiments on the Resynthesis of Symbiosis in *Convoluta roscoffensis* with Different Flagellate Cultures', *Journal of the Marine Biological Association of the UK*, *48*, 465–79

Roughgavren, J. (1975) 'Evolution of Marine Symbiosis — a Simple Cost-Benefit Model', *Ecology*, *56*, 1201–8

Roughgarden, J. (1983) 'The Theory of Coevolution' in Futuyma, D.J. and M. Slatkin (eds), *Coevolution*, Sinauer Associates, Sunderland, Mass., pp. 33–64

Rudman, W.B. (1981) 'The Anatomy and Biology of Alcynarian-Feeding Aeolid Nudibranchs and Their Development of Symbiosis with Zooxanthellae', *Zoological Journal of the Linnean Society of London*, *76*, 219–62

Schlichter, D. (1982) 'Epidermal Nutrition of the Alcynarian *Heteroxenia fuscesens* Ehrbg.: Absorption of Dissolved Organic Material and Lost Endogenous Photosynthate', *Oecologia*, *53*, 40–9

Schoenberg, D.A. (1980a) 'An Ecological View of Specificity in Algal-Invertebrate Associations, With Reference to the Association of *Symbiodinium microadriaticum* and Coelenterates' in Schwemmler, W. and H.E.A. Schenk (eds), *Endocytobiology*, vol. 1, Walter de Gruyter, Berlin, pp. 145–54

Schoenberg, D.A. (1980b) 'Intraspecific Variation in a Zooxanthella' in Schwemmler, W. and H.E.A. Schenk, *Endocytobiology*, pp. 155–62

Schoenberg, D.A. and R.K. Trench (1976) 'Specificity of Symbioses Between Marine Cnidarians and Zooxanthellae' in Mackie, G.O. (ed.), *Coelenterate Ecology and Behavior*, Plenum Press, New York

Schoenberg, D.A. and R.K. Trench (1980a) 'Genetic Variation in *Symbiodinium (Gymnodinium) microadriaticum* Freudenthal, and Specificity in its Symbiosis with Marine Invertebrates. I. Isozyme and Soluble Protein Patterns of Axenic Cultures of *Symbiodinium microadriaticum*', *Proceedings of the Royal Society of London Series B*, *207*, 405–27

Schoenberg, R.K. and R.K. Trench (1980b) 'Genetic Variation in *Symbiodinium (Gymnodinium) microadriaticum* Freudenthal, and Specificity in its Symbiosis with Marine Invertebrates. II. Morphological Variation in *Symbiodinium microadriaticum*', *Proceedings of the Royal Society of London Series B*, *207*, 429–44

Schoenberg, D.A. and R.K. Trench (1980c) 'Genetic Variation in *Symbiodinium (Gymnodinium) microadriaticum* Freudenthal, and Specificity in its Symbiosis With Marine Invertebrates. III. Specificity and Infectivity of *Symbiodinium microadriaticum*', *Proceedings of the Royal Society of London Series B*, *207*, 445–60

Smith, D.C., L. Muscatine and D.H. Lewis (1969) 'Carbohydrate Movement from Autotrophs to Heterotrophs in Parasitic and Mutualistic Symbiosis', *Biological Review*, *44*, 17–90

Smith, D.C. (1975) 'Symbiosis and the Biology of the Lichenised Fungi', *Symposium of the Society for Experimental Biology*, *29*, 373–406

Smith, D.C. (1980) 'Principles of the Colonisation of Cells by Symbionts as Illustrated by Symbiotic Algae' in Schwemmler, W. and H.E.A. Schenk (eds), *Endocytobiology*, vol. 1, Walter de Gruyter, Berlin, pp. 317–32

Smith, D.C. (1981) 'The Role of Nutrient Exchange in Recognition Between Symbionts', *Ber. Deutsch. Bot. Ges.*, *94*, 517–28

Steele, R.D. (1975) 'Stages in the Life History of a Zooxanthella in Pellets Extruded by its Host *Aiptasia tagetes*', *Biological Bulletin*, *149*, 590–600

Steele, R.D. (1976) 'Light Intensity as a Factor in the Regulation of Density of Symbiotic Zooxanthellae in *Aiptasia tagetes* (Coelenterata, Anthozoa)', *Journal of Zoology*, *179*, 387–405

Taylor, D.L. (1969) 'On the Regulation of and Maintenance of Algal Numbers in Zooxanthellae-Coelenterate Symbiosis, with a Note on the Nutritional Relationship in *Anemona sulcate*', *Journal of the Marine Biological Association of the UK*, *49*, 1057–65

Taylor, D.L. (1971a) 'On the Symbiosis Between *Amphidinium klebsii* and *Amphiscolops langerhansi* (Turbellaria: Acoela)', *Journal of the Marine Biological Association of the UK*, *51*, 301–15

Taylor, D.L. (1971b) 'Ultrastructure of the "Zooxanthella" *Endodinium chattonii in situ*', *Journal of the Marine Biological Association of the UK*, *52*, 227–34

Taylor, D.L. (1973) 'The Cellular Interactions of Algal-Invertebrate Symbiosis', *Advances in Marine Biology*, *11*, 1–56

Taylor, D.L. (1974) 'Symbiotic Marine Algae: Taxonomy and Biological Fitness' in Vernberg, W.B. (ed.), *Symbiosis in the Sea*, University of South Carolina Press, Columbia, pp. 245–62

Trench, R.K. (1971a) 'The Physiology and Biochemistry of Zooxanthellae Symbiotic with Marine Coelenterates. II. The Liberation of Fixed ^{14}C by Zooxanthellae *in vitro*', *Proceedings of the Royal Society of London Series B177*, 237–50

Trench, R.K. (1971b) 'The Physiology and Biochemistry of Zooxanthellae Symbiotic with Marine Coelenterates. III. The Effect of Homogenates of Host Tissue on the Excretion of Photosynthetic Products *in vitro* by Zooxanthellae from Two Marine Coelenterates', *Proceedings of the Royal Society of London Series B*, *177*, 251–64

Trench, R.K. (1979) 'The Cell Biology of Plant-Animal Symbiosis', *Annual Review of Plant Physiology*, *30*, 485–531

Trench, R.K. (1983) 'Dinoflagellates in Non-Parasitic Symbioses' in Taylor, F.J.R. (ed), *Biology of Dinoflagellates*, Blackwell, New York (in press)

Wells, J.W. (1956) 'Scleractinia' in Moore, R.C. (ed.), *Treatise on Invertebrate Paleontology. Part F, Coelenterata*, University of Kansas Press, Lawrence, pp. 328–44

Wilkerson, F.P. and L. Muscatine (1984) 'Uptake and Assimilation of Dissolved Inorganic Nitrogen by a Symbiotic Sea Anemone', *Proceedings of the Royal Society of London Series B*, *221*, 71–86

Wilkerson, F.P., G.M. Parker and L. Muscatine (1983) 'Temporal Patterns of Cell Division in Natural Populations of Symbiotic Algae', *Limnology and Oceanography*, *28*, 1009–14

Withers, N.W., W.C. Kokke, W. Fenical and C. Djerassi (1982) 'Sterol Patterns of Zooxanthellae Isolated From Marine Invertebrates: Synthesis of Gorgosterol and Desmethylgorgosterol by Aposymbiotic Algae', *Proceedings of the National Academy of Science of the USA*, *79*, 3764–8

Yonge, C.M. (1957) 'Symbiosis' in Hedgpeth, J.W. (ed.), *Treatise on Marine Ecology and Paleoecology*, vol. 1 (Geological Society of America Memoir 67), pp. 429–42

8 CHEATING AND TAKING ADVANTAGE IN MUTUALISTIC ASSOCIATIONS

Jorge Soberon Mainero and Carlos Martinez del Rio

Many mutualistic associations are characterized by the tight coevolution of their member species. It is not surprising, in view of the myriad of beautiful examples of coadaptations, that the idea that these coadaptations evolved for the good of both partners (that is, for the the community), is common. For example, how could the perfect fit between the proboscis of the hawkmoth *Xantophan morgani* and the flower of *Angraecum sesquipedale* have evolved, except to benefit both plant and hawkmoth? The examples of good-to-the-community thinking (conscious or not) are many, both in older and more recent literature (Allee *et al.* 1949; Odum 1971; Wilson 1980, 1983). This view has culminated in the extreme example of good-for-the-Biosphere thinking (the Gaia hypothesis, Lovelock 1979; criticized by Dawkins 1982 and Van Valen 1983).

Darwin (1859: 200) recognized the problems of good-for-the-community thinking when he stated that: 'Natural selection cannot possibly produce any modification in any one species exclusively for the good of another species; though throughout nature one species incessantly takes advantage of, and profits by, the structures of another.' Recently, a more thorough understanding of the operation of natural selection has lead to the suggestion of mechanisms that may operate with higher level vehicles (*sensu* Dawkins 1982) such as groups, species and communities and that such mechanisms may be particularly important in the evolution of mutualism (Wilson 1980, 1983).

Despite the undeniable interest of group and community selection, in this chapter we shall stress gene and individual selection, the existence of conflicts and the possibility that individuals of a third species take advantage of mutualistic partnerships. The chapter will be divided into two sections; the first is devoted to conflicts that arise within the mutualism, because some individuals do not provide the expected benefit to the partner species. This problem will be analysed from the point of view of evolutionary stable strategies (ESS, Maynard Smith 1982). In the second section we address the problem of the conditions which permit a third species to take advantage of the existence of an established mutualism. Discussion is biased towards plant-pollinator associations; this reflects both authors' preferences and availability of data.

Cheating

Once a mutualistic relationship has arisen, the appearance of cheaters becomes highly probable (Boucher *et al.* 1982). A cheater is an individual of a partner species that receives the benefits of mutualism but does not reciprocate. For cheating to increase in frequency, cheaters must obtain an additional advantage over non-cheaters, such as the avoidance of the costs (energetic or otherwise) of mutualism.

It seems that cheating can be facilitated in mutualistic interactions which are based on interspecific reciprocal altruism, in which there is a delay between the offering and receipt of benefits. This is because a delay impedes the discrimination of cheaters and non-cheaters by the cheated partner. In plant-pollinator systems, for example, the fact that nectar is often concealed and produced long before each visit, permits the appearance of both nectarless flowers and nectar robbers. Some of the conditions that favour the evolution of intraspecific reciprocal altruism, such as long life of the partners, low dispersal rates and high mutual interdependence (Trivers 1971), work against cheating in mutualistic systems. If both partners are long lived and there is individual recognition, cheating is unlikely because repeated encounters between individuals are expected to be common. Conversely, asymmetries in the lengths of life of partner species can favour cheating: cheaters of the longer-lived species can exploit unexperienced individuals of the species with the shorter generation time. Similarly, cheaters of the shorter-lived species could benefit from a relatively high rate of individual turnover that makes difficult recognition of cheaters by longer-lived partners. Low dispersal rates may increase the chance of repeated interaction, individual recognition, and site-specific encounters, thus reducing the advantages of cheating.

Dawkins and Krebs (1979) have pointed out that cheating can be one of the driving forces of coevolution. The evolution of discrimination as a counteradaptation against cheating benefits non-cheaters and thus, in a subtle way, promote the tightening of a mutualistic relation. Many coevolved characteristics of mutualistic species can be interpreted in such light. Snow (1981) provides examples from plant-disperser in interactions. The chemical and mechanical defences of seeds of avian-dispersed fruits are probably counteradaptations to seed predators among the fruit-consuming species, and the nutritional properties of bird-dispersed fruits (and also of some nectars; see Baker and Baker 1975) are presumably the result of the discriminative capabilities of the dispersers.

Cheating in Nectar Production

The possibility of cheating by plant in plant-pollinator systems, arises if the pollinator cannot determine how much food the flower will provide until after it has visited and presumably pollinated it. A plant can be pollinated even if it produces no nectar provided that it is surrounded by nectar-producing neighbours (Heinrich 1979). If the energy saved can be diverted to other fitness — contributing activities, the cheating plant will be at a selective advantage.

The relative advantage of offering a particular reward is dependent on the distribution of rewards encountered by the pollinators in the population of plants; plants are, in the game — theoretical lore of Maynard Smith (1982), 'playing the field'. The pollinators condition the optimal strategy of the plant mainly by being more or less discriminatory. In the following we discuss the evolution of nectar rewards in an ESS context; our discussion owes a lot to two previous papers by Pyke (1979, 1981) and to Feinsinger's work on the distribution of nectar rewards in a community of hummingbird-pollinated plants (Feinsinger 1978)

Heinrich (1975a) proposed as a principle that the optimal reward that an obligated outcrossing plant should offer is one that maximizes outcrossing while minimizing reward costs. Although it is not possible to optimize both variables simultaneously (Pyke 1981), the principle is useful because it recognizes the existence of a trade-off between the gain in fitness obtained through outcrossing and the costs of rewarding pollinators.

The outcrossing fitness gains basically come through two components: the transmission and the receipt of pollen (Kodric-Brown and Brown 1979; Snow 1982), both of which are positively correlated with the frequency of visits to a plant and to the number of visits per plant (Feinsinger 1978; Pyke 1979, 1981). The movements of a foraging pollinator within and between plants are the outcome of a series of animal decisions that are dependent, among other things, on the spatial distribution of flowers and plants (whether they are clumped, random or homogeneously distributed), the distribution of rewards among the flowers (Heinrich 1979; Gass and Montgomerie 1979) and the distribution of pollinators on flowers (Pleasants 1981; Martinez del Rio and Eguiarte, in review). The gains in fitness due to outcrossing are not a simple function of the reward offered per flower but also of the discriminatory and special memory capabilities of the pollinators, and of the distribution of rewards within and between flowers, plants and clumps.

The energetic cost of producing nectar is relatively easy to estimate, as nectar is basically a sugar and water solution (Baker 1975), with variable but usually minute quantities of aminoacids (Baker and Baker 1975; Baker 1977). If nectar composition and concentration are constant (which sometimes is not the case; see Corbet 1978; Real 1981b; Willson and Bertin 1979) the cost is simply proportional to the rate at which it is produced (Pyke 1981). A difficulty arises in calculating the net benefit of nectar production by subtracting the costs of nectar production from the benefits of outcrossing, because the 'currencies' are different and it is necessary to make assumptions on 'exchange' values (Schoener 1971; Mitchell 1981; Pyke *et al.* 1977; Pyke 1981). The cost of nectar production can be extremely different for plants with photosynthetic flower parts than for plants without them (Bazzaz *et al.* 1979). In the following discussion we assume that in principle an exchange function can be found to convert costs and benefits so as to treat them in the same units.

From the elements outlined above a graphical model can be constructed to obtain the optimal quantity of nectar that has to be offered to the pollinators, which will be the quantity that maximizes the difference between the gains brought by outcrossing and the costs of nectar production (Pyke 1981).

The simplest possible assumption about the costs of nectar production is that they increase linearly with the quantity of nectar produced per flower (Figure 8.1). However the form of the benefit function depends on more conditions. In Figure 8.1 the curves of benefit are shown for two extreme distributions of nectar rewards in the population. In drawing them it was supposed that the pollinators behave in a way that maximizes their rate of energy intake and that they possess a certain degree of spatial memory.

Curve 8.1a shows the increase in benefit due to increased nectar production when nectar rewards are abundant in the population.

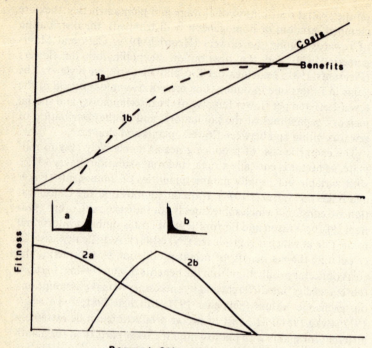

Figure 8.1: Optimal Nectar Rewards. Curve la shows the increase in benefits through increases in nectar rewards for a plant living in a nectar-rich environment in which most plants offer large quantities of nectar (dark frequency distribution a). The lower curve, 2a, is the standarized difference between costs and benefits. Curve 1b shows the increase in benefits in a nectar-poor environment (dark frequency distribution b) and the corresponding standarized curve 2b, of differences between costs and benefits.

Since nectarless and low-producing plants receive high frequencies of visits through the conditioning of pollinators on their high-rewarding neighbours, they are favoured by selection, since they receive the benefits of pollination at very low expense. A population of high-rewarding plants can be invaded by low-rewarding phenotypes and is therefore not stable in an evolutionary sense. In the same graph a benefit curve is shown for a nectar-poor environment (8.1b) in which the nectar reward distribution is skewed towards nectarless flowers. If the pollinators can remember the location of nectar-rich plants and avoid poor ones, the frequency of visits will increase sharply beyond a certain threshold of adequate nectar rewards, and tend asymptotically to a saturation level. The optimal quantity of nectar will be situated well to the right of the mean quantity offered, suggesting that the frequency of plants that offer richer rewards will increase; thus a pure 'low-reward' strategy is not evolutionary stable either. It is to be expected that, in due time, an equilibrium distribution centred around an optimal mean of nectar production that is stable to invasion (an ESS), will be achieved by the population (Pyke 1981).

A basic supposition of this result is that the pollinators possess a relatively good spatial memory and behave in ways that maximize their rates of energy intake. Although there is good evidence for the fact that some pollinators, such as hummingbirds and bumblebees, are at least approximately optimal foragers (Pyke 1978: Pleasants 1981; Gass 1978; Hainsworth and Wolf 1976), hard data on the spatial memory of nectar-feeding animals are extremely scant, although the visiting of discrete clumps of plants in definite sequences (trap-lining), which certainly requires this capability, is believed to be quite common (Janzen 1971; Stiles 1975; Feinsinger and Chaplin 1975; Heithaus *et al.* 1975).

In addition to good memory, if a pollinator is to be discriminatory and visit only the relatively richest plants, then flowers must be abundant relative to pollinators. If flowers are scarce, travel times among rich plants might be so high as to make it profitable to visit poorer but commoner plants (MacArthur and Pianka 1966; Krebs 1978); if in contrast flowers are abundant relative to pollinators, they can afford to be discriminatory and visit only rich plants. It can be suggested that relative scarcity of plants will tend to reduce the optimal mean quantity of nectar offered per visit. Feinsinger (1978) arrived at the same conclusions following a slightly different reasoning.

Among the pressures that influence the rewards offered to pollinators are: the presence of robbers that deplete nectar from flowers without otherwise damaging them, and the existence of flower mimics (*aprovechados*, see next section). Both shift the distribution of floral rewards to the left, towards low rewards, favouring plants that offer rich rewards if pollinators are discriminatory. It can be suggested that nectar production will be higher in heavily robbed populations and in the presence of flower mimics, other things being equal, than in populations without robbers or mimics.

The precise form of the benefit curves depends on the discriminatory capabilities of the pollinators. Animals with keen memory and well-developed perceptual skills will, under similar conditions, be more discriminatory and will favour higher-yielding plants than less capable animals. Thus if plants and flowers are visited more or less randomly, the possibility of cheating by individual plants is increased. A good example is provided by dioecious plants such as some Caricaceae, in which one sex (usually the female) does not give nectar but nevertheless receives enough pollen (Baker 1979; Bawa 1980; Bullock and Bawa 1981). Baker (1979) mentions that of all the flower visitors observed in an orchard of *Carica papaya*, only crepuscular hawkmoths (Sphingidae), which probably approach the plants using olfactory clues (Brantjes 1978), visited the nectarless pistillate flowers. Diurnal hummingbirds, bees, flies and butterflies collected nectar from staminate flowers but largely ignored female flowers.

The existence of large variances in nectar reward is a well-documented fact. This variance can be the result of the removal of nectar by pollinators and robbers (Hainsworth and Wolf 1972; Heinrich 1975a; Zimmerman 1981) or it can be due to intra- and inter-plant variation in nectar production (Heinrich 1975a; Feinsinger 1978; Herrera and Soriguer 1983; Feinsinger 1983); in the latter case the resulting distribution of nectar rewards has been christened the 'bonanza-blank' pattern (Feinsinger 1978). Although bonanza-blank patterns can arise from variation in the micro-environment (Fahn 1949) or from variation of physiological conditions of plants or parts (Shuel 1955, 1957), they can also conceivably arise as adaptive responses of the plants to their pollinators' visit patterns. Feinsinger (1978) pointed out that relative scarcity of plants relative to pollinators and local spatial and temporal unpredictability of nectar resources are necessary conditions for bonanza-blank patterns to be beneficial to plants. It should also be pointed out that patterns with great variances are to be expected more in shrubs and trees with many flowers than in herbaceous plants with few. For a plant with many flowers to produce too much nectar in each flower can result in satiation of its pollinators, establishment of territories, and long intraplant foraging bouts that promote self-fertilization and reduce gene flow (Heinrich and Raven 1972; Heinrich 1975a; Carpenter 1976; Frankie and Baker 1974);

under these conditions a bonanza-blank pattern can increase inter-plant visits while maintaining the pollinators conditioned through an 'intermittent reinforcement schedule' (Skinner 1938, cited in Feinsinger 1978). For herbaceous plants with few flowers a bonanza-blank pattern is an unlikely result of natural selection because it is evolutionary unstable; if pollinators are discriminatory, nectar-rich plants will be chosen and eventually an equilibrium distribution of rewards will be achieved around an optimal mean; if pollinators are absolutely undiscriminatory nectarless plants will benefit because they will be pollinated at no expense and the plant population will shift from mutualism to deceit. Cheating in plants may have long-term effects on the population of pollinators, and through these, on the genetic structure and population dynamics of the plant. If the population of pollinators is seriously reduced, autogamous plants are benefited and inbreeding increases, with the deleterious effects that it sometimes brings. Individual selection thus might have long-term deleterious effects on the population.

Nectar Robbers

Cheating pollinators seem to be more rare than cheating plants; this may be a result of specialization. Nectar larceny is the evolutionary outcome of plants specializing on a particular class of pollinators. The filtering mechanisms of flowers (*sensu* Stiles 1975) evolved to avoid wasting rewards on inefficient and time-consuming pollinators; the concealment of rewards in long spurs, long corolla tubes or hidden nectaries resulted in an increased specialization of true pollinators but also in the improvement of the robbing techniques of the excluded nectar exploiters (Kodric-Brown and Brown 1979). This evolutionary divide that separates true pollinators from nectar robbers has had the consequence of committing true pollinators to efficient pollen transfer and nectar robbers to a life of crime. Of course the involvement of pollinators is not a moral one but is rather the result of functional and behavioural constraints (Brian 1957); it is easier and energetically cheaper for a coevolved pollinator to behave correctly.

The fact that flower units evolved for the benefit of plants and not of pollinators is evidenced by the observation that some potential pollinators become secondary nectar robbers, taking nectar through holes made by primary nectar robbers, when pierced flowers are available (Rust 1979). All pollinators are potential nectar robbers and behave as such if it increases their rate and efficiency of nectar intake, but it is remarkable that few true pollinators misbehave. Skutch (1954) states that primary robbing is a pretty rare behaviour in tropical hummingbirds and it is rarely found that an species of animal behave both as pollinator and robber in the same species of plant (McDade and Kinsman 1980; Koeman-Kwak 1973), although it is common to find that a pollinator of one plant species is a robber of another (McDade and Kinsman 1980; Inouye 1983).

Taking Advantage

An organism will be called an *aprovechado* (Spanish for 'one who takes advantage') of a given mutualistic association if it derives benefits from the existence of the association without providing any benefits in return (Figure 8.2).

The idea of 'taking advantage' has a connotation of causing some amount of damage to the mutualists. Probably many examples discussed below undergo periods in which the *aprovechado* behaves as a commensal, but there is no doubt that in several cases there is a loss in fitness for at least one of the mutualists due to the presence of the *aprovechado*. Since hard data on the subject are scarce, we shall assume for convenience that the net result of having an *aprovechado* is negative for both mutualistic partners. We do not consider as *aprovechados* those organisms that simply parasitize or prey upon any or both mutualists regardless of the particular features of the mutualism (signals, rewards, behaviour, etc.). Therefore lichen herbivores, seed or fruit predators, etc. are not *aprovechados*.

In this section, instead of looking for ESS for the *aprovechados* we address the more general problems of the circumstances under which an *aprovechado* can invade a mutualistic association. We will consider formally only two mutualistic partners, although in reality these may be two groups of species. All the examples we will mention can be divided in two broad classes, according to the effects on the *aprovechado*: those associations in which the *aprovechado* benefits from *both* mutualists (henceforth called type I) and those in which the *aprovechado* receives a negative effect from one of the mutualists (type II, Figure 8.2b). Both types can be represented by the following model:

$$M_1 = M_1 \cdot F_1(M_1, M_2, A)$$
$$M_2 = M_2 \cdot F_2(M_1, M_2, A)$$
$$A = A \cdot F_3(M_1, M_2, A) \tag{8.1}$$

Type I

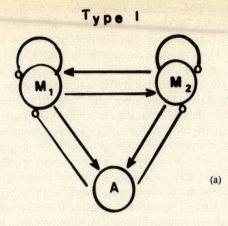

(a)

Type II

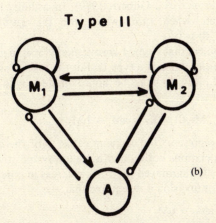

(b)

Figure 8.2: Possible Interactions Between Two Mutualists and an Aprovechado. The effect of M_2 on A is negative in type II. Effects of A on M_1 and M_2 have been depicted as negative, but may be also neutral (see text).

in which M_1 and M_2 represent the population densities of the mutualists and A the sensity of the *aprovechado*. Such three-species
systems may exibit very complicated dynamics, and intuition
derived from two-species models is seldom sufficient to understand
the behaviour of some of the intricate equilibrium solutions that
may arise (Hirsch and Smale 1974; May and Leonard 1975). However, as argued by Vandermeer and Boucher (1978), stability of
feasible equilibria may be entirely irrelevant to persistence in the real
world. Instead of looking for stable equilibria, what is needed is to
find regions in the three-species space with the property that once
any trajectory enters the region, it remains there (these regions are
called invariate sets). Inside these regions stable equilibria, limit
cycles or strange attractors may be found but this is irrelevant from
the point of view of persistence. Proving the existence of invariate
sets is in general a formidable mathematical problem. However, it is
possible to show that given certain conditions, invasibility by all
three species guarantees persistence of the system. The conditions
are: (1) the origin is invasible by all the three species, (2) every two-
species equilibrium point is globally stable, (3) every species in the
absence of the other two has a positive carrying capacity. Essentially
these mean that the interactions are facultative and weak, which is
probably a fairly common case in non-symbiotic mutualisms
(Boucher *et al*. 1982; Gilbert, 1980). Invasibility requires positive
growth rate when rare, which for the *aprovechado* means
$F_3(M_1, M_2, 0) > 0$.

The problem that a type I *aprovechado* faces in invading are very
different from those of a type II. For the sake of simplicity assume
that $F_3(M_1, M_2, 0)$ may be approximated by a Lotka-Volterra
expression:

$$F_3(M_1, M_2, 0) = K_3 + aM_1 + bM_2 \qquad (8.2)$$

where K_3 represents the carrying capacity of the *aprovechado*. In
type I associations both a and b are always positive and therefore
invasibility is guaranteed ($F_3 > 0$), whereas in type II, b is negative
and thus for invasion it is required that:

$$K_3 + aM_1 > bM_2 \qquad (8.3)$$

which may not be fulfilled. This very simple observation has some
interesting consequences. In the first place, one should expect type I
association to be more common that type II because establishment
of the former requires less stringent conditions. Secondly, there

appears to be more room for specialization and coevolution of the association in type I than in type II. Indeed, taking K_3 to represent the degree of specialization of the *aprovechado*, K_3 near zero means a very obligate relationship with the mutualists while $K_3 \gg 0$ means a facultative relationship (Soberon and Martinez del Rio 1981). Thus in type II a high degree of specialization ($K_3 = 0$) will make fulfilment of (8.3) difficult. Besides, if the negative effect of M_2 on A corresponds to exploitative competition for some resource, this should put a selective premium on divergence and niche shifts. In short, a type I *aprovechado* is engaged in 'arms race' (Dawkins 1982) coevolution with both mutualists. On the other hand, a type II is likely to be exposed to pressures to shift and 'loosen' the association. We now discuss some examples in order to flesh up these ideas.

Type I Aprovechados

Batesian Mimicry. There are mutualisms in which several unpalatable or poisonous species display the same set of signals to potential predators, thus sharing the costs of 'teaching' the predators to avoid that set of signals (Mullerian mimicry). The *aprovechados* of the mutualism are the Batesian mimics which display the same signals but are perfectly palatable to predators (see Rettenmeyer 1970; Papageorgis 1975; Vane-Wright 1976). This is a type I *aprovechado* because it derives benefits from both models.

Evolution of this kind of mimicry probably begins with non-zero values of a and b due to mistakes committed by the predators that operate the system. Once admitted into the association, the *aprovechado* would evolve towards the morph of the models. The best documented examples of Batesian mimicry come from butterflies and from syrphid flies and bees (Wickler 1968; Rettenmeyer 1970). There is no doubt that some *aprovechados* benefit from the presence of their Mullerian mimics (Rettenmeyer 1970), and in many cases the *aprovechados* have evolved a striking degree of resemblance to their models. It is much less clear, however, whether the models are damaged by the *aprovechados*. This probably depends on diversity and memory of predators, frequency of the *aprovechado* and phenology. It would also be advantageous for the *aprovechado* to appear with some delay after the models in order to cope with already trained predators. Superficially this shift may appear as a case of competition for 'enemy free space' (Atsatt 1981), but the real mechanism is probably the one outlined above (Rothschild 1963; Llorente and Garces 1983).

Reproductive Mimicry in Plants. Reproductive (food source) mimicry (Wiens 1978) occurs when a nectarless plant resembles a rewarding one and pollinators trained to visit the latter spend some time visiting the mimic, which is the *aprovechado* of the association. There are many cases reported of this kind of *aprovechado* (Thien and Marks 1972; Proctor and Yeo 1973; Wiens 1978; Brown and Kodric-Brown 1979; Heinrich 1979; Dafni and Ivry 1981; Dressler 1981) but these are largely anecdotal and several have been contested (Byerzychudek 1981; Williamson and Black 1981; Boyden 1982). Working with the orchid *Orchis israelitica*, a proposed mimic of the lily *Bellevalia flexuosa*, Dafni and Ivry (1981) have found a much higher capsule production rate in orchids growing in the presence of the lilies than in orchids without them. Unfortunately, these authors do not give details of the presence of other plant species, density of model and mimic in the different treatments or density of pollinators. Therefore, their results may be explained by a variety of non-mimetic mechanisms. Byerzychudek (1981) compared pollinia removal rates in pure and mixed stands of the milkweed *Asclepias curassavica* and its proposed mimic, the orchid *Epidendrum radicans*. She expected higher removal rates in *E. radicans* mixed with *A. curassavica*, but the differences found were not significant. However, pollinators in the area were mainly secondary-growth butterflies which are vagile and probably not very 'trainable', so mistake visits could be expected to be high, making deceit and mimicry indistinguishable. For the purpose of this chapter, deceitful plants mean non-mimetic, attractive, nectarless plants. In another example, Brown and Kodric-Brown (1979) reported a nectarless population of *Lobelia cardinalis* and suggested that it may be a mimic of neighbouring *Ipomopsis agregata* and *Penstemon barbatus*. However, it is quite possible that instead of being a case of reproductive mimicry, *Lobelia* is receiving only a random rate of visits due to untrained pollinators (Williamson and Black 1981).

A proper test of the hypothesis of reproductive mimicry would require comparing the pollination rates in pure and mixed stands of models and mimics due to resident, 'intelligent' pollinators like hummingbirds, bumblebees, certain long lived butterflies (*Heliconius* sp.), etc. Experimental set-ups similar to Real's (1981a) can be useful. The theoretical curves of visits should have a shape similar to those in Figure 8.3. The intensity of the selective pressure for resemblance to the model is related to the increases in visits, and therefore (to an extent) in pollination rate, that a mimic receives

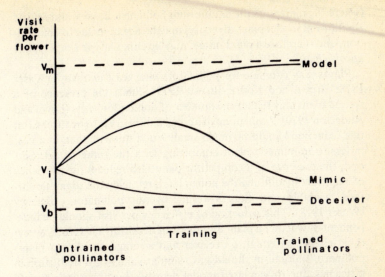

Figure 8.3 Visitation Rates to Different Kinds of Flowers by Untrained and Trained Pollinators. The upper curve is the visitation rate to a rewarding plant. The middle curve corresponds to a mimetic plant and the lower one to a non-mimetic, nectarless plant. V_i represents the visitation rate due to naïve pollinators. It has to be weighed by the frequencies of plants and the genetic preferences of pollinators. V_m is the maximum visit rate, determined by the functional response of a well trained pollinator and V_b is a basal rate due to errors, re-sampling, etc.

relative to a deceitful plant. Deceitful plants tend to get very low visit rates (Melampy and Hayworth 1980; Ackerman 1981) and a true mimic should receive a higher rate (Heinrich 1979; Boyden 1982). It would be very interesting to have more quantitative data on the subject.

Although conclusive data is unavailable, the anecdotal evidence strongly suggests the existence of reproductive mimicry in plants. This type of *aprovechado* presents an important difference with the preceding Batesian mimics, because a plant reproductive mimic can also compete with its model for the very pollinators that the latter is training. That is, the sign of b in:

$$K_3 + aM_1 + bM_2 \qquad (8.4)$$

(where M_1 represents the equilibrium pollinator density and M_2 the equilibrium model plant diversity) may be negative due to competition, and therefore a plant mimic may be either a type I or a type II *aprovechado*.

Plants may compete for pollinators in at least two ways (Waser 1978; Brown and Kodric-Brown 1979). First, the presence of a second plant may reduce the number of pollination visits (Levin and Anderson 1970). As mentioned above, if mimicry is to be successful, this reduction should be much smaller in a mimetic than in a nectarless, non-mimetic plant competing with the same model. Second, the presence of a competitor plant also reduces the efficiency of each visit by diluting the pollen load, reducing the stigmatic surface available and thus lowering overall pollinator efficiency (Waser 1978). This reduction of efficiency per visit should be independent of whether a plant is mimetic or deceitful. Therefore, everything else being equal, a deceiver and a mimic invading a plant-pollinator mutualism should have similar values of b (reduction in efficiency due to increased model density), but the mimic should get increased values of a (visitaticn rate) relative to a deceitful plant. If inequality (8.3) is fulfilled at the visit 'base line' (pollinator errors, resampling visits, naïve or young visitors, etc.; see Figure 8.3), a mimetic *aprovechado* should be able to invade with a selective premium on increased resemblance to the model (higher a; see Heinrich 1979). On the other hand, if $K_3 + aM_1 < bM_2$ the relationship corresponds to a type II *aprovechado* and will be discussed later.

Once the mimetic plant has invaded, if its resemblance to the model is high enough, its presence may have the effect of simulating the presence of cheaters, with a consequent change in the ESS for nectar production of the model plant. It follows that nectar production in a population with mimetic plants should be higher than in pure stands of the model. Both the systems *Lantana* — *Asclepias* — *Epidendrum* and *Ipomopsis* — *Penstemon* — *Lobelia* may be used to test this hypothesis provided local pollinators are discriminatory enough. Real (1981a) has proposed a similar hypothesis based on the idea that pollinators avoid plants with highly variable nectar rewards and therefore a variable (individual) plant should increase its average nectar production to keep the pollinators interested. Deciding between the two hypothesis would be very difficult.

Peckhammian Mimicry in Cleaning Fishes. Peckhammian (Wickler 1968), or antergic mimicry (Vane-Wright 1976) occurs when the operator in a mimetic relationship is attracted to a mimic that inflicts some kind of damage upon the former. Several examples are known of fish species that clean other 'client' species of ectoparasites, uneaten food particles, etc. This mutualistic system is often invaded by *aprovechados* that mimic the cleaners' coloration pattern and general aspect, but bite the client (Wickler 1968; Losey 1972; Gorlick *et al.* 1978). In this case, as in the preceding two, the system probably evolved due to 'errors' (Wickler 1968: 164) on the part of the operator of the mimetic relationship (the client) that opened the door to the *aprovechados*. Convergence towards the morph of the cleaner may be extremely close, as in the case of *Aspidontus teniatus* and its model *Labroides dimidiatus* (Wickler 1968). *A. teniatus* certainly benefits from the presence of both cleaner and client, but it is not clear whether its net effect on the mutualists is negative to both or only to the client. As in Batesian mimicry, if the *aprovechado* 'de-trains' the client, the cleaner is being negatively affected. However, certain client species do not appear to associate damage with the morphological pattern of the cleaner but rather with the particular cleaning spot (Wickler 1968). As in the other examples, carefully designed experiments are needed to clarify the status of this kind of *aprovechado*.

Peckhammian Mimicry in Predatory Insects. This is the case of a plant and its insect pollinators (the mutualistic system) and a predator of the insects which disguises itself with the colour and/or shape of the flower. Very much as with the plant mimics, evidence for this is largely anecdotal (Emerton 1902; Wickler 1968) and few field experiments have been performed. Nevertheless, it is probably true to say that such an *aprovechado* benefits from both pollinators and the plants that attract them. A further advantage is obtained when the *aprovechado* resembles the plant, because potential predators may be led astray, as is probably the case of the mantis *Hymenopus coronatus* (Wickler 1968). Clearly, pollinators are damaged by the existence of these *aprovechados*. Whether the plant gets a negative or a positive effect is less clear. Louda (1982) found that the presence of the spider *Peucetia viridans* on the inflorescences of *Haplopappus venetus* reduced both pollination and seed set. However, release of viable, undamaged seeds was higher on inflorescences with spiders because seed predation was reduced. This positive net effect by the spider depends critically on relative timing of spider and plant phenologies (Louda 1982) and may shift from positive to negative and *vice versa*. As Louda (1982) stresses, such conflicting, alternating selective pressures may prevent the fine tuning of plant-*aprovechado* coevolution. Nevertheless, since both mutualists provide benefits to the *aprovechado*, one should expect an easy invasion and perhaps further evolution towards better disguised *aprovechados*. A spectacular example of this is the mantis *H. coronatus* (Wickler 1968). Less clear examples may be found among Thomisid spiders (Emerton 1902) and Phymatid bugs (Wright 1981).

Dispersal Mimicry in Plants. It is interesting to mention a last example of type I *aprovechado* which is still poorly documented. This is the case of plants (mainly legumes) whose seeds resemble the fruits of other species and obtain the benefits of seed dispersal without providing fruity tissue (McKey 1975; Wiens 1978). Experiments to substantiate this as true mimicry are scarce. It seems reasonable to suppose that the mimics benefit from obtaining 'free' transport, but this is by no means sure. Circumstantial evidence for the hypothesis of mimicry is good (McKey 1975), but as in preceding examples, field experiments are required.

Type II Aprovechados

Reproductive Mimicry in Plants. As was discussed before, a mimetic plant may compete with its model for pollinators, and the balance between benefits due to training of pollinators and detriments due to competition for them may be negative, rendering invasion difficult. Several possibilities arise. The first is that the mimetic population adopts some form of autogamy, wind pollination, etc. Of course, in this case the selective pressure for the evolution of mimicry (higher pollination rates than non-mimetic, nectarless plants) disappears and mimics and therefore *aprovechados* are not to be expected. Another possibility is that the mimetic plant shifts its phenology to appear somewhat than its model. Some pollinators will remain faithful to their search images for long enough to make this a plausible mechanism (Waser 1978; Heinrich 1979). A mimetic plant may thus obtain the benefits of trained pollinators without paying the costs of competition (Boyden 1982, Heinrich 1975c). These shifts in phenology may turn the *aprovechado* into a type II (benefiting from both mutualists) and at the same time reduce considerably whatever negative effect the 'de-training' of pollinators may have on the models. Of course, the effect on pollinators remains negative. A third possibility is that the *aprovechado* remains in space and time as a true mimic but reducing its death rate and increasing its pollination efficiency (Soberon and Martinez del Rio 1981). Significantly, many perennial and highly efficient orchids seem to behave as *aprovechados* or deceivers (Dressler 1981).

Flower Nectar Robbers. Probably the best documented example of type II *aprovechados* are flower nectar robbers (Inouye, 1983). Interspecific nectar robbers are very clear cases of type II because they derive benefits from flowers while competing with the pollinators for nectar. The effect of the *aprovechado* on the plant can be negative due to damages to floral structures and to severe depletion of nectar standing crop, as has been reported for *Aphelandra golfodulcensis* and *Justicia aurea* in Costa Rica (McDade and Kinsman 1980). Roubik (1982) was able to document a significant decrease in fruit and seed set in a population of *Pavonia dasypetala* in Panama due to the activities of the robber bee *Trigona ferricauda*. This was due to the fact that the robbers attacked the hummingbird pollinators and reduced their per-patch flower visit rate (Roubik 1982). Further evidence of widespread negative effects on plants is the different protective mechanisms against robbers (Proctor and Yeo 1973; Bolten and Feinsinger 1978; Guerrant and Fiedler 1981; Inouye 1983).

On the other hand, Koeman-Kwak (1973) found that some bumblebee robbers of *Pedicularis palustris* (Scrophulariacea) actually did pollinate the flowers. Hawkins (1961) reported an increased pollination rate due to the activity of nectar robbers. Heinrich and Raven (1972) suggested that reduced foraging times in robbed flowers may increase visit rate (see also Soberon and Martinez del Rio 1981). However, such effect is probably rare (Roubik 1982).

The competitive effects of robbers on pollinators is less well documented. Roubik (1982) demonstrated that the robber *Trigona ferricauda* chased away the hummingbird *Phaetornis superciliosus* from flower patches, although perhaps not without the latter having extracted some nectar. Colwell *et al.* (1974) documented a case of competition between *Diglossa plumbea* and hummingbirds for the nectar of *Centropogon valerii* in Costa Rica.

Very much as with established flower mimics, nectar robbers may simulate the presence of plant cheaters and consequently, an increased nectar production is to be expected.

Nectar Robbing of Aphids and Lepidopteran Larvae. Mutualisms between honeydew-producing insects and attending ants are well known (Hinton 1951; Way 1963; Hocking 1975, Boucher *et al.* 1982). Honeydew insect foragers that do not protect the aphids or larvae benefit from the presence of the honeydew producer and may compete with the ants, and thus are type II *aprovechados*. An example of this is the African butterfly *Megalopalpus zymna* which has been observed feeding on secretions of homopterans (Owen 1976) or lepidopteran larvae (Gilbert 1976). Some bee species may also adopt this behaviour (Brian 1957, cited in Inouye 1983). Experimental studies on this kind of system would be very interesting.

Discussion

At the beginning of this section, two hypotheses were proposed based on the conditions for invasibility: that type I *aprovechados* should be more common, and that they should show more traits specifically evolved for the association with the mutualists than type II. Both hypotheses are supported by the examples cited above. Not only there are some spectacular examples of well-adapted type I *aprovechados* (some Batesian and cleaning and perhaps Peckhammian mimics) but also type I *aprovechados* appear to be fairly common (several orchids, many butterflies, moths, syrphid flies and other Batesian mimics, perhaps several Thomisid spiders and Phymatid bugs, etc.). It is very interesting to notice that all cases of type I *aprovechado* involve a mimetic relation of some kind, and are based on errors committed by one of the mutualistic partners (Batesian mimics are the exception). Selective pressures on the *aprovechado* should act to increase this error rate. Many non-symbiotic mutualisms depend on a flow of information between a species advertising some reward and another providing some service, and therefore such a prevalence of mimetic *aprovechados* is to be expected. Dawkins (1982) has called the use of the muscular power of one species (or individual) by another 'manipulation'. Some mutualists are cases of 'good' manipulators (for example, flowers that manipulate their pollinators to act as flying penises, but provide a reward) and their mimetic *aprovechados* may be considered 'bad' manipulators.

Despite the above, non-mimetic type I *aprovechados* should exist. Perhaps there are *aprovechados* of truly mutualistic phoretic associations (Wilson 1980) and unless the carrier 'selects' its phoretic associates or *vice versa*, it is difficult to see how mimicry could evolve in these cases.

Type II *aprovechados* appear to be less common than type I, nectar robbers being the only common examples. Although they certainly present adaptations to their way of life (long legs and a particular body posture, Gilbert 1976; toothed mandibles, Inouye 1983) there are no reports of the kind of very fine tuning that some type I *aprovechados* show. As a matter of fact, nectar robbers are often regarded as opportunistic, generalized nectar feeders (Inouye 1983).

Most examples of *aprovechados* have been considered as parasites (*sensu lato*) by the original authors. Since not all parasites, not all manipulators and not all cheaters are *aprovechados*, in this chapter we have used the word as a label to stress the special features that arise when a third species takes advantage of a mutualistic partnership. It is quite certain that further observations and experiments will add to and modify the small catalogue of examples we provide. In particular, it is important to document whether some proposed mimetic patterns are truly mimetic. This can only be done by experiment.

It is perhaps ironic to find a chapter devoted to 'cheating' and 'taking advantage' in a book devoted to mutually beneficial interspecific effects. Prevalent evolutionary schools emphasize the individual or genic level (Dawkins 1982) rather than the group or community levels (Wilson 1980). It is thus not surprising that conflicts of interest appear when biological relations are analysed in such a light. In the words of Dawkins (1978: 70): '. . . we must expect lies and deceit, and selfish exploitation of communication to arise whenever the interests of the genes of different individuals diverge'. However, it is not difficult to imagine how individual (or gene) selection can lead to such changes in a population of *aprovechados* or cheaters (low numbers, inbreeding) that its extinction may follow. For example, it is well known that, in models of predator-prey interactions, increasing the predator's searching efficiency (which is individually selected) may decrease its prey equilibrium size to very low levels (Hassel and Moran 1976) with the consequent risk of extinction for specific predators. Soberon and Martinez del Rio (1981) provide an example in which individual selection for optimal nectar production in a plant-pollination system can result in the appearance of unstable equilibria that may bring the plant population to extinction. In both examples extinction is made possible by the obligacy of at least one of the partners. Highly facultative mutualisms are unlikely to have such long term penalties (Boucher *et al*. 1982).

In this chapter we have stressed individual selection and its negative effects on mutualistic systems, despite our belief that certain, still poorly understood, properties of groups may be of particular importance in the evolution of mutualistic associations. As Boucher *et al*. (1982) have said, mutualistic relationship have appealed to people who think in terms of benefit of groups and societies. Their study may provide a natural ground to integrate group and individual selection.

References

Akerman, J.D. (1981) 'Pollination Biology of *Calypso bulbosa* var. *occidentalis*: A Food Deception System', *Madrono, 28*, 101–10

Aiee, W.C., A. Emerson, O. Park, R. Park and K.P. Schmidt (1949) *Principles of Animal Ecology*, W.B. Saunders and Co., Philadelphia

Atsatt, P.R. (1981) 'Lycaenid Butterflies: Selection for Enemy Free Space, *American Naturalist, 118*, 638–54

Baker, H.G. (1975) 'Sugar Concentration in Hummingbird Flowers', *Biotropica, 7*, 37–41

Baker, H.G. (1977) 'Chemical Aspects of the Pollination of Woody Plants in the Tropics' in Tomlinson, P.B. and M. Zimmerman (eds), *Tropical Trees as Living Systems*, Cambridge University Press, Cambridge

Baker, H.G. (1979) ' "Mistake" Pollination as a Reproductive System With Special Reference to the Caricaceae' in Burley, J. and B.T. Styles (eds), *Tropical Trees: Variation, Breeding and Conservation*, Linnean Society Symposium series No 2, Academic Press, London

Baker, H.G. and I. Baker (1975) 'Studies of Nectar Constitution and Plant-Pollinator Coevolution' in Gilbert, L.E. and P.H. Raven (eds), *Coevolution of Animals and Plants*, University of Texas Press, Austin

Bawa, K.S. (1980) 'Mimicry of Male by Female Flowers and Intrasexual Competition for Pollinators in *Jacharatia dolichaula* (D. Smith) Woodson (Caricaceae)', *Evolution, 34*, 467–74

Bazzaz, F.A., R.W. Carlson and J.L. Harper (1979) 'Contribution to Reproductive Effort by Photosynthesis of Flowers and Fruits', *Nature, 279*, 554–5

Bolten, A.B. and P. Feinsinger (1978) 'Why Do Hummingbird Flowers Secrete Dilute Nectar', *Biotropica, 10(4)*, 307–9

Boucher, D.H., S. James and K.H. Keeler (1982) 'The Ecology of Mutualism', *Annual Review of Ecology and Systematics, 13*, 315–47

Boyden, T. (1980a) 'Floral Mimicry by *Epidendrum ibaguense* (Orchidaceae) in Panama', *Evolution, 34*, 135–6

Boyden, T. (1982) 'The Pollination Biology of *Calypso bulbosa* var. *americana* (Orchidaceae): Initial Deception of Bumblebee Visitors', *Oecologia, 55*, 178–84

Brantjes, N.B.M. (1978) 'Sensory Responses to Flowers in Night Flying Moths' in Richards, A.J. (ed.), *The Pollination of Flowers by Insects*, Linnean Society Symposium series No 6, Academic Press, London

Brian, A.D. (1957) 'Differences in the Flowers Visited by Four Species of Bumble-bees and Their Causes', *Journal of Animal Ecology, 26*, 71–98

Brown, J.H. and A. Kodric-Brown (1979) 'Convergence Competition and Mimicry in a Temperate Community of Hummingbird-Pollinated Flowers', *Ecology, 60*, 1022–35

Bullock, S.H. and K.S. Bawa (1981) 'Sexual Dimorphism and the Annual Flowering Pattern in *Jacaratia dolichaula* (D. Smith) Woodson (Caricaceae) in a Costa Rica Rain Forest', *Ecology, 62*, 1494–1504

Byerzychudek, P. (1981) '*Asclepias, Lantana*, and *Epidendrum*: A Floral Mimicry Complex?', *Biotropica, 13*, 54–8

Caspenter, L., (1976) 'Plant-Pollinator Interactions in Hawaii: Pollination, Energetics of *Metrosideos collin* (Myrtaceae), *Ecology, 57(6)*, 1125–44

Colwell, R.K. B. Betts, P. Bunnell, F. Carpenter and P. Feinsinger (1974) 'Competition for the Nectar of *Centropogon valerii* by the Hummingbird *Colibri thalassinus* and the Flower Piercer *Diglossa plumbea* and Its Evolutionary Implications', *Condor , 76*, 447–84

Corbet, S.A. (1978) 'Bee Visits and the Nectar of *Echium vulgare* L. and *Synapsis alba* L.', *Ecological Entomology, 3*, 25–37

Dafni, A. and Y. Ivry (1981) 'Floral Mimicry Between *Orchis israelita* Baumann and Dafni (Orchidaceae) and *Bellerolia flexuosa* Boiss (Liliaceae)', *Oecologia, 49*, 229–32

Darwin, C. (1859) *On The Origin Of Species by Means of Natural Selection.*

Dawkins, R. (1978) *The Selfish Gene*, Granada Publishing Company, 224pp.

Dawkins, R. (1982) *The Extended Phenotype*, W.H. Freeman and Company, San Francisco

Dawkins, R. and J.R. Krebs (1979) 'Arms Races Between and Within Species', *Proceedings of the Royal Society of London B., 205*, 489–511

Dressler, R. (1981) *The Orchids, Natural History and Classification*, Harvard University Press, Boston

Emerton, J.H. (1902) *The Common Spiders of the United States*, Republished by Dover Publications, New York

Fahn, A. (1949) 'Studies in the Ecology of Nectar Secretion', *Palestine Journal of Botany, Jerusalem Series, 4*, 207–24

Feinsinger, P. (1978) 'Ecological Interactions Between Plants and Hummingbirds in a Succesional Tropical Community', *Ecological Monographs, 48*, 269–87

Feinsinger, P. (1983) 'Variable Nectar Secretion in a *Heliconia* Species Pollinated by Hermit Hummingbirds, *Biotropica, 15*, 48–52

Feinsinger, P. and S.P. Chaplin (1975) 'On the Relationship Between Wing-Disc Loading and Foraging Strategy in Hummingbirds', *American Naturalist, 109*, 217–24

Frankie, G.W. and H.G. Baker (1974) 'The Importance of Pollinator Behavior in the Reproductive Biology of Tropical Trees', *Anales del Instituto de Biologia, Universidad National Autonoma de Mexico, Serie de Botanica, 1*, 1–40

Gass, C.L. (1978) 'Experimental Studies of Foraging in Complex Laboratory Environments, *American Zoologist, 18*, 617–76

Gass, C.L. and R.D. Montgomerie (1979) 'Hummingbird Foraging Behavior: Decision Making and Energy Regulation' in Kamil, A.C. and T.D. Sargent (eds), *Foraging Behavior: Ecological, Ethological and Psychological Approaches*, Garland STPM Press, New York/London

Gilbert, L.E. (1976) 'Adult Resources in Butterflies: African Lycaenid *Megalopalpus* Feeds on Larval Nectary', *Biotropica, 8(4)*, 282–3

Gilbert, L.E. (1980) 'Coevolution of Animals and Plants: A 1979 postscript' in Gilbert, L.E. and Raven P.H. (eds.), *Coevolution of Animals and Plants*, University of Texas Press, Austin, revised edition

Gorlick, D.L., P.D. Atkin and G.S. Losey (1978) 'Cleaning Stations as Water Holes,

Garbage Dumps and Sites for The Evolution of Reciprocal Altruism', *American Naturalist*, *112*, 341–3

Guerrant, E.O. and P. Fiedler (1981) 'Flower Defences Against Nectar-Pilferage by Ants', *Biotropica*, *13(2)*, 25–33

Hainsworth, F.R. and L.L. Wolf (1972) 'Energetics of Nectar Extraction in a Small, High Altitude, Tropical Hummingbird, *Selasphorus flamula*', *Journal of Comparative Physiology*, *80*, 377–87

Hainsworth, F.R. and L.L. Wolf (1976) 'Nectar Characteristics and Food Selection by Hummingbirds', *Oecologia*, *25*, 101–14

Hawkins, R.P. (1961) 'Observations on the Pollination of Red Clover by Bees: I, The Yield of Seeds in Relation to Number and Kinds of Pollinators', *Annals of Applied Biology*, *49*, 55–65

Heinrich, B. (1975a) 'Energetics of Pollination', *Annual Review of Ecology and Systematics*, *6*, 139–69

Heinrich, B. (1975b) 'The Role Of Energetics in Bumblebee-Flower Interactions' in Gilbert, L.E. and P.H. Raven (eds), *Coevolution of Animals and Plants*, Texas University press, Austin

Heinrich, B. (1975c) 'Bee Flowers: a Hypothesis on Flower Variety and Blooming Times', *Evolution*, *29*, 325–34

Heinrich, B. (1979) *Bumblebee Economics*, Harvard University Press, Cambridge, Mass.

Heinrich, B. and P.H. Raven (1972) 'Energetics and Pollination Ecology', *Science*, *176*, 597–602

Heithaus, E.R., T.H. Fleming and P.A. Opler (1975) 'Foraging Patterns and Resource Utilization in Seven Species of Bats in Seasonal Tropical Forest', *Ecology*, *54*, 841–54

Herrera, C.M. and R.C. Soriguer (1983) 'Inter and Intra-Floral Heterogeneity of Nectar Production in *Helleborus foetidus* L. (Ranunculaceae)', *Botanical Journal of the Linnean Society*, *86*, 253–60

Hinton, H.E. (1951) 'Myrmecophilous Lycaneidae and Other Lepidoptera — a Summary', *Transactions of The South London Natural History Society, 1949–1950*, 111–75

Hirsch, M.W. and S. Smale (1974) *Differential Equations, Dynamical Systems and Linear Algebra*, Academic Press Inc., New York

Hocking, B. (1980) 'Ant-Plant Mutualism: Evolution and Energy' in Gilbert, J.E and Raven, P.H., (eds), *Coevolution of Animals and Plants*, University of Texas Press, Austin

Inouye, D.W. (1983) 'The Ecology of Nectar Robbing' in Bentley, B. and T. Elias (eds), *The Biology of Nectaries*, Columbia University Press, New York

Janzen, D. (1971) 'Euglossine Bees as Long Distance Pollinators of Tropical Plants', *Science*, *171*, 203–5

Kodric-Brown, A. and J. Brown (1979) 'Competition Between Distantly Related Taxa in the Coevolution of Plants and Pollinators', *American Zoologist*, *19*, 115–27

Koeman-Kwak, M. (1973) 'The Pollination of *Pedicularis pallustris* by Nectar Thieves (Short-Tongued Bumblebees)', *Acta Botanica Neederlandica*, *22(6)*, 608–15

Krebs, J.R (1978) 'Optimal Foraging: Decision Rules for Predators' in Krebs, J.R. and N.B. Davies (eds), *Behavioural Ecology: An Evolutionary Approach*, Blackwell Scientific Publications, Oxford

Levin, D.A. and Anderson W.W. (1970) 'Competition For Pollinators Between Simultaneously Flowering Species, *American Naturalist*, *104*, 455–67

Llorente, J. and A. Garces (1983) 'Notas Sobre *Dismorphia amphiona lupita* (Lamas) (Lepidoptera, Pieridae) y Observaciones Sobre Algunos Complejos Mimeticos en Mexico', *Revista de la Sociedad Mexicana de Lepidopterologia*,

8(2), 27–39

Losey, G.C. (1972) 'The Ecological Importance of Cleaning Symbiosis', *Copeia*, *1972*, 820–3

Louda, M. (1982) 'Inflorescence Spiders: a Cost/Benefit Analysis for The Host Plant, *Haplopappus venetus* Blake (Asreraceae)', *Oecologia, 55*, 185–91

Lovelock, J.E. (1979) *Gaia: A New Look at Life on Earth*, Oxford University Press, Oxford

MacArthur, R.H. and Pianka, E.R. (1966) 'On the Optimal Use of a Patchy Environment', *American Naturalist, 100*, 603–9

McDae, L.A., and Kinsman, S. (1980) 'The Impact of Floral Parasitism in Two Neotropical Hummingbird-pollinated Plant Species, *Evolution, 34*, 944–58

McKey, D. (1975) 'The Ecology of Seed Dispersal Systems' in Gilbert L. (ed.), *Coevolution of Animals and Plants*, University of Texas Press, Austin

May, R. and W. Leonard (1975) 'Nonlinear Aspects of Competition Between Three Species', *SIAM Journal of Applied Mathematics, 29*, 243–53

Maynard Smith, J. (1982) *Evolution and The Theory of The Games*, Cambridge University Press, London/New York

Melampy, M. and A.M. Hayworth (1980) 'Seed Production and Pollen Vectors in Several Nectarless Plants', *Evolution, 34*, 1144–54

Mitchell, R. (1981) 'Insect Behaviour, Resource Exploitation and Fitness', *Annual Review of Entomology, 26*, 373–96

Odum, E.P. (1971) *Fundamentals of Ecology*, W.B. Saunders and Company, Philadelphia

Owen, D.F. (1971) *Tropical Butterflies*, Clarendon Press, Oxford

Papageorgis, C. (1975) 'Mimicry in Neotropical Butterflies', *American Scientist, 63*, 522–32

Pleasants, J.M. (1981) 'Bumblebee Responses to Variations in Nectar Availability', *Ecology, 62*, 1648–61

Proctor, M. and P. Yeo (1973) *The Pollination of Flowers*, The New Naturalist, Collins, London

Pyke, G.H. (1979) 'Optimal Foraging in Nectar Feeding Animals and Coevolution With their Plants' in Kamil, A.C. and T.D. Sargent (eds), *Foraging Behavior: Ecological, Ethological and Psychological Approaches*, Garland STPM Press, New York/London

Pyke, G.H. (1980) 'Optimal Foraging in Bumblebees: Calculation of Net Rate of Energy Intake and Optimal Patch Choice', *Theoretical Population Biology, 19*, 232–40

Pyke, G.H. (1981) 'Optimal Nectar Production in a Hummingbird Pollinated Plant', *Theoretical Population Biology, 20*, 326–43

Pyke, G.H., H.R. Pulliam and E.L. Charnov (1977) 'Optimal Foraging: A Selective Review of Theory and Tests', *Quarterly Review of Biology, 52*, 137–54

Real, L.A. (1981a) 'Uncertainty and Pollinator-Plant Interactions: The Foraging Behavior of Bees and Wasps on Artificial Flowers', *Ecology, 62*, 20–6

Real, L.A. (1981b) 'Nectar Availability and Bee Foraging on *Ipomoea* (Convolvulaceae)', *Biotropica, 13*, 64–9

Rettenmeyer, C.R. (1970) 'Insect Mimicry', *Annual Review of Entomology, 15*, 43–74

Rothschild, M. (1963) 'Is the Buff Ermine (*Spilosoma lutea* Huf.) a Mimic of the White Ermine (*Spilosoma lubricipeda* L.)?', *Proceedings of The Royal Entomological Society, 38*, 159–64

Roubik, D. (1982) 'The Ecological Impact of Nectar Robbing Bees and Pollinating Hummingbirds on a Tropical Shrub', *Ecology, 63*, 354–60

Rust, R.W. (1979) 'Pollination of *Impatiens capensis*: Pollinators and Nectar Robbers', *Journal of the Kansas Entomological Society, 52*, 297–308

Schoener, T.W. (1971) 'Theory of Feeding Strategies', *Annual Review of Ecology and Systematics*, *2*, 369–404

Shuel, R.W. (1955) 'Nectar Secretion in Relation to Nitrogen Supply, Nutritional Status and Growth of the Plant', *Canadian Journal of Agricultural Science*, *35*, 124–38

Shuel, R.W. (1957) 'Some Aspects of the Relation Between Nectar Secretion and Nitrogen, Phosphorus and Potassium Nutrition', *Canadian Journal of Plant Science*, *37*, 220–36

Skutch, A.F. (1954) *Life Histories of Central American Birds*, vol. 1, Pacific Coast Avifauna Series no. 31, Berkeley, California

Snow, A. (1982) 'Pollination Intensity and Potential Seed Set in *Passiflora vitifolia*', *Oecologia*, *55*, 231–7

Snow, D. (1981) 'Coevolution of Birds and Plants' in Greenwood, P.H. (ed.), *The Evolving Biosphere*, The British Museum, Cambridge University Press, Cambridge

Soberon, J. and C. Martinez del Rio (1981) 'The Dynamics of a Plant-Pollinator Interaction', *Journal of Theoretical Biology*, *91*, 363–78

Stiles, F.G. (1975) 'Ecology, Flowering Phenology and Hummingbird Pollination of Some Costa Rican *Heliconia* Species', *Ecology*, *56*, 285–301

Stiles, F.G. (1981) 'Geographical Aspects of Bird-Flower Coevolution With Particular Reference to Central America', *Annals of the Missouri Botanical Garden*, *68*, 323–51

Thien, L.B. and B.G. Marcks (1972) 'The Floral Biology of *Arethusa bulbosa*, *Calopogon tuberosus* and *Pogonia ophioglossoides* (Orchidaceae)', *Canadian Journal of Botany*, *50*, 2319–25

Trivers, R.L. (1971) 'The Evolution of Reciprocal Altruism', *Quarterly Review of Biology*, *46*, 35–57

Vane-Wright, R.I. (1976) 'A Unified Classification of Mimetic Resemblances', *Biological Journal of the Linnean Society*, *8*, 25–56

Vandermeer, J.H. and D.H. Boucher (1978) 'Varieties of Mutualistic Interactions in Population Models', *Journal of Theoretical Biology*, *74*, 549–58

Van Valen, L. (1983) 'How Pervasive is Coevolution?' in Nitecki, M. (ed.), *Coevolution*, University of Chicago Press

Waser, N.M. (1978) 'Competition for Hummingbird Pollination and Sequential Flowering in Two Colorado Wild Flowers', *Ecology*, *59*, 934–44

Way, M. (1954) 'Studies on The Association of the Ant *Oecophyla longinoda* (Latr.) (Formicidae) With The Scale Insect *Saissetia zanzibarensis* William (Coccidae)', *Bulletin of Entomological Research*, *45*, 113–34

Way, M. (1963) 'Mutualism Between Ants and Honeydew Producing Homoptera', *Annual Review of Entomology*, *8*, 307–44

Wickler, W. (1968) *Mimicry in Plants and Animals*, World University Library, McGraw Hill Company, New York

Wiens, D. (1978) 'Mimicry in Plants', *Evolutionary Biology*, *11*, 365–403

Williamson, G.B. and E.M. Black (1981) 'Mimicry in Hummingbird Pollinated Plants', *Ecology*, *62*, 494–6

Willson, M.F. and R.I. Bertin (1979) 'Flower Visitors, Nectar Production and Inflorescence Size of *Asclepias syriaca*', *Canadian Journal of Botany*, *57*, 1380–8

Wilson, D.S. (1980) *Natural Selection in Populations and Ecosystems*, Series in Evolutionary Biology, The Benjamin Cummins Publishing Co.

Wilson, D.S. (1983) 'The Effect of Population Structure on the Evolution of Mutualism', *American Naturalist*, *121*, 851–70

Wright, D.M. (1981) 'Historical and Biological Observations of Lepidoptera Captured by Ambush Bugs (Hemiptera)', *Journal of the Lepidopterist's Society*, *35*, 120–3

Zimmerman, N. (1981) 'Patchiness in the Dispersion of Nectar Resources: Probable Causes', *Oecologia*, *49*, 154–7

9 COMPETITION IN MUTUALISTIC SYSTEMS

John F. Addicott

Introduction

There are three major interfaces between competition and mutualism. The first involves determining the existence, distribution and relative frequency of competition, predation, mutualism, and commensalism in different ecosystems, and the relative impact of these interactions on the distribution and abundance of species populations. A lack of information about mutualism (see Colwell and Fuentes 1975; Risch and Boucher 1976) hampered previous assessments of the distribution and relative importance of mutualism (for example, May 1976; Pianka 1974; Williamson 1972), but the surge of interest in mutualism during the last ten years is now providing the basis for making reasonable comparisons.

A second interface involves the existence of benefits in mutualistic systems based upon modifications by one species of the competitive interactions encountered by another. This can occur through niche differentiation (Davidson and Morton 1981a; Friedmann and Kern 1956; Handel 1978), deterring competition (Osman and Haugeness 1981; Wright 1973), increasing competition (Janzen 1969; Messina 1981; Quinlan and Cherrett 1978), feeding on competitors (Barbehenn 1969; Springett 1968), and competing with competitors (Davidson 1980; Lawlor 1979).

The third interface is the focus for this chapter: competition that occurs for the resources or services provided by mutualists. I will develop this topic around three different problems: detecting whether there is limitation by mutualists, and determining whether intra- and/or interspecific competition is occurring for mutualists. I will also consider three related topics: competition with non-mutualists for mutualistic resources; mutualisms for mutualists; and the importance of scales of observation in the detection of competition or mutualism for mutualists.

Evidence for limitation and competition can be direct, from manipulative experiments, indirect, from natural experiments, and secondary, from patterns expected as a result of limitation or competition (Connell 1975; Waser 1983). There are now numerous studies that provide direct evidence for competition (Connell 1975, 1983; Schoener 1983), but few of the studies of competition for mutualists fall in this category. I will emphasize direct evidence where it is available, but I will of necessity have to rely heavily upon both indirect and secondary evidence.

The most difficult task in the study of competition for mutualists may be to determine not whether competition occurs but whether it is based upon the resources or services provided by mutualists. Competition may occur simultaneously for a number of limiting resources, and I know of no unique protocol for determining which of a number of possible limiting factors may be the basis for competition. Furthermore, mutualistic resources may be limiting, but there need not be competition for them. Therefore, to establish competition for mutualists there must be careful observation and/or control of other potentially limiting factors. There are also some kinds of competition that occur without mutualists being limiting.

Limitation by Mutualists

By definition all species are limited by their mutualists, because complete removal of the mutualists lowers population growth rate or individual fitness of its partner. However, a more useful view of limitation is the following: at field densities does a small change in the density of one species lead to corresponding changes in individual fitness or *per capita* population growth rate of the other? A mutualist must limit its partner at some absolute or relative densities, but it need not do so at all densities.

I will treat the empirical evidence for limitation by mutualists in three sections. I will present examples of indirect evidence for limitation, in which natural variation in the density or activity of one species is associated with changes in the fitness or growth rate of another. I will discuss direct evidence for pollinator limitation of plant reproduction, temporal and spatial variation of pollinator limitation, and interactions between pollinator and resource limitation. I will consider in detail the evidence for pollinator limitation in the obligate and highly specific interaction between yuccas and yucca moths.

Indirect Evidence for Mutualist Limitation

Variation in the number of ants on both local and regional scales provides indirect evidence for limitation by ants of plants with extrafloral nectaries (EFNs) and homopterans tended by ants. Inouye and Taylor (1979) examined the effects of tending by ants on net seed production of *Helianthella quinquenervis*, a plant with EFNs. Ants significantly lowered the amount of seed predation by the larvae of tephritid flies. However, the number of ants active on a given flower stalk declined with increasing distance for ant nests, and the proportion of seeds damaged was inversely related to the number of ants tending a given stalk. Therefore, since natural variation in ant activity is positively correlated with predispersal seed survivorship, ants are probably limiting to seed production in *H. quinquenervis*.

On a larger scale, ant activity at sugar baits varied significantly among habitats in northwestern Costa Rica (Bentley 1976). Ant activity was positively correlated with the growth of cultivated beans which had been supplied with artificial EFNs. In locations with the highest ant activity, plants with artificial EFNs produced significantly higher dry weights than plants lacking the EFNs. It is not clear what factors account for changes in ant activity from one habitat to another (Bentley 1976). O'Dowd and Catchpole (1983) were unable to show a beneficial effect of ants tending EFNs on *Helichrysum* spp. in Australia, but the number of ants tending was very low. Their results probably reflect a pattern of ant limitation at a regional level.

Geographic variation in pollinator activity can account for differences in reproductive output of the plants they pollinate. For example, Cruden *et al.* (1976) hypothesize that plants pollinated by moths may reach their upper elevational limits where temperature restricts the activity of moths. Per cent fruit set in *Yucca glauca* declined with increasing elevation along the eastern slope of the Colorado Rockies (Cruden *et al.* 1976). The moth *Tegeticula yuccasella* pollinates *Y. glauca*, and is non-thermoregulatory. Therefore, its activity should be correlated with temperature and elevation. Although resource availability could also correlate negatively with elevation, Cruden *et al.* (1976) speculate that the decline in reproductive output of *Y. glauca* is a function of decreased moth activity.

Direct Evidence for Pollinator Limitation

The studies discussed above all rely on indirect evidence of mutualist limitation, but the observed patterns may reflect differences in factors other than mutualist availability. Direct evidence from manipulative experiments is readily available in the study of pollinator limitation of plant reproduction. Data are required on seed production under three conditions: exclusion of pollinators, normal pollinator visitation rates, and unlimited availability of pollen (Bierzychudek 1981b). Bagging flowers excludes pollinators, and hand pollination provides unlimited pollen, but care must be taken as to the source of pollen (Price and Waser 1979). Differences between normal and exclusion treatments establish the existence of mutualism, while an increase in seed set with hand pollination establishes pollinator limitation.

The numerous studies of pollinator limitation provide a number of qualitatively different results, which are reviewed by Bierzychudek (1981b) and Stephenson (1981). In many cases the results show clearly that plant reproduction is limited by pollinators (for example, Bierzychudek 1981b; Augspurger 1981; Janzen *et al.* 1980, Petersen *et al.* 1982; Snow 1982). In other cases the results are equally clear in the direction of resource limitation (Motten *et al.* 1981; Motten 1983; Stephenson 1981). Other results are more complex.

Whole plants or individual flowers that bloom at different times may encounter different numbers or kinds of pollinators, and pollinator limitation may occur during one period but not another. Gross and Werner (1983) studied pollination of four species of *Solidago* by honey bees in southern Michigan, repeating hand pollination experiments at intervals throughout the flowering season. Pollinators limited seed production in *Solidago canadensis* at all times, in *S. graminifolia* only during the early part of its flowering period, and in *S. juncae* not at all. Similarly, Zimmerman (1980) studied pollination of *Polemonium foliossisimum* bumblebees in Colorado. During early flowering there was no correlation between a bee visitation index and average seed set. During mid to late flowering, correlations were positive, and hard pollination significantly increased seed set. These studies illustrate differences in pollinator limitation between time, as well as between closely related species.

More than one factor can limit seed or fruit production simultaneously. Pollinators may be required for seed set, and hand pollination may increase seed set above normal field levels. However, the extent to which hand pollination increases seed set may be limited, thereby implicating resource limitation as well as pollinator limitation. For example, hand pollination of *Lithospermum caroliniense* increased seed production from 9 percent to just 17 percent (Weller 1980), and similarly hand pollination increased fruit production in *Asclepias tuberosa* from c. 2 per cent to 14.7 percent (Wyatt 1976). Fruit production in a number of species of *Yucca* illustrates the combined effects of both pollinator and resource limitation.

Pollinator Limitation of Yuccas

The interaction between yuccas and yucca moths provides a favourable system for examining resource and pollinator limitation of fruit production in plants, and I will examine this interaction in some detail. Moths of the genus *Tegeticula* are the only pollinators of yuccas (Powell and Mackie 1966), and each species of *Yucca* is associated with just one species of *Tegeticula*: *Yucca whipplei* is associated with *T. maculata*, *Y. brevifolia* with *T. synthetica*, and all other yuccas with *T. yuccasella* (Davis 1967, but see Miles 1983). Specificity makes pollinator limitation more likely, since other pollinators cannot compensate for changes in the relative numbers of moths to flowers. Similarly, yuccas will not compete for moths with other plants or with other species of *Yucca*, because a different population of moths is associated with each sympatric species of *Yucca* (Miles 1983). Therefore, this system lacks some of the complexity found in other pollination systems.

Evidence on pollinator/resource limitation in yuccas comes primarily from the work of Aker and Udovic (Aker 1982; Aker and Udovic 1981; Udovic 1981; Udovic and Aker 1981) on *Y. whipplei* in southern California, and from my own studies on various yuccas in the southwestern United States (see below). Aker and Udovic provide five lines of evidence that strongly support the resource limitation hypothesis. First, the number of mature fruits produced per plant increased with plant size (basal rosette, inflorescence height, or number of flowers) both within sites (Aker 1982; Udovic and Aker 1981) and between sites (Udovic 1981). However, fruit production per flower did not increase with plant size (Aker 1982), in contradiction to the pattern predicted by some forms of the pollinator limitation hypothesis (Schaffer and Schaffer 1977, 1979; Udovic 1981). Second, fruit production per flower was generally less than 10 per cent, and was a function of both low rates of fruit initiation and fruit abortion rates of 47 per cent to 74 per cent.

Third, they performed a hierarchical multiple regression of the number of mature fruit (MF) on measures of plant size and the number of initiated fruit (IF), entering plant size into the regression model before IF, MF was positively correlated with both IF and plant size. Although both plant size and IF were positively correlated with MF, plant size explained 1.9x to 4.8x more of the variation in MF than did IF, and IF explained only 11 per cent to 22 per cent of the variation in MF.

Fourth, a similar hierarchical regression of the number of aborted fruit (AF) on IF and plant size showed that AF was positively correlated with IF and negatively correlated with measures of plant size. For plants of a given size, the greater IF, the greater the proportion of fruit which aborted. If plants regulate the number of mature fruits based upon plant size, then the partial regression coefficient of AF on IF should be greater than the overall proportion of aborted fruit to initiated fruit (Aker 1982, Udovic and Aker 1981). Overall abortion rates varied from 47 per cent to 67 per cent, and were significantly different from the partial regression coefficients, which varied from .74 to .86 (Aker 1982).

Finally, the abortion rate of fruit from flowers initiated late, at the top of inflorescences, was higher than from flowers initiated early, at the bottom of inflorescences (Aker and Udovic 1981). Altogether this evidence leads to the conclusion that most plants at their study sites 'initiated more fruits than they could support, that larger plants could support more fruits, and that fruit abortion helped to reduce the number of fruits to a sustainable level' (Udovic and Aker 1981), and supports the hypothesis of resource limitation in *Y. whipplei*.

However, there is also some evidence for pollinator limitation. IF correlated positively with indices of pollinator visitation within sites and between sites (Udovic 1981). Twelve of 85 plants initiated a lower percentage of fruit than the average percentage of mature fruit, indicating that these plants received fewer visits than necessary for full fruit production (Udovic and Aker 1981). Also, a number of plants, which bloomed well after most other individuals in the population, received few visits and set relatively little fruit (Udovic and Aker 1981). There are also other observations of possible pollinator limitation in *Y. whipplei* (Powell and Mackie 1966, Addicott, personal observations).

But, considering both sets of evidence, Udovic and Aker (1981) conclude that 'the amount of resources available to support fruit development is considerably more important in determining the ultimate number of fruits matured'. However, the genus *Yucca* is diverse, comprising 20–40 species, depending upon one's taxonomic preferences. As Aker and Udovic point out (Aker 1982; Udovic 1981; Udovic and Aker 1981), the importance of pollinator limitation can vary among species, populations, years and individuals. I have therefore undertaken to assess the relative importance of pollinator and resource limitation in a broader sample of yuccas from the southwestern United States.

During 1983 I collected data on fruit production and fruit abortion form a total of 616 inflorescences at 21 sites in the upper Colorado River basin in Arizona, Utah and Colorado. Collections were made from 23 different populations of yuccas, representing 8 different species: *Y. angustissima, Y. baccata, Y. baileyi, Y. elata, Y. gilbertiana, Y. glauca, Y. kanabensis* and *Y. neomexicana*. For each inflorescence I counted the number of mature fruit, the number of fruit which initiated development but subsequently aborted, and the number of flowers which did not initiate fruit. I used the same criteria for distinguishing these categories as Udovic and Aker (1981) used.

There is considerable variation in the reproductive parameters for yuccas between species and within species (Table 9.1), as well as between individuals within populations. For individual plants the proportion of flowers which initiated fruit ranged from 0 per cent to 71.4 per cent, mature fruit production ranged from 0 per cent to 66.3 per cent, and abortion rates ranged from 0 per cent to 100 per cent. There is more variation at both an individual and population level than Aker and Udovic found for *Y. whipplei*. Also, in comparison with *Y. whipplei*, many of the populations in my survey have higher proportions of mature fruit and lower abortion rates. Twelve of 23 populations have abortion rates of less than 30 per cent, and seven of these were less than 20 per cent (Table 9.1). Of the eight species in my survey, *Y. elata* most closely matches *Y. whipplei*, in that *Y. elata* has relatively low fruit production and high rates of fruit abortion.

There are four lines of evidence indicating the relative importance of pollinator rather than resource limitation for most populations. First, a high proportion of flowering individuals initiating no fruit is an indicator of pollinator limitation. For five of the eight species there was no such indication, as the proportion of inflorescences

Table 9.1: Average Number of Flowers per Plant, Proportion of Flowers Initiating Fruit and Maturing Fruit per Plant, and Number of Plants Initiating No Fruit, in 1983

Species	Study site	Sample size	# Flowers	Fruit initiation	Fruit maturation	# Plants initiating no fruit
Y. angustissima	Gateway, Co.	39	56.6	.235	.203	0
	Grandview Pt., Ut.	35	77.8	.185	.151	1
	Peach Spr., Az.	22	94.7	.235	.180	0
	Seligman, Az.	21	92.5	.139	.112	2
	Squaw Flats, Ut.	61	54.8	.180	.152	2
Y. baccata	Gateway, Co.	14	95.2	.057	.050	2
	LaSal Mtns., Ut.	17	50.5	.222	.138	0
	Peach Spr., Az.	29	67.2	.032	.018	12
Y. baileyi	Black Mesa, Az.	21	48.1	.049	.038	13
	Marble Cyn., Az.	16	57.8	.138	.078	5
	Moab, Ut.	13	62.9	.284	.183	0
	Glen Cyn. City, Ut.	62	54.9	.106	.065	19
Y. elata	Cottonwood, Az.	10	154.8	.232	.074	0
	Wickenburg, Az.	13	661.3	.218	.102	0
	Cedar City, Ut.	20	225.1	.273	.189	0
	LaVerkin, Ut.	10	105.9	.396	.126	0
	Zion NP, Ut.	31	95.0	.195	.099	1
	Dewey, Az.	10	167.4	.297	.115	0
Y. gilbertiana	Cedar City, Ut.	33	97.1	.277	.193	0
Y. glauca	Poncha Spr., Co.	36	52.6	.129	.087	3
	Saguache, Co.	42	40.3	.226	.152	0
Y. kanabensis	Kanab, Ut.	31	101.1	.107	.074	0
Y. neomexicana	Montrose, Co.	30	123.4	.085	.076	6

initiating no fruit varied from 0 per cent to 4 per cent (Table 9.1). However, for *Y. baccata* and *Y. baileyi* the values for all populations combined were 23.3 per cent and 33.0 per cent, respectively, suggesting pollinator limitation in at least some of the populations of these species. A high proportion of plants not initiating fruit could be the result of incongruence between flight and flowering seasons, or differential attraction of pollinators among plants within a population. The former is more likely, as two pieces of information cast doubt upon the hypothesis of differential attraction. Plants that initiated no fruit did not have significantly fewer flowers than plants that did initiate fruit, and for all plants the proportion of initiated fruit was not positively correlated with the number of flowers on a plant.

The next two lines of evidence depend upon the same hierarchical regression procedure which Aker (1982) used. I regressed MF on the number of flowers per plant (FL) and on IF. Only plants that initiated some fruit were included in the analysis. As with *Y. whipplei* MF was positively correlated with both FL and IF. However, unlike *Y. whipplei* IF explained more of the variation in MF in 12 out of 23 populations, accounting for between 44.7 per cent and 91.8 per cent of the variation (Table 9.2). Only in *Y. elata* did FL consistently explain more of the variation in MF than did IF.

I also tested whether the partial regression coefficient of AF on IF was significantly greater than the overall ratio of AF to IF for a given population (Table 9.2). This test was significant for only 8 of 23 populations. Three were populations of *Y. elata*, and two others were the populations of *Y. angustissima* and *Y. baileyi* for which IF explained very little of the variation in MF. The magnitude of the partial regression coefficient indicates the extent to which fruit abortion compensates for excess initiation of fruit. Values close to 1.0 indicate perfect compensation. Except for *Y. elata* which generally had values greater than 0.8, the values were low, 12 of 23 being less than 0.5.

Finally, qualitative observations on the relationship between fruit set and the relative abundance of moths support the hypothesis of limitation by pollinators. For example, *Y. harrimaniae* blooms regularly in Colorado National Monument, near Grand Junction, Colorado. However, I have seen few or no moths during flowering, and fruit set is correspondingly low.

Therefore, there is strong support for the pollinator limitation hypothesis in many of the population surveyed. The major excep-

Table 9.2: Results of Hierarchical Multiple Regressions of MF (or AF) on FL and IF. Significance values are given for the partial correlations of MF on FL and on IF, and for the comparison of the partial regression coefficient of AF on IF with the population rate of fruit abortion (* – $p < .05$; ** – $p < .01$; *** – $p < .001$)

Species	Study site	Sample size	Proportion of variation in MF due to FL	Proportion of variation in MF due to IF	Fruit abortion rate	Partial regression coeff. of AF on IF
Y. angustissima	Gateway, Co.	39	.342***	.532***	.137	.250
	Grandview Pt., Ut.	34	.379***	.480***	.148	.220
	Peach Spr., Az.	22	.711***	.169***	.190	.557***
	Seligman, Az.	19	.093	.803***	.261	.222
	Squaw Flats, Ut.	61	.476***	.391***	.141	.186
Y. baccata	Gateway, Co.	12	.059	.856***	.164	.169
	LaSal Mt., Ut.	17	.411**	.220*	.329	.537
	Peach Spr., Az.	17	.000	.501***	.347	.559
Y. baileyi	Black Mesa, Az.	8	.498**	.391**	.256	.326
	Marble Cyn., Az.	11	.033	.806***	.429	.479
	Moab, Ut.	13	.412*	.195*	.304	.681*
Y. elata	Glen Cyn. City, Ut.	43	.214**	.447***	.340	.380
	Cottonwood, Az.	10	.073	.064	.591	.809
	Wickenburg, Az.	13	.471**	.059	.456	.868**
	Cedar City, Ut.	20	.725***	.203***	.291	.335
	LaVerkin, Ut.	10	.364	.046	.676	.884
	Zion NP, Ut.	30	.500***	.023	.413	.921***
	Dewey, Az.	10	.659**	.023	.550	1.084**
Y. gilbertiana	Cedar City, Ut.	33	.091	.574***	.317	.324
Y. glauca	Poncha Spr., Co.	33	.011	.669***	.192	.561***
	Saguache, Co.	42	.200**	.600***	.252	.473***
Y. kanabensis	Kanab, Ut.	31	.054	.599***	.255	.428*
Y. neomexicana	Montrose, Co.	24	.039	.918***	.122	.020

tions were *Y. elata* in general, and *Y. angustissima* at Peach Springs, Arizona, and *Y. baileyi* at Moab, Utah. *Yucca elata* is similar to *Y. whipplei* in having large, branched inflorescence with many hundreds of flowers. The other yuccas are all smaller, with inflorescences averaging 100 flowers or less (Table 9.1).

In an obligate and highly specific pollination system, such as the yucca-yucca moth interaction, pollinator limitation of fruit production in individual plants could be accentuated for three reasons. First, there is considerable variation in the density of flowering individuals from year to year, particularly for populations of *Y. angustissima, Y. baileyi, Y. harrimaniae* and *Y. schidigera*. If the factors which initiate flowering are not consistent with factors which initiate the emergence of adult moths, then considerable year to year variation could exist in the number of moths per flower. In less specific and more facultative pollination systems, one pollinator may compensate for relatively low numbers of another (Inouye 1978). Second, the flight and flowering seasons may not be synchronized within a season. I believe that synchronization may be the exception rather than the rule for the yuccas of the southwestern United States. However, it will be difficult to assess this, because moths cannot be censused independently of flowers. Third, some plants may differentially attract pollinators, achieve high initial fruit set and be resource limited, while less attractive plants may become pollinator limited. This hypothesis is essentially one of intraspecific competition for pollinators.

Intraspecific Competition for Mutualists

A demonstration of mutualists as a limiting resource is practically essential for demonstrating competition for mutualists (but see Waser 1983). However, at densities normally occurring in the field, a species need not be limited by its mutualists, and even if it does, competition will not necessarily occur. A direct demonstration of intraspecific, consumptive or pre-emptive competition for mutualists must show that in addition to mutualists being limiting, that as the density of a species increases, its *per capita* population growth rate or the fitness of individuals declines, and that this density dependence is at least a function of the availability of mutualists. There are few studies in which there is a direct demonstration of all three criteria. In most cases direct information is incomplete, or the case for competition is based upon indirect or secondary information.

Pre-emptive intraspecific competition may occur among trees for dispersal agents, particularly for highly specific dispersal agents (Howe and Estabrook 1977). Trees with large seed crops or trees close to conspecifics that are also fruiting may saturate the numerical and functional responses of the dispersal agents, so that fruit removal rates are lower than with smaller crops or lower tree density (Howe and Estabrook 1977). In northwestern Costa Rica the proportion of the seed crop of *Casearia corymbosa* removed by seed dispersing birds is greatest for intermediate fruit crop sizes per tree, trees with large and small crops having smaller proportion of the seed crop removed (Howe and Vande Kerckhove 1979). In 1979 there was no effect of fruit crop size on the proportion of fruits removed from *Virola surinamensis* by birds and monkeys on Barro Colorado Island (Howe and Vande Kerckhove 1980). However, in a subsequent study fruit removal rates for this tree were shown to be a decreasing function of the number of neighbours fruiting within a radius of 50 m (Manasse and Howe 1983).

In ant-homopteran mutualisms there is strong, but indirect evidence for pre-emptive intraspecific competition among different populations of homopterans for the services of ants. The foraging range of *Formica obscuripes* and *Dolichoderus taschenbergi* became larger when there were fewer aphids to tend on jack pine in Manitoba (Bradley and Hinks 1968). Therefore, aphid populations farther from the nest are likely to be poorly tended or abandoned, and newly founded colonies at the periphery of the foraging range are less likely to be tended when other, closer colonies already exist. As the number of populations of aphids on fireweed in Colorado increases at a given site, fewer newly founded populations become tended (Addicott, unpublished). Given the importance of ants for these aphids (Addicott 1979), this provides indirect support for the hypothesis that aphids compete intraspecifically for ants.

The conditions required for competition are not always met. For example, Fritz (1982) finds that the number of ants tending membracids (*Vanduzea arquata*) on black locust increases with the density and age of the membracids, essentially as a function of the rate of honeydew production, and there is no indication that membracids compete for the services of ants. In the high arctic Hocking (1968) found that there was generally an excess of nectar in flowers, indicating that pollinating bees and flies were probably not in competition for this resource. The many plants that are resource rather than pollination limited will not compete intraspecifically for pollination, but they might compete interspecifically (see below).

It is tempting to assume that a demonstration of limitation by mutualists suffices as a demonstration of competition for mutualists (Schemske *et al.* 1978; Zimmerman 1980). However, this is not the case. For example, energetic constraints on foraging in ants could restrict the number of tending ants per aphid, and this number could be limiting to aphid fitness. With an increase in density of aphids, more ants would forage, but the number of ants per aphids might remain constant and no competition would result. This is seen in ants which tend membracids on black locust (Fritz 1982), where the number of ants tending increases with the rate of honeydew production, which is a function of instar and generation. Another reason to reject limitation as a demonstration of competition is that instead of competing for mutualists, individuals or species can be mutualists for their mutualists (see below).

Establishing that observed density dependence is a function of the availability of mutualists, as opposed to some other potentially limited resource, can be the most difficult task. Simply observing decreased fitness of a mutualistic species as its density increases does not mean that there is competition for mutualists. Density dependence can arise intraspecifically in mutualistic systems in many ways (see Addicott 1979, 1985), often in response to factors unrelated to mutualism. The work of Zimmerman (1980) shows very clearly that seed set in *Pollimonium foliossisimum* is limited by pollinators, but in the absence of differential set among individuals or manipulations of plant density, competition for pollinators is not demonstrated. Manupulations of plant density would be required and would have to control for other potentially limiting factors, such as water or soil nutrients. This could be done either by repeating hand pollination to control statistically for changes in resource availability, or by using potted plants which do not interact except through pollinators (Waser 1978a; Zimmerman, personal communication).

Intraspecific competition can also occur through mechanisms involving direct, usually aggressive, encounters among individuals. Encounter competition can occur in the absence of limiting resources. However, if encounters prevent access to mutualistic resources, those resources may effectively become limiting, and it is reasonable to consider cases of competition for mutualists through encounter competition.

Intraspecific competition for mutualists occurs in symbiotic mutualisms in which aggressive, territorial behaviour of the symbiont limits the number of symbionts per host. With symbiotic interactions, the basis for the aggressive interactions are relatively clear, as those individuals which are successful are able to obtain the resources or services which their host provides. Therefore, any competition which occurs is presumably for mutualists. For example, the polychaete worm, *Arctonoe pulchra* is a symbiont of the limpet, *Megathura crenulata*. In the field there is usually just one worm per occupied host, and the worms are intraspecifically aggressive (Dimock 1974). The worms move from heavily occupied hosts to unoccupied hosts, leaving a non-random, regular dispersion pattern (Dimock 1974). There is a similar interaction among conspecific crabs (*Trapezia* spp.) and shrimp (*Alpheus* spp.), that are symbiotic on pocilloporid corals. The crabs and shrimp feed on the mucus produced by the corals, and they deter predators from feeding on their host corals (Glynn 1976, 1980). Within a species aggressive interactions lead to no more than a single male-female pair of crabs per coral head (Preston 1973).

Where mutualists are not symbiotic, it is not as easy to infer that the ultimate basis for aggressive interactions is the resources or services provided by the mutualist. Within an exclusive use area there may be many other resources important to the species. Therefore, our ability to infer intraspecific competition for mutualists in territorial systems depends upon our ability to infer the importance of the resources provided by the mutualists relative to other resources. For many ants which tend homopterans of EFNs, for nectar feeding birds, and for nectar and pollen feeding bees which exhibit intraspecific territoriality (for example, Bradley 1973; Feinsinger 1976; Johnson and Hubbell 1975) the resources provided by their mutualists appear to be important relative to other kinds of resources, and it is reasonable to infer intraspecific competition for mutualists.

Interspecific Competition for Mutualists

As with intraspecific competition for mutualists, there are a number of qualitatively different mechanisms of competition, including encounter, consumption, pre-emption, and one mechanism specific to pollination systems. Competition for mutualists among different species also takes on a number of added complexities. The simplest form of competition for mutualists involves two species and a single mutualist. However, when two or more species interact with two or more of the same mutualist species, new patterns can arise. Differences in the quality of resources provided by mutualists can add a new dimension to competition, and there may be simultaneous or sequential competition among both sets of species for the services of the other.

Encounter Competition

Indirect evidence for interspecific competition for mutualists based upon aggressive encounters is relatively common. The ant mosaic is a product of encounter competition (Leston 1978, Bradley 1973), and in many systems the most important resources within areas of exclusive use are the resources provided by mutualists: honeydew from homopterans or extrafloral nectar from plants. Among pollinators aggressive interactions are common. Hummingbirds display interspecific territoriality, which has a pronounced effect upon the fitness, distribution and abundance of other species of hummingbirds (for example, Feinsinger 1976; Kodric-Brown and Brown 1978). Bees show interspecific competition by aggressive interactions (for example, Roubik 1978; Johnson and Hubbell 1975; Hubbell and Johnson 1977). Aggressive behaviour may also be expressed among guild members from very different taxonomic groups. For example, hummingbirds exhibit aggressive behaviour to a variety of insect species (Carpenter 1979; Boyden 1978; Primack and Howe 1975), and in some cases the reverse occurs (Roubik 1982).

Consumptive Competition

Competition may arise through the differential abilities of species to exploit the resources provided by their mutualists (for example, Bond and Brown 1979; Brown *et al.* 1981; Heinrich 1976b; Inouye 1978). Brown *et al.* (1981) studied the effects of different environmental conditions on the potential for interspecific competition among hummingbirds and bees for the nectar of *Ribes pinetorum* and *Chilopsis linearis* in southeastern Arizona. At high elevations *R. pinetorum* is visited by both hummingbirds and bees. Bees can forage only during the warmer parts of the day, but their foraging lowers nectar levels in *R. pinetorum*. This displaces hummingbird foraging to the cooler parts of the day. At lower elevations the same principles hold, but the pattern is different. Higher temperatures throughout the day allow bees to forage on *C. linearis* during all daylight hours, and they can again deplete nectar to levels insufficient for hummingbird foraging. When bee numbers are low, hummingbirds can forage on *C. linearis* during early morning hours, but with higher bee numbers there are no times when nectar levels are high enough for hummingbirds.

Inouye (1978) experimentally demonstrated a similar pattern for two species of bumblebees (*Bombus appositus* and *B. flavifrons*) feeding on the nectar of two flowers (*Delphinium barbeyi* and *Aconitum columbianum*) in montane meadows in Colorado. Under normal circumstances *D. barbeyi* is visited primarily by *B. appositus* while *A. columbianum* is visited primarily by *B. flavifrons*. However, in a series of experiments in which the density of first one and then the other bumblebee was lowered by capturing it at its preferred plant, Inouye showed that each species of bumblebee would forage on the other species of plant in much greater numbers that it did before. He argues that this is not due to aggressive interactions, but instead results from differential abilities to utilize nectar from each plant.

Pre-emptive Competition

If the number of mutualists is limited in a given area, then two or more species could compete for that mutualist by pre-emption: while individuals of one species are receiving the services of the mutualist, individuals of the other species may not receive that same service. This phenomenon is shown clearly in ant-homopteran and ant-EFN mutualism, where homopterans or plants may compete for the services of ants. Buckley (1983) describes a system in Australia in which *Acacia* sp. with EFNs competes with the membracid treehopper, *Sextius virescens*, for tending by an *Irodomyrmex* ant. Ants tend both EFNs and membracids on acacia. The membracids occurred in sufficient numbers and concentrations to attract ants away from the EFNs. This need not have been detrimental to *Acacia*, for ants tending homopterans may still have a beneficial effect on the host plants (Messina 1981; Fritz 1983). However, in Buckley's system the membracids were less well dispersed over the host plants than were the EFNs and consequently the ant defence against herbivores of acacia was less effective when membracids were present.

Competition among myrmecochorous plants for the services of ants is likely to occur through pre-emption. Davidson and Morton (1981b) present a pattern which could be the result of interspecific competition among two species of myrmecochorous plants for the services of ants. In Australia the ant *Rhytidoponera* sp. disperses *Sclerolaena diacantha* to its mounds. This plant occurs in greater densities on the mound than off the mounds, and in greater densities when it is the only myrmecochorous plant available to the ants. This could be the result of sharing dispersal services of the ants, but it could also result from competition on ant mounds with other myrmecochorous plants.

I have indirect evidence for the existence of interspecific competition among aphids for the services of ants (Addicott 1978). Early in the season *Formica integroides* tended only *Aphis farinosa* on willow, even though there were large numbers of *A. varians* on fireweed. When the densities of aphids on willow declined, only then did these ants begin to tend aphids on fireweed. Similarly, aphids on fireweed performed less well when they were in close proximity to shrubs on which other species of aphids were being tended, presumably as the result of sharing a limited supply of ants.

Brown and Kodric-Brown (1979) question whether pre-emptive competition will occur among plants for pollinators. They argue that 'competition for pollinators does not use up the limited resource and make it absolutely unavailable to other individuals' and consequently 'complete interspecific overlap in utilization of pollinators need not result in competitive exclusion or even significant reductions in population density among competing plant species'. That is, the fact that one plant is visited by a bird, does not preclude another plant from being visited by that same bird at a later time. However, whether or not there will be competition among the plants will depend upon the extent to which plant reproduction is limited by the number of pollinator visits, and the extent to which the number of visits to a plant is affected by the number of conspecific and heterospecific neighbours which it has.

Competition through Interspecific Pollen Transfer

The greatest amount of work on interspecific competition for pollinators has been conducted on pollination systems. Waser (1978a, 1978b, 1983) has argued that plants compete for pollination in two distinctly different ways. One is through differential attraction or sharing of a limited number of pollinators (see above). The second involves interspecific transfer of pollen, which would result in either the loss of pollen for conspecific pollination or the clogging of stigmas with heterospecific pollen (Feinsinger 1978; Waser 1978a). As with other forms of interference, competition by interspecific pollen transfer can occur even when pollinators are not limiting (Waser 1983).

Waser (1978a) has direct, experimental evidence for this form of competition between two species of montane plants in Colorado. *Delphinium nelsoni* and *Ipomopsis aggregata* share hummingbird and bumblebee pollinators. *Delphinium nelsoni* blooms before *I. aggregata*, but there is a period of overlap in flowering, and interspecific transfer of pollen occurs. During the period of overlap, the seed set for each species was lower than during non-overlap periods (Waser 1978a). By using potted plants in pure and mixed stands, Waser (1978a) showed that plants in mixed stands had lower seed set than plants in pure stands even though grown at the same time under the same conditions, thereby demonstrating competition.

Secondary Evidence for Competition — Phenology

Waser's study is one of the few studies in which there is a direct, experimental demonstration of competition. Much of the controversy surrounding the study of competition stems from the use of secondary evidence: correspondence between observed and predicted patterns of niche differences based upon competition theory. There is a variety of evidence for morphological, behavioural, and phenological differences among plants or among pollinators which could be interpreted as the result of competition for mutualists (for example Inouye 1980; Mosquin 1971; Reader 1975; Stiles 1977; Waser 1983).

The credibility of the hypothesis that competition for mutualists (pollinators or dispersal agents) could select for staggered flowering or fruiting times depends upon two kinds of evidence. First, are there models which predict that competition for pollinators or dispersal agents could cause staggered floweriing times? Second, are patterns of flowering or fruiting phenology more uniformly distributed than would be expected by chance, as predicted by the competition hypothesis? Waser (1978b) has adequately addressed the first question in a simulation model of competition for pollinators, based upon loss of pollen, loss of receptive stigmatic surfaces and loss of visits. He also argues (Waser 1978a) that competition for pollination based upon pre-emption would be relatively ineffective, because only one species, the poorer competitor, would be under much selection.

The more controversial problem is whether phenological patterns are actually non-random. As with other questions that depend upon a 'null' hypothesis, the choice of a 'null' model, the choice of analysis procedures, and the choice of data sets to analyse, influences conclusions (Cole 1981; Pleasants 1980; Rabinowitz *et al.* 1981; Poole and Rathcke 1979; Stiles 1977, 1979). Interpretation of secondary evidence from phenological patterns will remain controversial.

The shape of the flowering curve may also respond to competition for pollinators (Thomson 1980). Since competition for pollination is frequency dependent (Levin and Anderson 1970; Straw 1972) and since individual pollinators retain floral preferences (Free 1968), individual plants that flower with many other conspecifics should suffer less competition. This should lead to flowering curves with positive skewness (Thomson 1980). For the animal pollinated plants of montane meadows in Colorado, he finds that most species have flowering curves with positive skewness, and that where two species share pollinators sequentially, the second species tends to have a more positively skewed flowering curve than the second. To determine whether this or any other pattern predicted from competition theory is actually the result of competition for pollination will require manipulative experiments that will be difficult to perform.

Complexity of Competition in Multispecies Systems

When two species share just one mutualist, competition between them will be a relatively simple process. However, each species can a associate with two or more mutualists that they share, as occurs with most pollination systems (Schemske 1983), seed dispersal systems (Beattie 1983), and ant-EFN or ant-homopteran systems (Schemske 1982; Addicott 1979). Under these conditions competition for mutualists becomes considerably more complex.

One level of complexity arises from differences among mutualists. If there are significant differences in the resources or services provided by different mutualist species, and if one species differentially attracts the better mutualist, then the other species would probably achieve lower fitness and suffer competition. That there are significant differences in the quality of mutualists is well documented for pollination systems (for example, Primack and Silander 1975; but see Motten *et al.* 1981) and ant-homopteran systems

(Addicott 1978, 1979; Burns 1973). However, the consequences of such differences are just beginning to be explored.

Another level of complexity arises from the possibility of simultaneous or sequential competition among members of both sets of species for the services or resources of the others. In most studies, competition is examined in just one direction, such as plants competing for pollination or pollinators competing for plants. However, it is possible for both sets of mutualists to compete within themselves for the services of the other, either simultaneously or sequentially. Heinrich (1976) argues that plants could compete for pollinators at the same time that pollinators might compete for plants. Culver and Beattie (1978) speculate that during early spring ants could compete for seeds, whereas later in the summer plants might compete for the services of ants. I know of no study which has actually measured both kinds of competition at the same time.

Parasites of Mutualism

There are many situations in which the resources or services that are normally utilized by a mutualist are also utilized by species that do not provide reciprocal benefits. These species are called parasites of mutualism. Examples include bees and birds which utilize nectar but provide no pollination (Colwell *et al.* 1974), ants which do not protect their host plants (Janzen 1975), or fish which do not provide any cleaning service (Springer and Smith-Vaniz 1972). Soberon-Mainero and Martinez del Rio (Chapter 8, this volume) provide an extensive discussion of parasites of mutualism. The importance of parasites of mutualism in the present context is that they are potentially in competition with mutualists for the same resources and services (for example, Colwell *et al.* 1974; McDade and Kinsman 1980; Roubik 1982; Willmer and Corbet 1981).

Two examples illustrate this phenomenon. McDade and Kinsman (1980) studied the interaction between two hummingbird pollinated plants (*Aphelandra golfodulcensis* and *Justica aurea*) and their pollinators and nectar robbers in Costa Rica. Nectar robbing by birds, bees and ants lowered both the availability of nectar and its subsequent production. For *Justica aurea* they suggest that the level of nectar robbing causes the legitimate pollinators to cease visiting these plants. Roubik (1982) studied the consequences of nectar robbing by *Trigona ferricauda* on a forest-edge shrub, *Pavonia dasypetala* in Panama. This plant is normally pollinated by the hummingbird, *Phaethornis superciliosus*. However, bees aggressively defend the flowers against visits by hummingbirds. This lowers the number of visits by hummingbirds from about 1.8 visits per flower per hour to about 0.9 in robbed flowers. It is unclear what the impact of this is on the hummingbirds, but a reasonable assumption is that the birds must utilize less suitable plants as nectar sources.

Mutualism for Mutualists

I have argued above that limitation of a species by its mutualists does not necessarily imply that it competes either intra- or interspecifically for its mutualists. One reason for this is that rather than competing for mutualists, individuals of two or more species might facilitate the maintenance of their mutualists, resulting in mutualism for mutualists. At some densities, increases (or decreases) in the density of one or more species may increase the availability of mutualists, thereby increasing (or decreasing) their own fitness. More succintly, species involved in mutualism can encounter both positive and negative density-dependent interactions for mutualists, and simply knowing that mutualists are limiting does not indicate which (if either) will occur.

There are three reasons for considering mutualism for mutualists in this chapter. It is another possible outcome of experiments designed to test for competition for mutualists. In many systems both positive and negative density dependence is observed, depending upon density. And, an evolutionary consequence of competition for mutualists can be mutualist for mutualists.

Intraspecific Mutualism for Mutualists

Plants that occur in higher densities, and plants that flower or fruit at the same time as their conspecific may be at an advantage relative to plants in either temporal or spatial isolation. Advantages are seen in both pollination and seed dispersal interactions. For example, Augspurger (1981) showed that plants which bloomed in either spatial or temporal isolation from other plants had much lower pollination success, as well as higher seed predation rates. Platt *et al.*

(1974) showed that clumped plants had both higher pollination success and lower seed predation. Silander (1979) found that plants at the closest nearest neighbour distances had the higher pollinator success.

Seed dispersal systems show the same phenomenon. Manasse and Howe's (1983) observations on *Virola*, showed that the highest seed removal rates were on plants with intermediate numbers of neighbours within neighbourhoods of 50 m. In this system it is advantageous to be isolated in both space and time, but not too isolated. Moore and Willson (1982) demonstrated that fruit removal rates by thrushes and mimids on *Lindera benzoin* are greatest on isolated rather than clumped plants. Although all fruit were eventually removed from all plants, they speculate that faster removal may be advantageous if it leads to more successful dispersal or lower seed predation. The mutualistic seed dispersal system between various species of corvids and pines also shows the advantages of synchrony (Ligon 1978; Hutchins and Lanner 1982).

Interspecific Mutualism for Mutualists

Although species may compete for the services of mutualists, the sequential or simultaneous presence of these species may help to maintain their mutualist. Snow (1965) suggested the possibility of sequential mutualism among trees dependent upon frugivorous birds for the dispersal of their fruit. Competition among 22 species of *Miconia* on Trinidad for dispersal agents may have selected for fruiting seasons spaced throughout the year. However, a staggering of fruiting seasons of individual species would lead to a more constant fruit supply for birds throughout the year and provide conditions for maintaining sedentary populations of fruit eating birds (Snow 1965). Grant (1966) developed a similar argument for simultaneous mutualism among hummingbird pollinated plants in California. She argued that although competition might select for different pollen placement on birds to avoid interspecific pollen transfer, species with similar morphology might increase the density of flowers available to birds and allowing the birds to remain.

The best demonstration of sequential mutualism for mutualists comes from the work of Waser and Real (1979). *Delphinium nelsoni* and *Ipomopsis aggregata* compete for pollination (Waser 1978a), yet Waser and Real (1979) demonstrated that these two plants are sequential mutualists. In two of four years the number of flowers of the early blooming *D. nelsoni* was depressed below normal levels, and the number of hummingbird pollinators declined. The seed set of the later flowering *I. aggregata* was lower in these two years, compared to the years with normal *D. nelsoni* flowering and normal hummingbird numbers. Although these two species appear to be sequential mutualists, Waser and Real (1979) caution that such mutualisms are 'an unanticipated outcome, rather than a cause, of divergence in flowering time of sympatric species'. In other words, competition for pollination can lead to divergence in flowering time, which can in turn lead to effective mutualism between sequentially flowering species. A correlate of the sequential mutualism for pollinators is that the second or subsequent species in a sequence could produce less nectar and still attract and maintain its pollinator (Heinrich 1975).

Simultaneous mutualism for mutualists results when increases in the combined densities of two of more species allow their mutualist to remain in a given location, rather than leaving for another location. However, the benefits of simultaneous presence are frequency dependent (Bobisub and Neuhaus 1975; Straw 1972). Two rare species which are treated identically by pollinators could benefit by their mutual presence. However, as their frequencies increase the benefit would decline and competition might result. This illustrates the fine line and interconnections between mutualism and competition.

Evidence supporting this idea varies. Bierzychudek (1981a) attempted to determine whether the 'Müllerian' mimics, *Asclepias curassavica* and *Lantana camara* in Costa Rica, actually received greater pollinator visitation when they occurred together; her results were negative, but this may be a function of scale of measurement (Bierzuchudek 1981a). Atsatt (1981) suggested that some Australian lycaenids and homopterans could be simultaneous mutualists. He observed lycaenids ovipositing primarily where homopterans were being tended by ants, so that the lycaenids benefited from the presence of homopterans. Only when lycaenids become too numerous and begin to draw off visits from homopterans would lycaenids and homopterans begin to come into competition. Schemske (1981) observed pollination of two species of *Costus* by the same pollinator, *Euglossa imperialis*. These plants are rare, understory, tropical herbs, which are pollinated only by this bee. He observed no evidence for competition among these plants, and concluded that they represent a case of convergence for pollinator sharing among rare species.

Whether competition or simultaneous mutualism exists may depend not just on the relative densities of species, but also on their interspersion (Thomson 1982). If plants are separated into large clumps, then flower constancy and movement patterns of pollinators could make these plants competitors. Similarly, if the plants are completely interspersed, then there will be a high number of interspecific movements, and competition through interspecific pollen transfer will occur. Only when plants show an intermediate degree of interspersion will they potentially become mutualists.

An entirely different class of mutualism for mutualists depends upon species participating in two or more different kinds of mutualism simultaneously. For example, Cluett and Boucher (1983) suggested that for plants with both mycorrhizae and nitrogen fixing bacteria, there might be competition between the two types of mutualists for their host plant. However, their data indicate just the opposite. Where plants are infected with mycorrhizae, the nitrogen fixing bacteria did better. These kinds of interactions in complex mutualistic systems deserve further attention.

Concluding Remarks

There have been many studies which have directly or indirectly inferred the existence of competition for mutualists. However, two major reviews of competition (Schoener 1983; Connell 1983) show that competition for mutualists has provided very little information about competition, relative to studies of other kinds of systems. Schoener (1983) surveyed 164 experimental field studies of competition, and of these only five would appear to involve systems in which the basis of competition is either a service or resource provided by a mutualist. Of the 72 studies surved by Connell (1983), only two possibly involved competition for mutualists.

This reflects an almost complete absence of manipulative, field experiments designed to demonstrate the existence of competition for mutualists. There are numerous studies in which the relatively weak evidence of patterns or natural experiments (Connell 1975, Waser 1983) suggests the existence of competition for mutualists, but studies using the stricter criteria based upon manipulation of density and resources (Reynoldson and Bellamy 1970; Connell 1975) are rarely attempted. However, mutualistic systems have great potential for the study of competition, and manipulative approaches should be adopted more widely.

By definition mutualists provide resources or services which are important. Although it is tempting to assume that mutualists are limiting, this need not be the case, as is shown particularly well in pollination systems. Further attention should be given to the conditions under which limitation does or does not occur, how much mutualist densities can vary without a species being limited, and particularly spatial and temporal variation in limitation.

With respect to developing our understanding of competition for mutualists, the greatest opportunity exists in complex systems where two or more species each interact with a common set of mutualists. Differences in the quality of mutualists, and the potential for simultaneous or sequential competition in both directions will make these systems difficult, but rewarding, to analyse.

Finally, one of the greatest challenges facing students of mutualism is how to make the jump from individual level effects to population level consequences (Addicott 1985). In most studies of mutualism, in general, and competition for mutualists, specifically, we generally measure one or a few correlates of fitness at an individual level. Differences in the relative success of individuals will alter the genetic composition of the population, but will it alter the dynamics or density of the population?

References

Addicott, J.F. (1978) 'Competition for Mutualists: Aphids and Ants', *Canadian Journal of Zoology, 56*, 2093–6

Addicott, J.F. (1979) 'A Multispecies Aphid-Ant Association: Density Dependence and Species-Specific Effects', *Canadian Journal of Zoology, 57*, 558–69

Addicott, J.F. (1985) 'On the Population Consequences of Mutualism' in Case, T.J. and J. Diamond (eds), *Ecological Communities*, Harper and Row, New York (in press)

Aker, C.L. (1982) 'Regulation of Flower, Fruit and Seed Production by a Monocarpic Perennial, *Yucca whipplei*', *Journal of Ecology, 70*, 357–72

Aker, C.L. and D. Udovic (1981) 'Oviposition and Pollination Behavior of the Yucca Moth, *Tegeticula maculata* (Lepidoptera: Prodoxidae), and its Relation to the Reproductive Biology of *Yucca whipplei* (Agavacae)', *Oecologia, 49*, 96–101

Atsatt, P.R. (1981) 'Ant-Dependent Food Plant Selection by the Mistletoe Butterfly *Ogyis amaryllis* (Lycaenidae)', *Oecologia, 48*, 60–3

Augspurger, C.K. (1981) 'Reproductive Synchrony of a Tropical Shrub: Experimental Studies on Effects of Pollinators and Seed Predators on *Hybanthus prunifolius* (Violaceae)', *Ecology, 62*, 775–88

Barbehenn, K.R. (1969) 'Host-Parasite Relationship and Species Diversity in Mammals: An Hypothesis', *Biotropica, 1*, 29–35

Beattie, A.J. (1983) 'Distribution of Ant-Dispersed Plants', *Sonderbd. Naturwiss. Ver. Hamburg, 7*, 249–70

Bentley, B.L. (1976) 'Plants Bearing Extrafloral Nectaries and the Associated Ant Community: Interhabitat Differences in the Reduction of Herbivore Damage', *Ecology, 57*, 815–20

Bierzychudek, P. (1981a) '*Asclepias, Lantana* and *Epidendrum*: a Floral Mimicry Complex?', *Biotropica, 13 (Supplement)*, 54–8

Bierzychudek, P. (1981b) 'Pollinator Limitation of Plant Reproductive Effort', *American Naturalist, 117*, 838–40

Bobisud, L.E. and R.J. Neuhaus (1975) 'Pollinator Constancy and Survival of Rare Species', *Oecologia, 21*, 263–72

Bond, H.W. and W.L. Brown (1979) 'The Exploitation of Floral Nectar in *Eucalyptus incrassata* by Honeyeaters and Honeybees', *Oecologia, 44*, 105–11

Boyden, T.C. (1978) 'Territorial Defence Against Hummingbirds and Insects by Tropical Hummingbirds', *Condor, 80*, 216–21

Bradley, G.A. (1973) 'Interference Between Nest Populations of *Formica obscuripes* and *Dolichoderus taschenbergi* (Hymenoptera: Formicidae)', *Canadian Entomologist, 105*, 1525–8

Bradley, G.A. and J.D. Hinks (1968) 'Ants, Aphids, and Jack Pine in Manitoba', *Canadian Entomologist, 100*, 40–50

Brown, J.H. and A. Kodrick-Brown (1979) 'Convergence, Competition, and Mimicry in a Temperate Community of Hummingbird Pollinated Flowers', *Ecology, 60*, 1022–35

Brown, J.H., A. Kodric-Brown, T.G. Whitman and H.W. Bond (1981) 'Competition Between Hummingbirds and Insects for the Nectar of Two Species of Shrubs', *Southwestern Naturalist, 26*, 133–45

Buckley, R. (1983) 'Interaction Between Ants and Membracid Bugs Decreases Growth and Seed Set of Host Plant Bearing Extrafloral Nectaries', *Oecologia, 58*, 132–6

Burns, D.P. (1973) 'The Foraging and Tending Behavior of *Dolichoderus taschenbergi* (Hymenoptera: Formicidae)', *Canadian Entomologist, 105*, 97–104

Carpenter, F.L. (1979) 'Competition Between Hummingbirds and Insects for Nectar', *American Zoologist, 19*, 1105–14

Cluett, H.C. and D.H. Boucher (1983) 'Indirect Mutualism in the Legume-*Rhizobium*-Mycorrhizal Fungus Interaction', *Oecologia, 59*, 405–8

Cole, B.J. (1981) 'Overlap, Regularity, and Flowering Phenologies', *American Naturalist, 117*, 993–7

Colwell, R.K., B.J. Betts, P. Bunnell, F.L. Carpenter and P. Feinsinger (1974) 'Competition for the Nectar of *Centropogon valerii* by the Hummingbird *Colibri thalassinus* and the Flower-Piercer *Diglossa plumbea*, and its Evolutionary Implications', *Condor, 76*, 447–52

Colwell, R.K. and E.R. Fuentes (1975) 'Experimental Studies of the Niche', *Annual Review of Ecology and Systematics, 6*, 281–310

Connell, J.H. (1975) 'Some Mechanisms Producing Structure in Natural Communities: a Model and Evidence from Field Experiments' in Cody, M.L. and J.M. Diamond (eds), *Ecology and Evolution of Communities*, Belknap Press, Cambridge, Mass., pp. 460–90

Connell, J.H. (1983) 'On the Prevalence and Relative Importance of Interspecific Competition: Evidence from Field Experiments', *American Naturalist, 122*, 661–96

Cruden, R.W., S. Kinsman II, R.E. Stockhouse and Y.B. Linhart (1976) 'Pollination, Fecundity, and the Distribution of Moth-Flowered Plants', *Biotropica, 8*, 204–10

Culver, D.C. and A.J. Beattie (1978) 'Myrmecochory in *Viola*: Dynamics of Seed-Ant Interactions in some West Virginia Species', *Journal of Ecology, 66*, 53–72

Davidson, D.W. (1980) 'Some Consequences of Diffuse Competition in a Desert Ant Community', *American Naturalist, 116*, 92–105

Davidson, D.W. and S.R. Morton (1981a) 'Myrmecochory in some Plants (F. Chenopodiaceae) of the Australian Arid Zone', *Oecologia, 50*, 357–66

Davidson, D.W. and S.R. Morton (1981b) 'Competition for Dispersal in Ant-Dispersed Plants', *Science, 213*, 1259–61

Davis, D.R. (1967) 'A Revision of the Moths of the Subfamily Prodoxinae (Lepidoptera: Incurvariidae)', *United States National Museum Bulletin, 255*, 1–170

Dimock, R.V. Jr (1974) 'Intraspecific Aggression and the Distribution of a Symbiotic Polychaete on its Host' in Vernberg, W.B. (ed.), *Symbiosis in the Sea*, University of South Carolina Press, Columbia, pp. 29–44

Feinsinger, P. (1976) 'Organization of a Tropical Guild of Nectarivorous Birds', *Ecological Monographs, 46*, 257–91

Feinsinger, P. (1978) 'Ecological Interactions Between Plants and Hummingbirds in a Successional Tropical Community', *Ecological Monographs, 48*, 269–87

Free, J.B. (1968) 'Dandelion as a Competitor to Fruit Trees for Bee Visits', *Journal of Applied Ecology, 5*, 169–78

Friedmann, H. and J. Kern (1956) '*Micrococcus cerolyticus*, nov. sp., an Aerobic Lypolytic Organism Isolated from the African Honey-Guide', *Canadian Journal of Microbiology, 2*, 515–7

Fritz, R.S. (1982) 'An Ant-Treehopper Mutualism: Effects of *Formica subsericea* on the Survival of *Vanduzea arquata*', *Ecological Entomology, 7*, 267–76

Fritz, R.S. (1983) 'Ant Protection of a Host Plant's Defoliator: Consequence of an Ant-Membracid Mutualism', *Ecology, 64*, 789–97

Glynn, P.W. (1976) 'Some Physical and Biological Determinants of Coral Community Structure in the Eastern Pacific', *Ecological Monographs, 46*, 431–56

Glynn, P.W. (1980) 'Defence by Symbiotic Crustacea of Host Corals Elicited by Chemical Cues from Predator', *Oecologia, 47*, 287–90

Grant, K.A. (1966) 'A Hypothesis Concerning the Prevalence of Red Coloration in California Hummingbird Flowers', *American Naturalist, 100*, 85–97

Gross, R.S. and P.A. Werner (1983) 'Relationships Among Flowering Phenology, Insect Visitors, and Seed-Set of Individuals: Experimental Studies on Four Co-Occuring Species of Goldenrod (*Solidago*: Compositae)', *Ecological Monographs, 53*, 95–117

Handel, S.N. (1978) 'The Competitive Relationship of 3 Woodland Sedges and its Bearing on the Evolution of Ant Dispersal of *Carex pedunculata*', *Evolution, 32*, 151–63

Heinrich, B. (1975) Bee Flowers: a Hypothesis on Flower Variety and Blooming Times', *Evolution, 29*, 325–34

Heinrich, B. (1976a) 'Flowering Phenologies — Bog, Woodland, and Disturbed Habitats', *Ecology, 57*, 890–9

Heinrich, B. (1976b) 'Resource Partitioning Among Some Eusocial Insects: Bumble-bees', Ecology, 57, 874–89

Hocking, B. (1968) 'Insect-Flower Associations in the High Arctic with Special Reference to Nectar', *Oikos, 19*, 359–87

Howe, H.F. and G.F. Estabrook (1977) 'On Intraspecific Competition for Avian Dispersers in Tropical Trees', *American Naturalist, 111*, 817–32

Howe, H.F. and G.A. Vande Kerckhove (1979) 'Fecundity and Seed Dispersal of a Tropical Tree', *Ecology, 60*, 180–9

Howe, H.F. and G.A. Vande Kerckhove (1980) 'Nutmeg Dispersal by Tropical Birds', *Science, 210*, 925–7

Hubbell, S.P. and L.K. Johnson (1977) 'Competition and Nest Spacing in a Tropical

Stingless Bee Community', *Ecology, 58*, 949–63

Hutchins, H.E. and R.M. Lanner (1982) 'The Central Role of Clark's Nutcracker in the Dispersal and Establishment of Whitebark Pine', *Oecologia, 55*, 192–201

Inouye, D.W. (1978) 'Resource Partitioning in Bumblebees: Experimental Studies of Foraging Behavior', *Ecology, 59*, 672–8

Inouye, D.W. (1980) 'The Effect of Proboscis and Corolla Tube Lengths on Patterns and Rates of Flower Visitation by Bumblebees', *Oecologia, 45*, 197–201

Inouye, D.W. and O.R. Taylor, Jr (1979) 'A Temperate Region Plant-Ant-Seed Predator System: Consequences of Extrafloral Nectar Secretion by *Helianthella quinquenervis*', Ecology, 60, 1–7

Janzen, D.H. (1969) 'Allelopathy by Myrmecophytes: The Ant *Azteca* as an Allelopathic Agent of *Cecropia*', *Ecology, 50*, 147–53

Janzen, D.H. (1975) '*Pseudomyrmex nigropilosa*: a Parasite of a Mutualism', *Science, 188*, 936–8

Janzen, D.H., P. DeVries, D.E. Gladstone, M.L. Higgins and T.M. Lewisohn (1980) 'Self- and Cross-Pollination of *Encyclia cordigera* (Orchidaceae) in Santa Rosa National Park, Costa Rica', *Biotropica, 12*, 72–4

Johnson, L.K. and S.P. Hubbell (1975) 'Contrasting Foraging Strategies and Coexistence of Two Bee Species on a Single Resource', *Ecology, 56*, 1398–406

Kodric-Brown, A. and J.H. Brown (1978) 'Influence of Economics, Interspecific Competition, and Sexual Dimorphism on Territoriality of Migrant Rufous Hummingbirds', *Ecology, 59*, 285–96

Lawlor, L.R. (1979) 'Direct and Indirect Effects of N-Species Competition', *Oecologia, 43*, 355–64

Leston, D. (1978) 'A Neotropical Ant Mosaic', *Annals of the Entomological Society of America, 71*, 649–53

Levin, D.A. and W.W. Anderson (1970) 'Competition for Pollinators between Simultaneously Flowering Species', *American Naturalist, 104*, 455–67

Ligon, J.D. (1978) 'Reproductive Interdependence of Pinon Jays and Pinon Pines', *Ecological Monographs, 48*, 111–26

McDade, L. and S. Kinsman (1980) 'The Impact of Floral Parasitism in Two Neotropical Hummingbird-Pollinated Plant Species', *Evolution, 34*, 944–58

Macior, L.W. (1971) 'Coevolution of Plants and Animals — Systematic Insights from Plant-Insect Interactions', *Taxon, 20*, 17–28

Manasse, R.S. and H.F. Howe (1983) 'Competition for Dispersal Agents Among Tropical Trees: Influences of Neighbors', *Oecologia, 59*, 185–90

May, R.M. (1976) 'Models for Two Interacting Populations' in May, R.M. (ed.), *Theoretical Ecology: Principles and Applications*, W.B. Saunders Co., Philadelphia, pp. 49–70

Messina, F.J. (1981) 'Plant Protection as a Consequence of an Ant-Membracid Mutualism: Interactions on Goldenrod (*Solidago* sp.)', *Ecology, 62*, 1433–40

Miles, N.J. (1983) 'Variation and Host Specificity in the Yucca Moth, *Tegeticula yuccasella* (Incurvariidae): A Morphometric Approach', *Journal of the Lepidopterists Society, 37*, 207–16

Moore, L.A. and M.F. Willson (1982) 'The Effect of Microhabitat, Spatial Distribution, and Display Size on Dispersal of *Lindera benzoin* by Avian Frugivores', *Canadian Journal of Botany, 60*, 557–60

Mosquin, T. (1971) 'Competition for Pollinators as a Stimulus for the Evolution of Flowering Time', *Oikos, 22*, 398–402

Motten, A.F. (1983) 'Reproduction of *Erythronium umbilicatum* (Liliaceae): Pollination Success and Pollinator Effectiveness', *Oecologia, 59*, 351–9

Motten, A.F., D.R. Campbell, D.E. Alexander and H.L. Miller (1981) 'Pollination Effectiveness of Specialist and Generalist Visitors to a North Carolina Population of *Claytonia virginica*', *Ecology, 62*, 1278–87

O'Dowd, D.J. and E.A. Catchpole (1983) 'Ants and Extrafloral Nectaries: No Evidence for Plant Protection in *Helichrysum* spp. — Ant Interactions', *Oecologia*, 59, 191–200

Osman, R.W. and J.A. Haugeness (1981) 'Mutualism Among Sessile Invertebrates: A Mediator of Competition and Predation', *Science, 211*, 846–8

Petersen, C., J.H. Brown and A. Kodric-Brown (1982) 'An Experimental Study of Floral Display and Fruit Set in *Chilopsis linearis* (Bignoniaceae)', *Oecologia, 55*, 7–11

Pianka, E.R. (1974) *Evolutionary Ecology*, Harper & Row Publishers, New York

Platt, W.J., G.R. Hill and S. Clark (1974) 'Seed Production in a Prairie Legume (*Astragalus canadensis* L.). Interactions Between Pollination, Predispersal Seed Predation, and Plant Density', *Oecologia, 17*, 55–63

Pleasants, J.M. (1980) 'Competition for Bumblebee Pollinators in Rocky Mountain Plant Community', *Ecology, 61*, 1446–59

Poole, R.W. and B.J. Rathcke (1979) 'Regularity, Randomness, and Aggregation in Flowering Phenologies', *Science, 302*, 470–1

Powell, J.A. and R.A. Mackie (1966) 'Biological Relationships of Moths and *Yucca whipplei* (Lepidoptera: Gelichiidae, Blastobasidae, Prodoxidae)', *University of California Publications in Entomology, 42*, 1–59

Preston, E.M. (1973) 'A Computer Simulation of Competition Among Five Sympatric Congeneric Species of Xanthid Crabs', *Ecology, 54*, 469–83

Price, M.V. and N.M. Waser (1979) 'Pollen Dispersal and Optimal Outcrossing in *Delphinium nelsoni*', *Nature, 277*, 294–7

Primack, R.B. and H.F. Howe (1975) 'Interference Competition Between a Hummingbird (*Amazilia tzacatl*) and Skipper Butterflies (Hesperiidae)', *Biotropica, 7*, 55–8

Primack, R.B. and J.A. Silander (1975) 'Measuring the Relative Importance of Different Pollinators to Plants', *Nature, 255*, 143–4

Quinlan, R.J. and J.M. Cherret (1978) 'Aspects of the Symbiosis of the Leaf-Cutting Ant *Acromyrmex octospinosus* (Reich) and its Fungus Food', *Ecological Entomology, 3*, 221–30

Rabinowitz, D., J.K. Rapp, V.L. Sork, B.J. Rathcke, G.A. Reese and J.C. Weaver (1981) 'Phenological Properties of Wind- and Insect-Pollinated Prairie Plants', *Ecology, 62*, 49–56

Reader, R.J. (1975) 'Competitive Relationships of Some Bog Ericads for Major Insect Pollinators', *Canadian Journal of Botany, 53*, 1300–5

Reynoldson, T.B. and L.S. Bellamy (1970) 'The Establishment of Interspecific Competition in Field Populations, with an Example of Competition in Action Between *Polycelis nigra* (Mull.) and *P. tenuis* (Ijima) (Turbellaria, Tricladida)' in Boer, P.J. and G.R. Gradwell (eds), *Dynamics of Populations*, Pudoc, Wageningen, pp. 282–97

Risch, S. and D.H. Boucher (1976) 'What Ecologists Look for', *Bulletin of the Ecological Society of America, 57(3)*, 8–9

Roubik, D.W. (1978) 'Competitive Interactions Between Neotropical Pollinators and Africanized Honey Bees', *Science, 201*, 1030–2

Roubik, D.W. (1982) 'The Ecological Impact of Nectar-Robbing Bees and Pollinating Hummingbirds on a Tropical Shrub', *Ecology, 63*, 354–60

Schaffer, W.M. and M.V. Schaffer (1977) 'The Adaptive Significance of Variation in Reproductive Habit in Agavaceae' in Stonehouse, B. and C.M. Perrins (eds), *Evolutionary Ecology*, MacMillan, London, pp. 261–76

Schaffer, W.M. and M.V. Schaffer (1979) 'The Adaptive Significance of Variations in Reproductive Habit in the Agavaceae II: Pollinator Foraging Behavior and Selection for Increased Reproductive Expenditure', *Ecology, 60*, 1051–69

Schemske, D.W. (1981) 'Floral Convergence and Pollinator Sharing in Two Bee Pollinated Tropical Herbs', *Ecology, 62*, 946–64

Schemske, D.W. (1982) 'Ecological Correlates of a Neotropical Mutualism: Ant

Assemblages at *Costus* Extrafloral Nectaries', *Ecology, 63*, 932–41

Schemske, D.W. (1983) 'Limits to Specialization and Coevolution in Plant-Animal Mutualisms' in Nitecki, M.H. (ed.), *Coevolution*, University of Chicago Press, Chicago, pp. 67–110

Schemske, D.W., M.F. Willson, M.M. Melampy, L.J. Miller, L. Verner, K.M. Schemske and L.B. Best (1978) 'Flowering Ecology of Some Spring Woodland Herbs', *Ecology, 59*, 351–66

Schoener, T.W. (1983) 'Field Experiments on Interspecific Competition', *American Naturalist, 122*, 240–85

Silander, J.A. Jr (1979) 'Density-Dependent Control of Reproductive Success in *Cassia biflora*', *Biotropica, 10*, 292–6

Snow, A.A. (1982) 'Pollination Intensity and Potential Seed Set in *Passiflora vitifolia*', *Oecologia, 55*, 231–7

Snow, D.W. (1965) 'A Possible Selective Factor in the Evolution of Fruiting Seasons in Tropical Forests', *Oikos, 15*, 274–81

Springett, B.P. (1968) 'Aspects of the Relationship Between Burying Beetles, *Necrophorus* spp. and the Mite, *Poecilochirus necrophori*, Vitz', *Journal of Animal Ecology, 37*, 417–24

Springer, V.G. and W.F. Smith-Vaniz (1972) 'Mimetic Relationships Involving Fishes of the Family Bleniidae', *Smithsonian Contributions to Zoology, 112*, 1–36

Stephenson, A.G. (1981) 'Flower and Fruit Abortion: Proximate Causes and Ultimate Functions', *Annual Review of Ecology and Systematics, 12*, 253–79

Stiles, F.G. (1977) 'Coadapted Competitors: the Flowering Seasons of Hummingbird-Pollinated Plants in a Tropical Forest', *Science, 198*, 1177–8

Stiles, F.G. (1979) 'Regularity, Randomness, and Aggregation in Flowering Times', *Science, 203*, 471

Straw, R.M. (1972) 'A Markov Model for Pollinator Constancy and Competition', *American Naturalist, 106*, 597–620

Thomson, J.D. (1980) 'Skewed Flowering Distributions and Pollinator Attraction', *Ecology, 61*, 572–9

Thomson, J.D. (1982) 'Patterns of Visitation by Animal Pollinators', *Oikos, 39*, 241–50

Udovic, D. (1981) 'Determinants of Fruit Set in *Yucca whipplei*: Reproductive Expenditure vs. Pollinator Availability', *Oecologia, 48*, 389–99

Udovic, D. and C. Aker (1981) 'Fruit Abortion and the Regulation of Fruit Number in *Yucca whipplei*', *Oecologia, 49*, 245–8

Waser, N.M. (1978a) 'Competition for Hummingbird Pollination and Sequential Flowering in Two Colorado Wildflowers', *Ecology, 59*, 934–44

Waser, N.M. (1978b) 'Interspecific Pollen Transfer and Competition Between Co-Occuring Plant Species', *Oecologia, 36*, 223–36

Waser, N.M. (1983) 'Competition for Pollination and Floral Character Differences Among Sympatric Plant Species: A Review of Evidence' in Jones, C.E. and R.J. Little (eds), *Handbook of Experimental Pollination Ecology*, Van Nostrand Reinhold, New York, pp. 277–93

Waser, N.M. and L.A. Real (1979) 'Effective Mutualism Between Sequentially Flowering Plant Species', *Nature, 281*, 670–2

Weller, S.G. (1980) 'Pollen Flow and Fecundity in Populations of *Lithospermum caroliniense*', *American Journal of Botany, 67*, 1334–41

Williamson, M. (1972) *The Analysis of Biological Populations*, Edward Arnold Ltd, London

Willmer, P.G. and S.A. Corbet (1981) 'Temporal and Microclimatic Partitioning of the Floral Resources of *Justicia aurea* amongst a Concourse of Pollen Vectors and Nectar Robbers', *Oecologia, 51*, 67–78

Wright, H.O. (1973) 'Effect of Commensal Hydroids on Hermit Crab Competition in the Littoral Zone of Texas', *Nature, 241*, 139–40

Wyatt, R. (1976) 'Pollination and Fruit-Set in *Asclepias*: a Reappraisal', *American Journal of Botany, 63*, 845–51

Zimmerman, M. (1980) 'Reproduction in *Polemonium*: Competition for Pollinators', *Ecology, 61*, 497–501

10 THE POPULATION DYNAMICS OF MUTUALISTIC SYSTEMS

Carole L. Wolin

Introduction

Despite the great diversity and widespread occurrence of mutualism, theory of the population dynamics of mutualistic systems has developed haltingly. Mutualism models formulated by early workers (Kostitzin 1934, 1937, Gause and Witt 1935) were largely ignored until recently (Risch and Boucher 1976). In the last ten years along with a renewed interest in mutualism (recent reviews include Boucher *et al.* 1982; Addicott 1984) there has been an elaboration of mathematical theory to describe these interactions. Many aspects of this interaction, though, have yet to be formalized in models. In this chapter I will review models of the population dynamics of mutualism. In addition, I will point out some of the directions which need further attention in models.

Since mutualism is a multifarious set of interactions, it cannot be described adequately by a single model. Accordingly, in this chapter models appropriate to the different classes of mutualism are considered. The structure of this chapter is as follows. Types of mutualism are defined and classified in the second section in terms of traits that relate to the population dynamics of mutualistic systems. The next three sections treat obligate, facultative, and obligate-facultative associations, respectively. I will show that simple two species models of facultative mutualism are more stable and persistent than those for obligate mutualism. Mechanisms which promote persistence in obligate systems are examined. The following section explores reward-based approaches to mutualism, including those appropriate for symbioses and protection mutualisms. By incorporating particular patterns of real mutualistic exchange, we can derive more specific models which, in some cases, result in different predictions about the behaviour of mutualistic systems than the more general formulations. The role of spatial and temporal heterogeneity in mutualistic systems is discussed in the following section. I will suggest that such heterogeneity needs to be incorporated into models and will provide new insight into species persistence in mutualistic systems.

The models discussed here will be evaluated in terms of their stability and persistence properties. Stability herein refers to Lyapunov stability. A system is locally stable if it returns to the same equilibrium following small perturbations. It is globally stable if it returns to the same equilibrium following arbitarily large perturbations. Persistence occurs if species do not go extinct; that is, they remain at positive densities.

Classification of Mutualism

Because of the diversity of mutualistic interactions, it is useful to bear in mind attributes which can affect population dynamics and are shared by or differ between mutualistic systems. These influence the relevant parameters and the number of species to include in models. Such attributes are summarized in Table 10.1 adapted from Wolin (1982). Mutualism is characterized here with respect to several variables. The first is type of benefit received. These include protection (reduction of predation, parasitism, herbivory or competition), provision of nutrients and/or shelter, dispersal, or some combination of rewards. For instance, for a plant, seed dispersal to a safe site not only accomplishes dispersal, but can also reduce seed predation and can increase nutrient availability (Howe and Smallwood 1982). As can be seen from Table 10.1, benefits typically are not identical for mutualistic partners. Another important characteristic is the degree of dependency of an interaction. Obligate mutualists cannot survive and/or reproduce without mutualistic interaction, whereas facultative mutualists can. Furthermore, the degree of specificity of mutualism varies from one interaction to another, ranging from one-to-one species specific association to association with a wide diversity of mutualist partners. Specialist is used in the table to mean associations restricted to one to several partner species, and generalist to mean association with partners of a greater number of species. Here, again, the two partners need not have identical attributes. The duration and intimacy of association also varies between interactions: some mutualists are symbiotic, that is, live together, while others are free-living. Symbiosis as defined by DeBary in 1879 refers to an intimate, constant relationship between dissimilar species that can be mutualistic, parasitic or

Table 10.1: Classification of Mutualism

Participant Reward; Examples	Partner 1 O/F[1]	G/S[2]	b/d[3]	S/N[4]	Participant Reward; Examples	Partner 2 O/F	G/S	b/d
Protection					*Nutrients &/or Shelter*			
Homoptera	O,F	G	d	N	ants	O,F	G	(b),d
extra-floral nectary &/or shelter plants	O,F	G,S	b,d	S,N	Hymenoptera	O,F	G,S	(b),d
butterflies	O,F	G,S	d	N	ants	O,F	G,S	(b),d
invertebrate bed formers	O,F	G	d	S	invertebrate grazers	O,F	G,S	(b),d
parasitoids	O	S	b,d	S	virus	O	S	b,(d)
Gamete or Zygote Dispersal					*Nutrients (usually)*			
flowering plants	O,F	G,S	b,(d)	S,N	pollinators	O,F	G,S	(b),d
flowering plants	O,F	G,S	b	N	seed dispersers	O	S	(b),d
Nutrients					*Nutrients &/or Sheltered Environment*			
protozoans	O,F	S	d	S	chlorella	O,F	G	(b),d
invertebrates	O,F	S	d	S	algae	O,F	G	(b),d
plants	O,F	G,S	d	S	mycorrhizae	O,F	G,S	(b),d
vertebrates	O,F	S	d	S	bacteria	O,F	G,S	b,d
insects	O	S	(b),d	S	fungi	O	S	(b),d
fungi	O	G	b,d	S	algae	O	S	b,d
Protection					*Protection*			
butterflies (Mullerian mimics)	F	G,S	d	N	butterflies	F	G,S	d

Notes: 1^O = obligate; F = facultative
2^G = generalist; S = specialist
3^b = increase in births as a result of mutualism; d = decrease in deaths; parentheses indicate indirect effects
4^S = symbiotic; N = nonsymbiotic.

Sources:

Association	References
	Addicott 1979; Bristow 1983; Way 1963
homoptera-ant	Bentley 1977; Hocking 1982; Janzen 1966; Ridley 1910
extrafloral nectary and/or shelter plant-Hymenoptera	Atsatt 1981; Ross 1966; Don Harvey, pers. comm.
butterflies-ants	Osman and Haugsness 1981; Paine and Suchanek 1983; Suchanek ms.
invertebrate bed formers-invertebrate grazers	Edson et al. 1981; Stoltz and Vinson 1979
parasitoid-virus	Feinsinger 1983; Jones and Little 1983; Proctor 1978; Schemske 1983
plant-pollinators	Howe and Smallwood 1982; Wheelwright and Orians 1982
plant-seed dispersers	Karakshian 1975; Karakshian and Siegel 1965
protozoans-Chlorella	Droop 1963; Muscatine et al. 1975; Trench et al. 1981
invertebrates-algae	Barrett 1983; Janos 1980; Malloch et al. 1980; Mosse 1963
plants-mycorrhizae	Hungate 1963; Howard 1967
vertebrates-bacteria	Barrett 1983; Cooke 1977
fungi-algae	Gilbert 1983; Rettenmeyer 1970
butterflies-butterflies	

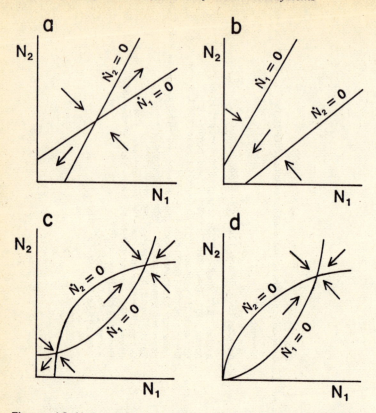

Figure 10.1: Isoclines for Two-species Models of Obligate Mutualism. a and b illustrate the isoclines for pairs of species each described by equation 10.1. c is one of the outcomes when $c_1 < 0$ in equation 10.4. d depicts pairs of species each described by equation 10.4 with $c_1 = 0$.

commensal (Henry 1966). In all mutualism birth rate is enhanced and/or death rate decreased as a result of mutualism, whether it be directly or indirectly (Wolin and Lawlor 1984) and this is the final attribute on which these interactions are classified.

Another important consideration is how these attributes covary. Pollination systems, for instance, can be obligate or facultative. While obligate associations between pollinators and plants are usually not symbiotic and need not be highly species-specific, the few symbiotic associations that exist, such as those between fig and fig-wasp or yucca and yucca moth, are all obligate, specialist associations (Janzen 1979; Powell and Mackie 1966). Further observations on the covariance of attributes will be included in discussing directions which need further exploration in models.

Obligate Mutualism

Obligate mutualism has typically been modelled using first-order differential equations, incorporating density-dependent self-regulation. Density-dependent self-regulation is used here to refer to population growth that is negatively density-dependent, and not to imply a particular mechanism of density-dependence. Models for obligate mutualism have been expressed in a general form by Freedman *et al.* (1983), and in specific forms by others (Dean 1983, May 1976, Vandermeer and Boucher 1978, Wells 1983), as summarized in Table 10.2. Obligate mutualists adequately described by these two species models are implicitly one-to-one species specialists. Some of the outcomes of such models are depicted in Figure 10.1. Simple two-species models of obligate mutualism have qualitatively different behaviour than those for facultative mutualism: they usually are unstable and predict species extinctions (Freedman *et al.* 1983, Vandermeer and Boucher 1978).

Modified Lotka-Volterra competition equations (Equation 10.1, Table 10.2) describe obligate mutualism when both species effectively have negative carrying capacities (Vandermeer and Boucher 1978). The negative carrying capacity is a mathematical artifact from the fact that a positive density of the mutualist partner is required for population increase (i.e. $dN_i/dt = 0$ must have a positive intercept on the N_j axis, herein called a positive threshold density). There are two possible outcomes with this model. The linear isoclines will intersect at a single equilibrium that is an unstable saddle point (Figure 10.1a) when $a_{12}a_{21}/a_{11}a_{22} > 1$. Extinction occurs below critical densities, and population explosion above them. When $a_{12}a_{21}/a_{11}a_{22} \leq 1$, there is no feasible equilibrium (i.e. one at positive species densities) and both species go extinct (isoclines parallel or intersecting as in Figure 10.1b). Though generally not considered, if there is no threshold density and the isoclines thus intersect at the origin, species persistence is now possible. Parameter values determine whether persistence with unbounded population growth or extinction occur.

Table 10.2: Models of Obligate Mutualism

$$dN_1/dt = N_1 f_1(N_1, N_2)$$
$$dN_2/dt = N_2 f_2(N_1, N_2)$$

Equation	f_1; f_2 if of different form than f_1	Reference
10.1	$f_1 = r_1 - a_{11}N_1 + a_{12}N_2$ ($K_1 = r_1/a_{11} < 0$)	Vandermeer and Boucher 1978
10.2	$f_1 = bN_2/(aN_1 + N_2 + c) - d(1 + fN_1)$ $f_2 = mN_1/(rN_2 + N_1 + h) - gN_2$ (N_1 = pollinator; N_2 = plant)	Wells 1983
10.3	$f_1 = r(1 - aN_1/N_2)$ $f_2 = -d - lN_1/(N_1N_2 + N_2c + cD)$ (N_1 = pollinator; N_2 = plant)	May 1976
10.4	$f_1 = r(1 - N_1/\hat{N}_m(1 - e^{-(aN_2 + c_1)/\hat{N}_m)})$ ($c_1 \leq 0$) (Adapted with $N_1 = Y$; $N_2 = X$; $K_y = \hat{N}_m$ = maximum value for $\hat{N}_1$; $k_y = \hat{N}_1$, the equilibrium density)	Dean 1983

The unrealistic result that populations can grow without bound (Figure 10.1a) is eliminated if *per capita* benefit of mutualism decreases with increasing recipient density or increasing benefactor density, the result being that mutualism cannot override all density-dependent regulation. This results in curvilinear isoclines that asymptote at some recipient density such that further increase of the mutualist partner does not increase the equilibrium density of the recipient. Population explosion is checked if one or both species behave in this manner (Dean 1983, Freedman *et al.* 1983, May 1976, 1981, Wells 1983). The curvilinear isoclines (or the one curvilinear and one linear isocline) intersect in zero, one or two feasible equilibria. If they do not intersect, extinction occurs. If there is one point of intersection, then this unstable equilibrium point does not allow for unlimited population growth, but does give rise to extinction below critical densities. The case with two points of intersection (Figure 10.1c) has an unstable and a stable equilibrium; stable species coexistence occurs provided that species densities remain large enough. These types of models have been used to describe pollinator-plant interactions (May 1976; Well 1983) and mutualism involving reciprocal nutrient exchange (Dean 1983).

Rather than incorporating an effectively negative carrying capacity, the carrying capacity can be effectively zero in the absence of the mutualist partner. In other words the threshold effect is removed: if the mutualist partner is above zero density, then there is a positive growth rate. Dean's (1983) model (Equation 10.4, Table 10.2) describes such a mutualist when $c_1 = 0$. As shown in Figure 10.1d, there is always stable coexistence with this formulation.

Mutualists described by the preceding models, with the exception of those without positive threshold densities, are likely to live a precarious existence. Synchronized biology of mutualist partners, for example coupled dispersal mechanisms such as those found in insects and their endosymbionts (Koch 1967) or fungi and their insect tenders (Francke-Grossman 1967), lower threshold densities. Threshold densities can also be low in certain cases in which the mutualist partner is needed for only part of the life cycle. For example, an individual plant does not need its pollinator for survival. Furthermore, if it flowers over the course of many years, the plant's pollinator need not be present during any given year in order for the individual to reproduce successfully during its lifetime.

Other mechanisms besides low threshold densities may enable species to persist. The previous models only considered dynamics within a single patch, i.e. in closed systems. In an open system, it is possible that within patches the mutualistic system is unstable, but observed over all patches, species persist. Patch structure is likely to be a very important factor in obligate mutualistic systems, though this has not been adequately explored empirically and has not been modelled at all. The role of spatial and temporal heterogeneity in mutualistic systems is discussed further in the penultimate section.

There are other reasons why many obligate mutualists are inadequately described by the simple two-species models presented. For obligate associations between symbionts, the relative time scale of host versus guest reproduction and the relative duration (and presence) of free-living in addition to host-guest states determine whether another model of mutualism may be more appropriate. This topic is explored further in the antepenultimate section.

Obligate mutualists are frequently not one-to-one species specialists. Thus mutualism often involves more than two species. This is true, for instance, of obligate interactions between plants and their pollinators, plants and their seed dispersers, and food-shelter plants and their insect defenders. Even in cases where a high degree of species specificity is seen, mutualists frequently are not restricted to interaction with a single other species. To describe such interactions alternative models must be constructed that do not incorporate total dependence on a single mutualist partner. Such models would resemble those for facultative mutualism. A more complete description of the system would necessitate multispecies models. Based on results with multispecies competition and multispecies predator-prey systems (May 1973), I would not expect multispecies systems of obligate mutualists to be more stable than two-species systems. However, if patch structure is incorporated, or if some of the species in mutualistic networks are facultative rather than obligate mutualists, multispecies associations may promote species persistence.

Facultative Mutualism

Models of facultative mutualism are more stable than two species models of obligate mutualism, and in all cases predict species persistence. Since facultative mutualists can survive and reproduce without this interspecific interaction, models incorporating density-dependent self-regulation predict that in the absence of a mutualist partner, species return to a positive carrying capacity. Moreover, either of the mutualists can increase if at zero density, that is the system is mutually invasible.

General formulations of facultative mutualism (Albrecht *et al.* 1974, Christiansen and Fenchel 1977, Freedman *et al.* 1983) have specified that for the general model $dN_i/dt = N_i f_i(N_1,N_2)$, that $\partial f_i(N_1,N_2)/\partial N_i < 0$ and $\partial f_i(N_1,N_2)/\partial N_j > 0$; $i,j = 1,2$; $i \neq j$, that is, the change in *per capita* rate of increase with respect to recipient density is always negative, while the change in *per capita* rate of increase with respect to density of the mutualist partner is always positive. For mutualism of this sort, when there is a unique, feasible equilibrium it is always globally stable (Albrecht *et al.* 1974, Freedman *et al.* 1983, Goh 1979). These conditions hold for the specific equations for facultative mutualism of Addicott (1981), Gause and Witt (1935), Goh (1979), May (1973, 1981), Travis and Post (1979), Vandermeer and Boucher (1978), Whittaker (1975) and Wolin and Lawlor (1984) presented in Table 10.3, with the exceptions of equations 10.6 and 10.10. In equation 10.6 mutualism can override density-dependent regulaton, thus $\partial f_i/\partial N_i$ can be greater than zero when the density of the mutualist partner, N_j, is sufficiently large. In equation 10.10, the interspecific interaction is mutualistic at low recipient densities but detrimental at high recipient densities, thus $\partial f_i/\partial N_j < 0$ at high densities of N_i.

Table 10.3: Models of Facultative Mutualism

$$dN_1/dt = N_1 f_1(N_1, N_2)$$
$$dN_2/dt = N_2 f_2(N_1, N_2)$$

Equation	f_1*	References using equations of this form
10.5	$r_1 - a_{11}N_1 + a_{12}N_2$	Addicott 1981; Gause and Witt 1935; Travis and Post 1979; Vandermeer and Boucher 1978; Wolin and Lawlor 1984
10.6	$r_1 - r_1 N_1/(K + Ka_{12}N_2)$	Addicott 1981; May 1981; Whittaker 1975; Wolin and Lawlor 1984
10.7	$r_1 - b_1 N_1/(1 + a_{12}N_2) - d_1 N_1$	Wolin and Lawlor 1984
10.8	$r_1 - r_1 N_1/K_1 + a_{12}N_1 N_2$	Wolin and Lawlor 1984
10.9	$r_1 - r_1 N_1/K_1 + a_{12}N_2 \exp(-\propto_{12}N_1)$	Wolin and Lawlor 1984
10.10	$(r_1 + a_{12}N_2)(1 - N_1/K_1)$	Addicott 1981; Wolin and Lawlor 1984
10.11	$r_1(1 - N_1/\hat{N}_m(1 - e^{-(aN_2 + c_1)/\hat{N}_m}))$ $(c_1 > 0)$ (Adapted with $Y = N_1$; $X = N_2$; $K_y\hat{N}^m =$ maximum value for $\hat{N}_1$; $k_y = \hat{N}_1$)	Dean 1983

*a_{12} or $\propto_{12}$ is effect of N_2 on N_1; all parameters are positive

In the terminology of Wolin and Lawlor (1984) facultative mutualists are of three types: density-independent mutualists, high density mutualists and low density mutualists. In density-independent mutualism (equation 10.5, Table 10.3) the *per capita* benefit of mutualism for N_i (the recipient) is independent of its density. In high density mutualism (equations 10.6–10.8, 10.11) the *per capita* benefit of mutualism for N_i increases with its density, thus $\partial f_i/\partial N_j$ increases with increasing N_i. Furthermore, a sufficiently large N_j can override all density-dependent self-regulation in equations 10.6 and 10.8 although not in equations 10.7 or 10.11. In low density mutualism (equations 10.9–10.10) the *per capita* benefit of mutualism for N_i decreases with increasing density of N_j, thus $\partial f_i / \partial N_j$ decreases with increasing N_i.

For high density and density-independent mutualism, whether stability ensues depends on the relative magnitudes of density-dependent self-regulation and mutualistic benefit (May 1981; Travis and Post 1979; Wolin and Lawlor 1984). Pairs of mutualists each described by equations 10.5 or 10.6 have linear isoclines and either a single stable equilibrium (Figure 10.2a) or no feasible equilibrium, populations increasing without bound (Figure 10.2b). Pairs of mutualists each described by equations 10.7 or 10.11 always have a stable equilibrium (Figure 10.2c). Stability analyses show that, for pairs of density-independent mutualists, the system, when stable, is as stable as a pair of noninteracting species described by a logistic model. That is, return time to equilibrium following perturbation is equally rapid (Addicott 1981, Wolin and Lawlor 1984). A stable system with a pair of high density mutualists is less stable (slower return time to equilibrium) than a pair of noninteracting species (Addicott 1981, May 1981, Wolin and Lawlor 1984).

Low density mutualism, that is mutualism in which the *per capita* benefit for N_i from mutualism decreases with increasing density of N_i, can arise when mutualism cannot override density-dependent factors or when the mutualist partner has detrimental effects at high densities (Addicott 1981; Wolin and Lawlor 1984). Equations 10.9 (Wolin and Lawlor 1984) and 10.10 (Addicott 1981; Wolin and Lawlor 1984) describe mutualists of this sort. Equation 10.10 is distinct from the other models since at low recipient densities the interspecific interaction is mutualistic, whereas at high recipient densities it is detrimental for the recipient. Facultative low density mutualism can increase the equilibrium density over the carrying capacity, but as seen in Figure 10.2d (equation 10.10), it need not do so. Low density mutualism, however, does change the return time to equilibrium. This form of mutualism is always stable for pairs of mutualists. Moreover, return time to equilibrium is always faster for mutualist pairs described by equations 10.9 or 10.10 than it is for two noninteracting species (Addicott 1981; Wolin and Lawlor 1984). The association of a low density mutualist with a high density or density-independent mutualist is also more stable than that with two high density or density-independent mutualists (Addicott 1981).

For species with discrete generations, a difference equation model is more appropriate than the differential equations previously dis-

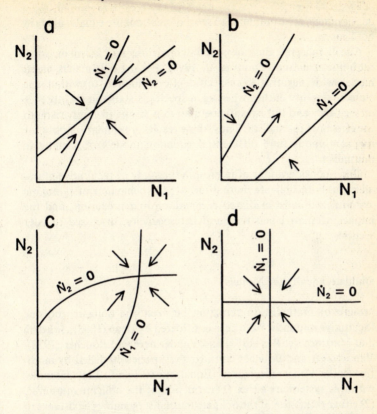

Figure 10.2: Isoclines for some Two Species Models of Facultative Mutualism. a and b diagram the isoclines for species which
are each described by equations 10.5 or 10.6. There is a stable
equilibrium in a and population explosion with no feasible
equilibrium in b. c depicts pairs of species described by equations 10.7 or 10.11. d is the isoclines for a pair of mutualists
each described by equation 10.10.

cussed. A discrete analogue to equation 10.5 is analysed for a two
species mutualistic system by Gilpin *et al.* (1982). Mutualism was
found to exhibit out-of-phase-limit-cycle oscillations at lower values
of the intrinsic rate of increase than the threshold value for oscillations in a single species system. It would be of interest to compare
this to the results for a discrete time model corresponding to a low

density facultative mutualist to see if oscillations are equally likely in both cases.

Another discrete time approach to modelling mutualism incorporates age structure (Travis *et al.* 1980). For a Leslie-matrix-based model with negative intraspecific effects limited to within age classes, stability ensues when each species is stable in the absence of the other and intraspecific regulation is strong in comparison to the effects of mutualism. Thus, these results are similar to those for the continuous time differential equation models of facultative mutualism.

Other considerations of interest in the study of facultative mutualism, as with obligate mutualism, are the behaviour of symbiotic systems, mutualist-mediated competition and predation, and the impact of spatial and temporal heterogeneity, discussed in later sections.

Obligate-Facultative Models

Models of mutualistic interaction between one obligate and one facultative mutualist have been considered by Dean (1983), Soberon and Martinez del Rio (1981) and Vandermeer and Boucher (1978). Vandermeer and Boucher consider two species modelled by modified Lotka-Volterra equations (Equations 10.1 and 10.5). Outcomes with this system are either (1) extinction of the obligate mutualist, (2) either extinction of the obligate mutualist or stable coexistence of the pair, depending on initial conditions (that is, the outcome is indeterminate), or (3) extinction of the obligate mutualist or unbounded population growth, depending on initial conditions. Similar cases arise if the model is modified to generate a curvilinear isocline for the facultative mutualist that asymptotes at some density of the facultative mutualist (Soberon and Martinez del Rio 1981) or curvilinear isoclines for both facultative and obligate mutualists such that the isocline for N_i asymptotes at some value of N_i (equations 10.4, 10.11) (Dean 1983). Because there is a limit to species densities, unbounded population growth is no longer possible and this outcome is replaced by one with a stable equilibrium. Additionally, certain parameter values result in a single stable equilibrium without threat of extinction of the obligate mutualist.

Reward-based Models

In addition to the basic distinction between obligate and facultative mutualism, it is of interest to consider type of reward in the formulation of models of mutualism. As seen in Table 10.1, the benefits exchanged between pairs of mutualists fall into several different classes. Reward-based approaches to modeling help insure that important aspects of the biology of real mutualists are incorporated into models. Such models are also more readily adaptable to empirical testing. One approach measures the effect of mutualistic interaction by analysing *per capita* births and deaths as a function of recipient density (Wolin and Lawlor 1984). Another approach in formulating more mechanistic models borrows from competition and predation studies (for example, Schoener 1976, 1978; Hassell 1978) in analyzing foraging behaviour and energy transfer. Some models of plant-pollinator mutualism have adopted this approach (May 1976; Soberon and Martinez del Rio 1981; Wells 1983). In addition to benefits, cost of mutualism can also be incorporated (Rai *et al.* 1983). By modifying the energetic, reward-based approach, Dean has described symbiotic nutrient-exchange mutualism (Dean 1983 and Chapter 11, this volume). It would seem that the energetic, reward-based approach could also be extended to some of the other types of mutualism.

Symbiosis

Mutualism in which the benefits exchanged are categorized in Table 10.1 as 'nutrients' for 'nutrient and/or sheltered environment' are all symbiotic associations. Symbiotic mutualisms such as the associations between algae, fungi or bacteria and their hosts, have some important differences from associations between nonsymbiotic mutualists. They are relatively long-lived (with respect to the organism's lifetime), intimate associations. A single guest frequently needs to interact with just a single individual of the host species during its lifetime. The host, likewise, frequently is colonized by a single individual guest, which may then replicate within the host. In contrast, in nonsymbiotic associations between plants and their pollinators, insect guards and nectar providers, and plants and seed dispersers, among others, an individual must interact with a number of individuals of the mutualist species.

Fifty years ago, a general framework for conceptualizing symbiotic mutualism was constructed by Kostitzin (1934, 1937).

This model tracks the numbers of free-living individuals (x_1, x_2) and symbiotic associations (x) of a mutualistic species pair. I have diagrammed what I believe are its essentials in Figure 10.3. Because of the difficulty of analysing this model, Kostitzin only considers two special cases. In one case the free state is not formed from the symbiotic one ($B_1 = B_2 = 0$, $v = v_1 = v_2$). A solution with just x_1 and x or just x_2 and x at positive values is stable for certain parameter values. There is a stable and an unstable equilibrium when all three entities are present. A large number of assumptions are made in the second special case ($v = c_1 = c_2 = \mu = E_1 = E_2 = d = 0$, $n_1 = v_1$, $n_2 = v_2$, $m_1 = u_1$, $m_2 = u_2$, $d_1 = e_1$, $d_2 = e_2$), but it is not clear which real mutualists this is meant to describe. For this case there is a stable equilibrium with x, x_1 and x_2 positive. Kostitzin also makes some preliminary analyses of the effects of age structure in models of symbiotic mutualism.

To make this framework more tractable to analyse, the models need to be designed for specific mutualisms, eliminating the unnecessary parameters. For instance, where the guest reproduces far more frequently than the host, guest dynamics can be subsumed within that of the host, because it is assumed that the guest more rapidly reaches equilibrium. Then, only two entities need to be tracked, the uninfected hosts (x_1) and infected hosts (x). This approach is similar to that taken in microparasite-host models (Anderson 1981, Anderson and May 1979). In other cases it is appropriate to include free-living stages of both partners as well as the symbiotic state, while eliminating other parameters such as births in the free-living state. I am in the process of analyzing these and other simple models that correspond to the life histories of various obligate and facultative symbiotic mutualists.

Protection as a Reward

Mutualism in which the benefit is decreased predation, herbivory, parasitism or competition (protection in Table 10.1) or increased predatory ability (Addicott 1984, Addicott and Freedman 1984) fundamentally involves three species. These three-species models can result in outcomes not predicted by the two-species models for mutualism (Rai *et al.* 1983). I will consider briefly the results from models in which the benefit for one of the mutualists is reduced predation or reduced competition. The effects of competition in mutualistic systems are treated in greater detail in the chapter by Addicott.

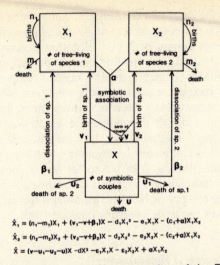

$$\dot{X}_1 = (n_1-m_1)X_1 + (v_1-v+\beta_1)X - d_1X_1{}^2 - e_1X_1X - (c_1+a)X_1X_2$$

$$\dot{X}_2 = (n_2-m_2)X_2 + (v_2-v+\beta_2)X - d_2X_2{}^2 - e_2X_2X - (c_2+a)X_1X_2$$

$$\dot{X} = (v-u_1-u_2-u)X - dX^2 - \varepsilon_1X_1X - \varepsilon_2X_2X + aX_1X_2$$

Figure 10.3: A Diagrammatic Interpretation of the Essentials of the Equations for Symbiotic Mutualism of Kostitzin. His equations (Kostitzin 1934), given below, include density-dependent effects as well as the processes diagrammed.

A slow-growing, predation-inhibiting mutualist can increase the local stability of a predator-prey system (Addicott and Freedman 1984). A three-species system in which the mutualist need not be slow-growing has been analysed by Rai *et al.* (1983). They found that, under conditions in which prey and its mutualist coexist in the absence of a predator, and predator and prey coexist in the absence of the mutualist, then it is possible for the mutualist to drive the predator extinct. Under conditions which allow a three species equilibrium, the equilibrium can be stable or unstable. When unstable, under some conditions stable limit cycles occur. An alternative way to conceptualize mutualist-mediated predation is as predator-mediated mutualism. Heithaus *et al.* (1983) find that an unstable system of facultative mutualism is stabilized by the addition of a predator.

The exact relation of mutualism and competition determines how mutualism affects competitor population dynamics and system stability.

A system composed of a slow-growing mutualist, a mutualist-competitor, and its competitor, with the benefit for the mutualist-competitor being decreased competition, can be more stable than a system with just two competitors (Addicott and Freedman 1984). In the three-species system with mutualist (not necessarily a slow-growing one), mutualist-competitor, and competitor, with the benefit to the mutualist-competitor being increased competitive ability, the mutualist-competitor persists under conditions which result in its extinction in the two-species competitive system. Species coexist under conditions in which competition alone would drive the mutualist-competitor extinct, and the mutualist-competitor is the competitively superior under conditions in which competition alone is indeterminate (Rai *et al.* 1983). Another type of competition that can occur in mutualistic systems is competition for mutualists. Addition of an obligate or facultative mutualist to a system with the competitive pair can be destabilizing (Hallam 1980).

Spatial and Temporal Heterogeneity

Mutualism may not benefit every individual in a population, just as other interspecific interactions need not affect every member of a population. In population models we deal with the average effect of the interspecific interaction. However, if the average effect varies with density or an environmental variable and this, in turn, varies over space and/or time, it is of interest to consider the effect of this variability on population dynamics. Levin (1976) has reviewed population dynamic models incorporating spatial and temporal heterogeneity.

This sort of heterogeneity does occur in mutualistic systems. For instance, in pollination systems, patch size and density are known to affect the pollination success of a plant (Antonovics and Levin 1980, Beattie 1976, Handel 1983, Platt *et al.* 1974, Silander 1978). Environmental variability can affect the benefits derived from mutualistic associations between *Paramecium bursaria* and *Chlorella* sp. (Weis 1969, Reisser 1981) and mycorrhizae and plants (Janos 1980). In the latter case, some plants which are facultative mutualists exploit nutrient poor habitats when mutualistically associated, but are restricted to richer habitats when not associated with mycorrhizae. Wilson (1980, 1983) discusses the importance of population substructure to mutualistic systems in general and to populations of phoretic mites mutualistically associated with burying beetles.

Under what conditions does spatial or temporal heterogeneity stabilize interactions or enhance species persistence, in mutualistic systems that are otherwise unstable and contain the threat of species extinctions? This question needs to be explored in models. However, the only population dynamic model, to date, to incorporate patch structure is that of DeAngelis *et al.* (1979). They construct an L-species, N-patch model with competitive and mutualistic interactions specified according to a Morishima matrix. This brings tractability to the problem of simultaneously analysing dynamics within patches and the effect of patch structure, though it places restrictions on which species are allowed to compete or interact mutualistically. They incorporate a Lotka-Volterra style of facultative mutualism into these models and find that stability can result. Further models using this approach, but with obligate rather than facultative mutualism, and formulations using alternative approaches such as that used by Slatkin (1974) are clearly needed to answer satisfactorily the question of what conditions promote species persistence.

Conclusion

The early prediction that models of mutualism are unstable and that this may explain the rareness of mutualism (May 1973) has, in the last ten years, been amended. Mutualism may be difficult to observe rather than rare. Evolutionary models are needed to address why mutualism is rare, if such is the case. Furthermore, models of population dynamics show that mutualism is not necessarily unstable. Morever, for facultative mutualism, instability does not imply lack of persistence, but rather unbounded population growth. Persistence mechanisms in obligate mutualistic systems may include spatial heterogeneity, multispecies associations and low densities of the mutualist partner allowing growth. The dynamics of simple two-species model systems of obligate and facultative mutualism are relatively well understood at this time. The elaboration of more mechanistic models and the exploration of the role of spatial and temporal heterogeneity in mutualistic systems are two promising areas for furthering our understanding of the population dynamics of mutualistic systems. Mutualism includes a diversity of patterns of interaction. Accordingly, it cannot be adequately described by a single model.

Acknowledgments

I am grateful to Tom Schoener, Cathy Toft, Dave Spiller and Steve Vogel for helpful comments on an earlier draft of this manuscript. I also thank Larry Gilbert and Doug Boucher for their encouragement. This work was facilitated by support from a Chancellor's Graduate Fellowship from the University of California at Davis.

References

Addicott, J.F. (1979) 'A Multispecies Aphid-Ant Association: Density Dependence and Species-Specific Effects', *Canadian Journal of Zoology, 57*, 558–69

Addicott, J.F. (1981) 'Stability Properties of Two-Species Models of Mutualism: Simulation Studies', *Oecologia, 49*, 42–9

Addicott, J.F. (1984) 'Mutualistic Interactions in Population and Community Processes' in Price, P.W., C.N. Slobokchikoff and W.S. Gaud (eds), *A New Ecology: Novel Approaches to Interactive Systems*, John Wiley and Sons, New York, pp. 437–55

Addicott, J.F. and H.I. Freedman (1984) 'On the Structure and Stability of Mutualistic Systems: Analysis of a Predator-Prey and Competition Models as Modified by the Action of a Slow-Growing Mutualist', *Theoretical Population Biology 26*: 320–39

Albrecht, F., H. Gratzke, A. Haddad and N. Wax (1974) 'The Dynamics of Two Interacting Populations', *Journal of Mathematical Analysis and Applications, 46*, 658–70

Anderson, R.M. (1981) 'Population Ecology of Infectious Disease Agents' in May, R.M. (ed.), *Theoretical Ecology: Principles and Application*, Sinauer, Sunderland, Mass., pp. 318–55

Anderson, R.M. and R.M. May (1979) 'Population Biology of Infectious Diseases. Part I', *Nature, 280*, 361–7

Antonovics, J. and D.A. Levin (1980) 'The Ecological and Genetic Consequences of Density-Dependent Regulation in Plants', *Annual Review of Ecology and Systematics, 13*, 315–47

Atsatt, P.R. (1981) 'Lycaenid Butterflies and Ants: Selection for Enemy-Free Space', *American Naturalist, 118*, 638–54

Baker, J.M. (1963) 'Ambrosia Beetles and their Fungi, with Particular Reference to *Platypus cylindrus* Fab.' in Nutman, P.S. and B. Mosse (eds), *Symbiotic Associations*, Cambridge University Press, Cambridge, pp. 232–65

Barrett, J.A. (1983) 'Plant-Fungus Symbioses' in Futuyma, D.J. and M. Slatkin (eds), *Coevolution*, Sinauer, Sunderland, Mass., pp. 137–60

Batra, L.R. (ed.) (1979) *Insect-Fungus Symbiosis: Nutrition, Mutualism and Commensalism*, John Wiley and Sons, New York

Beattie, A.J. (1976) 'Plant Dispersion, Pollination and Gene Flow in *Viola*', *Oecologia, 25*, 291–300

Bentley, B.L. (1977) 'Extrafloral Nectaries and Protection by Pugnacious Bodyguards', *Annual Review of Ecology and Systematics, 8*, 407–27

Boucher, D.H., S. James and K.H. Keeler (1982) 'The Ecology of Mutualism', *Annual Review of Ecology and Systematics, 13*, 315–47

Bristow, C.M. (1983) 'Treehoppers Transfer Parental Care to Ants: A New Benefit of Mutualism', *Science, 220*, 532–3

Christiansen, F.B. and T.M. Fenchel (1977) *Theories of Populations in Biological*

Communities, Springer-Verlag, Berlin

Dean, A.M. (1983) 'A Simple Model of Mutualism', *American Naturalist, 121*, 409–17

DeAngelis, D.L., C.C. Travis and W.M. Post (1979) 'Persistence and Stability of Seed-Dispersed Species in a Patchy Environment', *Theoretical Population Biology, 16*, 107–25

Droop, M.R. (1963) 'Algae and Invertebrates in Symbiosis' in Nutman, P.S. and B. Mosse (eds), *Symbiotic Associations*, Cambridge University Press, Cambridge, pp. 171–99

Edson, K.M., S.B. Vinson, D.B. Stoltz and M.D. Summers (1981) 'Virus in a Parasitoid Wasp: Suppression of the Immune Response in the Parasitoid's Host', *Science, 211*, 582–3

Feinsinger, P. (1983) 'Coevolution and Pollination' in Futuyma, D.J. and M. Slatkin (eds), *Coevolution*, Sinauer, Sunderland, Mass., pp. 282–310

Francke-Grossman, H. (1967) 'Ectosymbiosis in Wood-Inhabiting Insects' in Henry, S.M. (ed.), *Symbiosis*, vol. 2, Academic Press, New York, pp. 141–205

Freedman, H.I., J.F. Addicott and B. Ray (1983) 'Nonobligate and Obligate Models of Mutualism' in Freedman, H.I. and C. Strobeck (eds), *Population Biology Proceedings, Edmonton 1982, Notes in Biomathematics, 52*, 349–54

Gause, G.F. and A.A. Witt (1935) 'Behavior of Mixed Populations and the Problem of Natural Selection', *American Naturalist, 69*, 596–609

Gilbert, L.E. (1983) 'Coevolution and Mimicry' in Futuyma, D.J. and M. Slatkin (eds), *Coevolution*, Sinauer, Sunderland, Mass. pp. 263–81

Gilpin, M.E., T.J. Case and E.A. Bender (1982) 'Counterintuitive Oscillations in Systems of Competition and Mutualism', *American Naturalist, 119*, 584–8

Goh, B.S. (1979) 'Stability in Models of Mutualism', *American Naturalist, 113*, 261–75

Hallam, T.G. (1980) 'Effects of Cooperation on Competitive Systems', *Journal of Theoretical Biology, 82*, 415–23

Handel, S.N. (1983) 'Pollination Ecology, Plant Population Structure, and Gene Flow' in Real, L. (ed.), *Pollination Biology*, Academic Press, New York, pp. 163–211

Hartzell, A. (1967) 'Insect Ectosymbiosis' in Henry, S.M. (ed.), *Symbiosis*, vol. 2, Academic Press, New York, pp. 107–40

Hassell, M.P. (1978) *The Dynamics of Arthropod Predator-Prey Systems*, Princeton University Press, Princeton

Heithaus, E.R., D.C. Culver and A.J. Beattie (1980) 'Models of Ant-Plant Mutualisms', *American Naturalist 116*, 347–61

Henry, S.M. (1966) 'Foreword' in Henry, S.M. (ed.), *Symbiosis*, vol. 1, Academic Press, New York, pp. ix-xi

Hocking, B. (1982) 'Ant-Plant Mutualism: Evolution and Energy' in Gilbert, L.E. and P.H. Raven (eds), *Coevolution of Animals and Plants* (2nd edn), University of Texas Press, Austin, pp. 78–90

Howard, B.H. (1967) 'Intestinal Microorganisms of Vertebrates' in Henry, S.M. (ed.), *Symbiosis*, vol. 2, Academic Press, New York, pp. 314–84

Howe, H.F. and J. Smallwood (1982) 'Ecology of Seed Dispersal', *Annual Review of Ecology and Systematics, 13*, 201–8

Hungate, R.E. (1963) 'Symbiotic Associations: the Rumen Bacteria' in Nutman, P.S. and B. Mosse (eds), *Symbiotic Associations*, Cambridge University Press, Cambridge, pp. 266–97

Janos, D.P. (1980) 'Mycorrhizae Influence Tropical Succession', *Biotropica, 12 (Supplement)*, 56–64

Janzen, D.H. (1966) 'Coevolution of Mutualism Between Ants and Acacias in Central America', *Evolution, 20*, 249–75

Janzen, D.H. (1979) 'How to be a Fig', *Annual Review of Ecology and Systematics,*

10, 13–51

Jones, C.E. and R.J. Little (eds) (1983) *Handbook of Experimental Pollination Biology*, Van Nostrand Reinhold, New York

Karakshian, M.W. (1975) 'Symbiosis in *Paramecium bursuria*', in Jennings, D.H. and D.L. Lee (eds), *Symbiosis, Symposia of the Society for Experimental Biology, 29*, 145–74

Karakshian, S.J. and R.W. Siegel (1965) 'A Genetic Approach to Endocellular Symbiosis', *Experimental Parasitology, 17*, 103–22

Koch, A. (1966) 'Insects and their Endosymbionts', in Henry, S.M. (ed.), *Symbiosis*, vol. 1, Academic Press, New York, pp. 1–106

Kostitzin, V.A. (1934) *Symbiose, Parasitisme et Evolution (Etude Mathématique)*, Hermann et Cie., Paris

Kostitzin, V.A. (1937) *Biologie Mathématique*, Librairie Armand Colin, Paris

Lawlor, L.R. (1979) 'Direct and Indirect Effect of N-Species Competition', *Oecologia, 43*, 355–64

Levin, S.A. (1976) 'Population Dynamics Models in Heterogeneous Environments', *Annual Review of Ecology and Systematics, 7*, 287–310

Malloch, D.W., K.A. Pirozynski and P.H. Raven (1980) 'Ecological and Evolutionary Significance of Mycorrhizae Symbioses in Vascular Plants (A Review)', *Proceedings of the National Academy of Science of the US, 77*, 2113–8

May, R.M. (1973) *Stability and Complexity in Model Ecosystems*, Princeton University Press, Princeton

May, R.M. (1976) 'Mathematical Aspects of the Dynamics of Animal Populations' in Levin, S.A. (ed.), *Studies in Mathematical Biology*, American Mathematical Society, Providence, RI

May, R.M. (1981) 'Models for Two Interacting Populations' in May, R.M. (ed.), *Theoretical Ecology: Principles and Applications* (2nd edn), Sinauer, Sunderland, Mass., pp. 78–104

Mosse, B. (1963) 'Vesicular-Arbuscular Mycorrhizae: An Extreme Form of Fungal Adaptation' in Nutman, P.S. and B. Mosse (eds), *Symbiotic Associations*, Cambridge University Press, Cambridge, pp. 146–70

Muscatine, L., C.B. Cook, R.L. Pardy and R.R. Pool (1975) 'Uptake, Recognition and Maintenance of Symbiotic Chlorella by *Hydra viridis*' in Jennings, D.H. and D.L. Lee (eds), *Symbiosis, Symposia of the Society for Experimental Biology, 29*, 175–203

Osman, R.W. and J.A. Haugsness (1981) 'Mutualism Among Sessile Invertebrates: A Mediator of Competition and Predation, *Science, 211*, 846–8

Paine, R.T. and T.H. Suchanek (1983) 'Convergence of Ecological Processes Between Independently Evolved Competitive Dominants: A Tunicate-Mussel Comparison', *Evolution, 37*, 821–31

Platt, W.R., G.R. Hill and S. Clark (1974) 'Seed Production in a Prairie Legume', *Oecologia, 17*, 55–63

Powell, J.A. and R.A. Mackie (1966) 'Biological Interrelationships of Moths of *Yucca whipplei* (Lepidoptera: Gelchiidae, Blastobasidae, Pyroxidae)', *University of California Publications in Entomology, 42*

Proctor, M.C.F. (1978) 'Insect Pollination Syndromes in an Evolutionary and Ecosystemic Context', in Richards, A.J. (ed.), *The Pollination of Flowers by Insects*, Academic Press, New York, pp. 105–16

Rai, B., H.I. Freedman and J.F. Addicott (1983) 'Analysis of Three Species Models of Mutualism in Predator-Prey and Competitive Systems', *Mathematical Bioscience, 65*, 13–50

Reisser, W. (1981) 'Host-Symbiont Interaction in *Paramecium bursaria*: Physiological and Morphological Features and Their Evolutionary Significance', *Ber. Deutsch Bot. Ges, Bd., 94s*, 558–63

Rettenmeyer, C.W. (1970) 'Insect Mimicry', *Annual Review of Entomology, 15*,

43-74

Ridley, H.N. (1910) 'Symbiosis of Ants and Plants', *Annals of Botany, 24,* 457-83

Risch, S. and D. Boucher (1976) 'What Ecologists Look For', *Bulletin of the Ecological Society of America, 57,* 38-9

Ross, G.N. (1966) 'Life-History Studies on Mexican Butterflies. IV. The Ecology and Ethology of *Anatole rossi,* a Myrmecophilous Metalmark (Lepidoptera: Riodinidae)', *Annals of the Entomological Society of America, 59,* 985-1004

Schemske, D.W. (1983) 'Limits to the Specialization and Coevolution in Plant-Animal Mutualism' in Nitecki, M.H. (ed.), *Coevolution,* University of Chicago Press, Chicago, Illinois, pp. 67-109

Schoener, T.W. (1976) 'Alternatives to Lotka-Volterra Competition: Models of Intermediate Complexity', *Theoretical Population Biology, 10,* 309-33

Schoener, T.W. (1978) 'Effects of Density-Restricted Food Encounter on Some Single-Level Competition Models', *Theoretical Population Biology, 13,* 354-81

Silander, J.A. (1978) 'Density-Dependent Control of Reproductive Success in *Cassia biflora*', *Biotropica, 10,* 292-6

Slatkin, M. (1974) 'Competition and Regional Coexistence', *Ecology, 55,* 128-34

Soberon, J.M. and C. Martinez del Rio (1981) 'The Dynamics of a Plant-Pollinator Interaction', *Journal of Theoretical Biology, 91,* 363-78

Stoltz, D.B. and S.B. Vinson (1979) 'Viruses and Parasitism in Insects', *Advances in Virus Research 24,* 125-71

Suchanek, T.H. 'Mutualism in a Species-Rich Mussel Bed: Enhancement of Community Stability, *Ecology* (submitted)

Travis, C.C. and W.M. Post, III (1979) 'Dynamics and Comparative Statics of Mutualistic Communities', *Journal of Theoretical Biology, 78,* 553-71

Travis, C.C., W.M. Post, III, D.L. DeAngelis and J. Perkowski (1980) 'Analysis of Compensatory Leslie Matrix-Models for Competing Species', *Theoretical Population Biology, 18,* 16-30

Trench, R.K., N.J. Colley and W.K. Fitt (1981) 'Recognition Phenomena Between Marine Invertebrates and "Zooxanthellae": Uptake, Sequestration, Persistence', *Ber. Deutsch. Bot. Ges. Bd. 94, s,* 529-45

Vandermeer, J.H. (1980) 'Indirect Mutualism: Variation on a Theme by Stephen Levine', *American Naturalist, 116,* 441-48

Vandermeer, J.H. and D.H. Boucher (1978) 'Varieties of Mutualistic Interactions in Population Models', *Journal of Theoretical Biology, 74,* 549-58

Way, M.J. (1963) 'Mutualism Between Ants and Honeydew-Producing Homoptera', *Annual Review of Entomology, 8,* 307-44

Weis, D. (1969) 'Regulation of Host and Symbiont Population Size in *Paramecium bursaria*', *Experientia, 25,* 664-6

Wells, H. (1983) 'Population Equilibria and Stability in Plant-Animal Pollination Systems', *Journal of Theoretical Biology, 100,* 685-99

Wheelright, N.T. and G.H. Orians (1982) 'Seed Dispersal by Animals: Contrasts with Pollen Dispersal, Problems of Terminology and Constraints of Coevolution', *American Naturalist, 119,* 402-13

Whittaker, R.H. (1975) *Communities and Ecosystems,* MacMillan, New York

Wilson, D.S. (1980) *The Natural Selection of Populations and Communities,* Benjamin Cummings, Menlo Park, Ca.

Wilson, D.S. (1983) 'The Effect of Population Structure on the Evolution of Mutualism: a Field Test Involving Burying Beetles and Their Phoretic Mites', *American Naturalist, 121,* 851-70

Wolin, C.L. (1982) 'Models of Mutualism: Density Effects in Facultative Systems', Master's thesis, University of Texas, Austin

Wolin, C.L. and L.R. Lawlor (1984) 'Models of Facultative Mutualism: Density Effects', *American Naturalist 124,* 843-62

11 THE DYNAMICS OF MICROBIAL COMMENSALISMS AND MUTUALISMS

Antony M. Dean

Introduction

> Unfortunately, and in spite of a great deal of effort expended
> throughout many years, animal and plant interactions remain
> very poor models for an understanding of microbial competition.
> (D.L. Hartl 1984 personal communication)

An ecosystem is so complex, so difficult to comprehend, that any
attempt to understand the interactions of the component parts *in
situ* is frequently doomed to failure because of a lack of rigorous
controls. Under such circumstances the behaviour displayed by one
component may be ascribed to any number of phenomena. Consequently, if we wish to understand the mechanisms by which populations interact we must study them under simplified, controllable
laboratory conditions. These should be modelled for theoretical
insight, and under ideal circumstances the behaviour displayed
should be predictable under a variety of conditions imposed by the
experimentalist.

From such a perspective, mixed microbial cultures inhabiting
simple continuous culture devices are ideal model systems for the
study of many ecological phenomena. Unfortunately, population
biology has neglected this whole field of research for far too long,
and without good reason; for micro-organisms are not only economically and ecologically important, their world is every bit as
fascinating as that of higher forms of life that are the staple diet
of our researchers. Indeed, they can provide unique insights
unavailable from almost any other experimental approach.

Batch Cultures

Typically, the functional assay for commensalism or mutualism is
the observation that the growth of at least one of the partners is
dependent upon, or enhanced by, the presence of another. In studies

of microbial interactions, this conclusion is usually reached by the use of batch cultures. Frequently, the cause of a positive interaction is not detected, even though this technique has been used to establish its presence (for example, see Nurmikko 1956).

From an ecological standpoint, however, such an approach has two significant drawbacks. First, the possibility that other complicating effects are present is not excluded. Frequently, but by no means always, more intensive investigations do uncover complicating interactions: for example, two obligate mutualists may compete for a carbon source. Secondly, the choice of batch culture is unfortunate because the technique allows the environment to change rather drastically (for example, the depletion of once unlimited resources) with the result that such experiments are necessarily short term.

For such reasons, the dynamics of a system of interacting microbial populations are usually best studied in continuous culture, a technique which permits relatively long-term experiments to be carried out in considerably more stable environments. All of the work presented in this chapter relies heavily upon this method of culture, so that a brief description on the basic theory of the simplest apparatus, the chemostat, seems in order.

How a Chemostat Works

A simple chemostat is illustrated in Figure 11.1. A peristaltic pump feeds fresh media from a reservoir into the growth chamber at a steady rate: the excess culture of spent media, microbes and cellular debris overflows through a siphon so that a constant volume is maintained. Filtered, humidified air entering at the base of the growth chamber mixes and aerates the culture before escaping through a vent above the overflow siphon.

Temperature is usually controlled by immersing the growth chamber in a water bath, or by placing the whole apparatus in a warm room. Control of pH may be achieved by using suitably buffered media. Of course, different experiments require that there be variations upon this theme, and in order to ensure a more controlled chemical environment, auxiliary devices are sometimes employed to maintain correct pH, dissolved gas levels, ion concentrations, etc.

The whole system of tubes, reservoirs, fresh media and the chemostat itself may be autoclaved prior to inoculation; and with

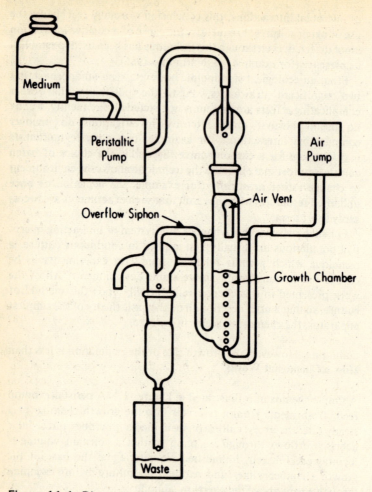

Figure 11.1: Diagram of a Simple Chemostat. See text for discussion of operation.

care, contamination prevented throughout the experiment. Samples are taken directly from the culture, or a few drops are collected from the overflow siphon. They may then be serially diluted and plated onto selective agar plates for counting (genetic markers can be used to distinguish strains), or a Coulter counter which can differentiate species of sufficiently different size may be used instead.

Basic Theory of the Chemostat

Since the more complex mathematics of mixed cultures is essentially an extension of the approach used by Monod (1942, 1950) to model the behaviour of a single population, the latter will be described first.

If the growth rate of a population of microbes is plotted against the concentration of a critical substrate, the result is commonly some sort of asymptotic curve. At low concentrations the substrate is limiting, whilst at high concentrations the population is growing at a maximum rate. A number of models have been developed to account for particular relationships (for example, those for dual nutrient limitation, substrate inhibition, etc.), but the Michaelis-Menten expression usually provides a good approximation and has the great advantage of simplicity. In this model, the rate of growth of cells, M_1, is related to the critical substrate concentration in the growth chamber, S_1, as follows:

$$M_1 = M_{max_1} [S_1/(k_{1.s_1} + S_1)] \tag{11.1}$$

where k is analogous to the Michalis constant, the subscript $1.s_1$ indicating that this applies to the population N_1 and the substrate S_1 respectively, and M_{max_1} is the maximal growth rate when S_1 is unlimiting. However, the growth rate of the population is less than in equation 11.1 because cells are continually washed out of the growth chamber. If D, known as the dilution rate, is the fractional rate of replacement of the culture by fresh media then the Monod equation describing the rate of the culture is:

$$dN_1/dt = (M_1 - D)N_1 \tag{11.2}$$

where N_1 is the density of the population. D^{-1}, known as the holding time, is the time for the feed to deliver a volume of medium equal to that of the volume of culture in the growth chamber. So long as $M_1 > D$ the population density should rise so that there will be more cells consuming less nutrients. Thus, the concentration of substrates in the growth chamber should fall: the Monod equation describing the rate of change of the critical substrate concentration is:

$$dS_1/dt = D(S_{1f} - S_1) - (1/Y_{1.s_1})M_1N_1 \tag{11.3}$$

where S_{1f} is the concentration of critical substrate in the feed (f stands for feed) and $Y_{1.s_1}$ is a yield constant, with the subscripts again representing N_1 growing on S_1. This equation assumes that the rate of consumption of the critical substrate is directly proportional to the rate of growth of the cells. Eventually an equilibrium will be reached because S_1 will be depleted to such a level that the rate of growth of the cells will exactly match the rate of washout, that is:

$$M_{max_1}[S_{1eq}/(k_{1.s_1} + S_{1eq})] = M_{1eq} = D \qquad (11.4)$$

Setting equation 11.3 equal to zero, then substituting in the right hand side of equation 11.4 and factoring out D yields:

$$N_{1eq} = Y_{1.s_1}(S_{1f} - S_{1eq}) \qquad (11.5)$$

S_{1eq} is typically so much smaller than S_{1f} (Dukhuizen and Hartl 1983) that the latter effectively determines the population density, whilst the dilution rate D determines the rate of cell division for a wide range of dilution rates. Thus, the two important parameters, N_{1eq} and M_{1eq}, are effectively under independent control (see Figure 11.2).

Unlike the Monod model, the logistic growth curve (the Verhulst-Pearl equation) expresses the equilibrium density (K) as a function of the intrinsic rate of growth (r);

$$dN/dt = rN - sN^2 \qquad (11.6)$$

or more familiarly as:

$$dN/dt = rN(K - N)/K \qquad (11.7)$$

where:

$$K = r/s$$

As is shown in Appendix One the Monod model may be reduced to a form similar to equation 11.7, namely:

$$dN_1/dt = [r_1N_1(K_1 - N_1)/K_1] \bullet [(k_{1.s1} + S_{1f})/(k_{1.s1} + S_1)] \quad (11.8)$$

where:

$$r_1 = M_{max_1}[S_{1f}/(k_{1.s_1} + S_{1f})] - D$$

and:

$$K_1 = Y_{1.s_1}(S_{1f} - S_{1eq})$$

If S_{1f} is very much smaller than $k_{1.s_1}$, then equation 11.8 reduces to that of the logistic. In practice, however, $k_{1.s_1}$ is also very small with the result that the dilution rate of a chemostat would have to be intolerably small. Note that 'density dependance' is interpreted in

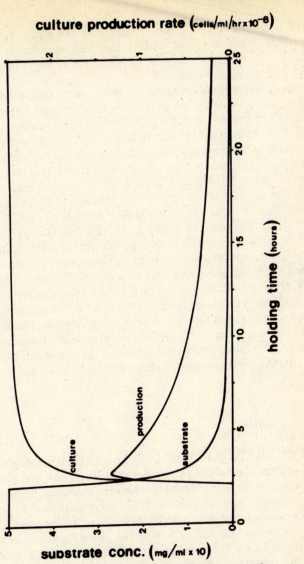

culture production rate (cells/ml/hr x 10⁻⁸)

Figure 11.2: The Theoretical Monod Relationships Between Holding Time and Equilibrium States of a Pure Chemostat Culture; $M_{max_1} = 0.5$, $S_{1f} = 0.5$, $k_{1.s_1} = 0.05$, $Y_{1.s_1} = 10-12$

terms of measurable mechanistic parameters and seen to be caused by nothing more than resource depletion. Also, note that the intrinsic rate of growth, r_1, is dependent both on the dilution rate and the initial substrate concentration S_{1f}.

Being expressed only in terms of r and K, the logistic cannot take into account changes in key environmental parameters affecting the density and growth rate of a population. These criticisms are not true of the Monod model, and whilst the logistic might well provide a suitable fit to a particular set of data, it would be unable to account for the resulting behaviour of the system if the dilution rate and/or the critical substrate concentration in the feed were altered.

In a more general vein, the logistic equation provides no insight whatsoever into the causes of density dependance, and for that matter, the intrinsic rate of growth is not so intrinsic as the name suggests. Since both parameters are basic to an understanding of population ecology, and since both may change as the environment changes, I have always been somewhat disturbed that so little effort has been expended to uncover their nature prior to a full-blooded experimental analysis of the interactions between a pair of species. Besides, it is not at all clear how the logistic equation should be modified to account for competition, mutualism, predation or other interactions. I shall return to this theme in the discussion, but I wish first to describe some aspects of microbial interactions and how they are modelled.

Commensalism

The approach taken by Monod may be readily extended to include two or more species and two or more nutrients, with the result that the ensuing ecological interactions between them may be modelled. For each and every component of the system there must be a differential equation that describes the impact of other components, such as in equation 11.2 and 11.3. However, even for relatively simple systems the dynamics may become mathematically quite intractable so that numerical integration is commonly used as a means of exploring temporal behaviour.

To discuss every model, let alone every facet and conceivable variation, is unnecessary. Instead, I shall only mention those models which form a basis for an understanding of microbial interactions together with those that are relevant to the discussion of data.

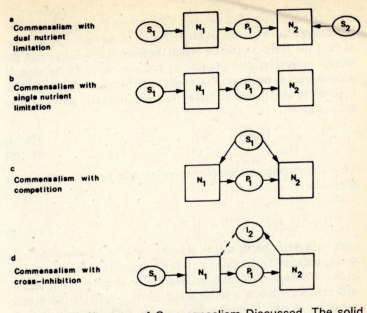

Figure 11.3: Patterns of Commensalism Discussed. The solid arrows indicate production and consumption of substances. The dashed arrow indicates inhibitory effects.

A simple model of commensalism is presented by Miura *et al.* (1980). Their system of crossfeeding is diagrammed in Figure 11.3a. Substrates S_1 and S_2 are introduced with the feed (the subscript f in the equations below designates the concentration of them in the reservoir), and P_1 is produced by the metabolic activities of N_1. N_1 grows on S_1, and N_2 grows on S_2 and also requires P_1. Thus N_2 is the commensal. Note that all other growth requirements for the populations are in excess, and so are not considered by the model.

The mass balance equations are as follow:

$$dN_1/dt = (M_1 - D)N_1 \tag{11.9}$$

$$dN_2/dt = (M_2 - D)N_2 \tag{11.10}$$

$$dS_1/dt = D(S_{1f} - S_1) - M_1N_1/Y_{1.s_1} \tag{11.11}$$

$$dS_2/dt = D(S_{2f} - S_2) - M_2N_2/Y_{2.s_2} \tag{11.12}$$

$$dP_1/dt = aM_1N_1 - M_2N_2/Y_{2.p_1} - DP_1 \tag{11.13}$$

Here, $Y_{1.s_1}$, $Y_{2.s_2}$, and $Y_{2.p_1}$ are simply the yield constants of N_1

growing on S_1 and N_2 growing on S_2 and on P_1 respectively. The constant a relates the rate of growth of N_1 to the rate of production of P_1. Thus, the model assumes that the rates of consumption of the various substrates and the rate of production of P_1 are directly proportional to the growth rates of the respective populations. The latter are related to the substrate concentrations in the growth chamber as follows:

$$M_1 = M_{max_1} [S_1/(k_{1.s_1} + S_1)] \tag{11.14}$$
$$M_2 = M_{max_2} [S_2/(k_{2.s_2} + S_2)] [P_1/(k_{2.p_1} + P_1)] \tag{11.15}$$

Thus the growth rate of N_2 can potentially be limited by either substrate.

If:

$$D > M_{max_1} [S_{1f}/(k_{1.s_1} + S_{1f})] \tag{11.16}$$

then both populations will be washed out of the growth chamber simply because the dilution rate will be greater than the growth rate of N_1. Should the inequality in equation 11.16 be reversed, then N_1 will eventually reach a stable steady state exactly in the manner predicted by the Monod model. Under such circumstances the commensal population, N_2, will survive as long as:

$$D < M_{max_2} [S_{2f}/(k_{2.s_2} + S_{2f})] [P_1/(k_{2.p_1} + P_1)] \tag{11.17}$$

If the inequality is reversed the commensal will be washed out of the system, and only N_1 will survive. Obviously, N_2 is an obligate commensal.

The three possible steady states are:
(a) Total washout
(b) Only N_1 survives
(c) Both N_1 and N_2 survive

All the above are stable nodes. Of course, (c) represents the commensalism. If the maximum growth rate (by maximum I refer to the initial conditions, not to M_{max}) of N_1 is always less than that of N_2, only two steady states, (a) and (c), exist. Clearly, the fact that an obligate commensalism may be successfully established in batch culture is no guarantee that it may also be in continuous culture.

If the P_1 produced by N_1 is unlimiting, the behaviour of the commensal N_2 will be decoupled from that of N_1 for small perturbations from steady state. Thus, near steady state, the behaviour of the system is akin to one of neutralism, despite the fact that N_2 is an obligate commensal. In the event that S_2 is always unlimiting (and so P_1 must always be limiting for N_2), then the behaviour would be more akin to what is normally considered commensalism because any fluctuation in N_1 will automatically have an effect on N_2.

The model reduces to that of Reilly (1974) if S_2 is unlimiting (see Figure 11.3b; equation 11.12 is now irrelevant) and the rate of consumption of S_1 is exactly equal to the rate of production of P_1. Such a situation may arise, for example, in the nitrogen cycle whereby nitrite is converted into nitrate. Under these circumstances:

$$a = 1/Y_{1.s_1} \qquad (11.18)$$

whence equations 11.13 and 11.15 become:

$$dP_1/dt = M_1N_1/Y_{1.s_1} - M_2N_2/Y_{2.p_1} - DP_1 \qquad (11.19)$$

and

$$M_2 = M_{max_2}[P_1/(k_{2.pl} + P_1) \qquad (11.20)$$

respectively.

Both Miura *et al.* (1980) and Reilly (1974) present stability analyses based on linearized equations from Taylor's expansions. In both cases all eigenvalues are real and negative demonstrating that the equilibrium is a stable node. However, Reilly notes that up to three aperiodic overshoots and undershoots are possible so long as the populations are not approaching a new steady state from an old one (as is the case if a step change in holding time occurs).

If a limited quantity of P_1 is introduced with the feed then, in addition to the three steady states already mentioned, a fourth will be possible, namely:

(d) Only N_2 survives

In (a) the dilution rate is greater than the maximum growth rate of both populations so that equation 11.16 is satisfied and the inequality of equation 11.17 is reversed. In (b), the dilution rate is greater than the maximum growth rate of N_2 so that the combined concentration of P_{1f} and the P_1 produced by N_1 is not sufficient to satisfy equation 11.17. In (d), the dilution rate is greater than the maximal growth rate of N_1 so that it is washed out, but the concentration of P_{1f} is sufficient to satisfy equation 11.17 and therefore N_2 can survive.

There are now several subtly different reasons for the survival of both populations in case (c). In the event that the P_1 produced by N_1 is always unlimiting for N_2, no effect should be observed upon the addition of P_1 to the feed. Another possibility is that P_1 produced from the activities of N_1 is limiting, so that the additional P_{1f} introduced with the feed will permit N_2 to increase in density. There are three possible effects which depend solely upon the concentration of P_{1f} added. First, the commensal may still be primarily limited by the P_1 produced by its partner so that the P_{1f} added is not sufficient to permit survival alone. Thus the commensalism is still obligate. Second, the concentration of P_{1f} added permits a facultative association in which the commensal detectably responds to the concentration of this nutrient. Thirdly, the P_{1f} may be added in excess so that the system will become one of neutralism, with only S_2 limiting the density of N_2. Yet another is that the combined concentrations of P_1 produced by N_1 and the P_{1f} introduced with the feed will prevent N_2 from being washed out of the system. In other words, the relationship is still obligate, but both sources of P_1 are now required for survival of the commensal.

If nothing else, the above demonstrates the complexity, subtlety and multiple outcomes of the simplest model of commensalism by varying just two environmental parameters, D and P_{1f}. At one extreme there is an apparent neutralism even though the relationship is obligate, and at another the relationship is purely facultative.

An Example of a Microbial Commensalism?

Lee *et al.* (1976) studied a commensalism important in the manufacture of Swiss cheese. Under anaerobic conditions, *Lactobacillus plantarum* metabolized glucose to lactate which was then oxydized further by *Propionibacterium shermanii* to propionic acid and carbon dioxide.

Although the commensalism studied was not complicated by other interactions, a number of modifications were introduced to the simplified model of Miura *et al.* (1980) because *L. plantarum* appeared to have a significant maintenance requirement. A maintenance requirement indicates that an organism must consume a nutrient without growing in order to stay viable. The necessary modifications were those of Marr *et al.* (1963) and Pirt (1966) for maintenance, and because this process is associated with the production of lactate by this species, the model of Luedeking and Piret (1959) was incorporated to describe the latter. Equations 11.9 and 11.10 still apply, 11.12 is irrelevant, but 11.11 and 11.13 become:

$$dS_1/dt = D(S_{1f} - S_1) - M_1N_1/Y_{1.s_1} - cN_1 \qquad (11.21)$$

and

$$dP_1/dt = aM_1N_1 + bN_1 - M_2N_2/Y_{2.p_1} - DP_1 \qquad (11.22)$$

where b and c are constants related to the maintenance of *L. plantarum*. Note that this process is independent of the rate of growth. The Monod relationships of equations 11.14 and 11.20 still apply.

Because the maximum rate of growth of *L. plantarum* is greater

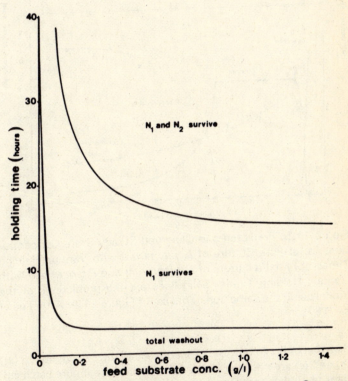

Figure 11.4: The Operating Diagramme Predicting the Outcome of the Theoretical Model of Commensalism of Lee et al. (1976). Data from pure cultures were used to estimate constants (see Lee *et al.* 1976), with the exception of two, b and c, estimated from a single mixed culture. After Lee *et al.* (1976)

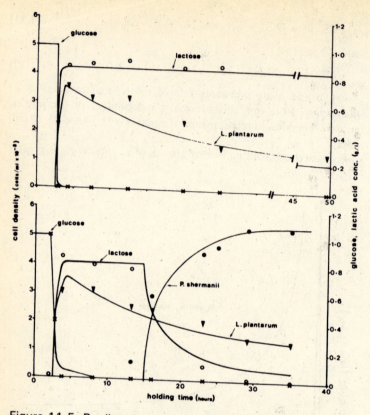

Figure 11.5: Predicted and Observed Steady State Concentrations in (a) Pure Culture of *L. plantarum* with Various Holding Times, (b) Mixed culture of *L. plantarum* and *P. shermanii* With Various Holding Times. Solid lines are the predictions of the model using the same constants as for Figure 11.4. After Lee *et al*. (1976).

than that of the commensal *P. shermanii*, all three steady states, (a), (b) and (c) are possible. Using data obtained from pure batch cultures (with the exception of the estimates of c and b, which proved inadequate under such conditions and so were obtained from mixed cultures), Lee *et al*. (1976) compared the model with the actual dynamics in a chemostat, Figures 11.4 and 11.5. The model successfully predicted the conditions under which the various steady states would occur when the glucose concentration in the reservoir was 1.0 g/litre, and was also in good qualitative agreement with the transient behaviour of the system.

When is a commensalism not a commensalism? When the commensal does not 'benefit'! As Lee *et al.* (1974) demonstrated, *P. shermanii* can utilize both the lactate produced by *L. plantarum* and the glucose provided in the feed. Thus, if the two species are inoculated into a medium containing glucose as the only source of energy they will inevitably compete. In fact, if the competition were to continue, *P. shermanii* would be eluted from the chemostat. But instead, *P. shermanii* switches from glucose as soon as *L. plantarum* produces lactate, so avoiding losing the competition and simultaneously generating what appears to be a commensalism. However, *P. shermanii* will reach significantly higher population densities if growing on the glucose in the absence of *L. plantarum*. Therefore, *P. shermanii* does not really 'benefit' at all from the presence of *L. plantarum*.

Commensalism With Competition

Not all experiments have displayed the simple dynamics that Lee *et al.* (1976) observed. Frequently a commensalistic interaction has been found to be compounded by other, unrelated ecological phenomena. For example, Shindala *et al.* (1965) studied the commensalistic dependence of the bacteria *Proteus vulgaris* on the yeast *Saccharomyces cerevisiae* in continuous culture. The commensalistic relationship could be disengaged by the addition of niacin, or a related compound, to the growth chamber of a steady state culture. The density of the bacteria immediately increased at the expense of the yeast, and only when the compound had been eluted was the original steady state reattained. Since the addition of niacin to a pure culture of yeast had no effect, there was clearly competition for some unknown substance in the original steady state culture (Meers 1973). In another study, Miura *et al.* (1978) discovered dampened oscillations in a system composed of *Pseudomonas oleovorans* and *Mycotorula japonica* growing on phenol and n-tetradecane. *P. oleovorans* could only assimilate phenol and excreted biotin as a product of metabolism. *M. japonica* could grow on both phenol and n-tetradecane but required the biotin for survival. Here, then, was a potential for competition, particularly if *M.*

japonica had a low affinity for the hydrophobic n-tetradecane.

None of the models of pure commensalism so far discussed can predict the existence of oscillations in any system variables. Miura *et al.* (1980) modified their model of pure commensalism so as to include competition. One of the two resulting models is diagrammed in Figure 11.3c. Equations 11.9, 11.10, 11.13 and 11.14 still apply, but equations 11.11 and 11.15 now become:

$$dS_1/dt = D(S_{1f} - S_1) - M_1N_1/Y_{1.s_1} - M_2N_2/Y_{2.s_1} \qquad (11.23)$$

$$M_2 = M_{max_2}[S_1/(k_{2.s_1} + S_1)][P_1/(k_{2.p_1} + P_1)] \qquad (11.24)$$

The model has the same three steady states as does that of Lee *et al.* (1976) yet quite different transient behaviours may now occur. For instance it predicts that the competitive interactions may result in dampened oscillations of both populations and both limiting substrates, thus qualitatively confirming the observations of Miura *et al.* (1978). The fact that Shindala *et al.* (1965) did not observe such behaviour is not surprising because the model predicts that a node can exist too; however, it has as yet to be rigorously tested.

MeGee *et al.* (1972) studied the interactions between the yeast *Saccharomyces cerevisiae* and the bacterium *Lactobacillus casei*. The bacterium required riboflavin, P_1, produced by the yeast. The interaction was one of commensalism compounded by competition for glucose, S_1, and by the fact that yeast has a large maintenance requirement which, if not met, would result in starvation. The model used is as follows:

$$dN_{1v}/dt = (M_1 - d_1 - D)N_{1v} \qquad (11.25)$$

$$dN_{1t}/dt = M_1N_{1v} - DN_{1t} \qquad (11.26)$$

$$dN_2/dt = (M_2 - D)N_2 \qquad (11.27)$$

$$dS_1/dt = D(S_{1f} - S_1) - M_1N_{1v}/Y_{1v.s_1}$$
$$\qquad - d_1S_1N_{1v}/Y_{1v.d} - M_2N_2/Y_{2.s_1} \qquad (11.28)$$

$$dP_1/dt = D(P_{1f} - P_1) + aM_1N_{1v} - M_2N_2/Y_{2.p_1} \qquad (11.29)$$

where N_{1v}, N_{1t} and N_2 are the densities of viable yeast, total yeast and the bacterium respectively, and M_1 and M_2 are described in equations 11.14 and 11.24 respectively. d_1 is the rate of death of viable cells so that:

$$d_1 = d_{max_1}/(1 + S_1/k_{1v.d}) \qquad (11.30)$$

Here, d_{max_1} is the maximum death rate of viable cells under total starvation ($S_1 = 0$) and $k_{1v.d}$ is simply a constant associated with this process. $Y_{1v.d}$ is a constant associated with the consumption of glucose for maintenance. Note that here, maintenance is a function of S_1 so that it is correlated with growth rate: the higher the growth rate the lower the maintenance. Again, the model predicts that three steady states:

(a) Total washout
(b) Only N_1 survives
(c) Both N_1 and N_2 survive

are possible, as well as with the fourth one

(d) Only N_2 survives

In (b), only the yeast survives because the levels of glucose and riboflavin are not sufficient to prevent the bacterium being washed out of the system. In (c), the steady state represents one of commensalism with competition. In (d), the bacterium is no longer dependent upon the yeast as a source of riboflavin, so that the ensuing pure competition results in washout of the yeast. In all of these cases, stability analyses show that the stable steady states may be either nodes or foci.

Estimates of the various parameters from pure batch cultures gave a reasonable fit to the data although some improvement was achieved when a few were recalculated, due to difficulties in estimating minute riboflavin concentrations and differences in pH between media. When no riboflavin was provided in the feed ($P_{1f} = 0$), the system was one of either commensalism with competition, case (c), or only the yeast survived at high dilution rates or low concentrations of glucose in the feed, case (b). The model accurately predicted the equilibrium densities of the populations for a range of dilution rates and glucose concentrations in the feed (Figure 11.6). Note that the difference between the total and viable counts of yeast increases dramatically with an increasing holding time, thus validating the inclusion of maintenance factors in the model.

For a range of concentrations of riboflavin provided in the feed the model again predicted the steady state values of the populations well. The yeast was completely washed out of the system when the concentration of riboflavin in the feed exceeded 12 mg/litre. This value, therefore, represents the boundary between a facultative commensalism with competition and pure competition. Thus, by simply changing the concentration of riboflavin in the feed, the behaviour of the bacterium is changed from being one of an obligate commensal, through facultative commensalism to being a superior competitor.

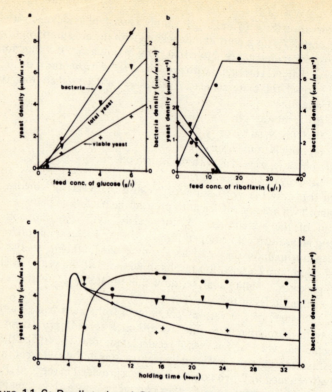

Figure 11.6: Predicted and Observed Steady State Concentrations (a) as a Function of Glucose Feed Concentration, (b) as a Function of Riboflavin Concentration and (c) as a Function of Holding Time. Solid lines are the predictions of the model for competitive commensalism, with constants estimated from pure cultures. After MeGee *et al*. (1972).

Pure Mutualism

Pure mutualism, as described by Meyer *et al*. (1975) for two populations, occurs if the growth rates of both are actually or potentially limited only by the concentrations of critical substrates produced by the partner (Figure 11.7a). The term 'pure' refers to the fact that the relationship is an obligate one in which there are no external con-

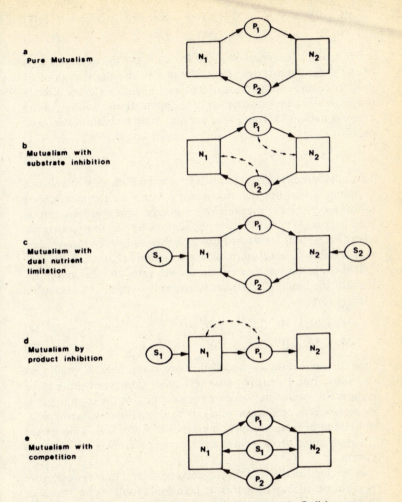

a
Pure Mutualism

b
Mutualism with
substrate inhibition

c
Mutualism with
dual nutrient
limitation

d
Mutualism by
product inhibition

e
Mutualism with
competition

Figure 11.7: Patterns of Mutualism Discussed. Solid arrows indicate the production and consumption of substances, broken arrows indicate the effects of inhibition.

straints imposed, such as limiting nutrients provided in the feed. The Monod equations describing the chemostat model are

$$dN_1/dt = (M_1 - D)N_1 \qquad (11.31)$$

$$dN_2/dt = (M_2 - D)N_2 \qquad (11.32)$$

$$dP_1/dt = a_1 M_1 N_1 - M_2 N_2 / Y_{2.p_1} - DP_1 \qquad (11.33)$$

$$dP_2/dt = a_2 M_2 N_2 - M_1 N_1 / Y_{1.p_2} - DP_2 \qquad (11.34)$$

where a_1 and a_2 are constants, and P_1 and P_2 are the concentrations of the critical substrates. Thus, the model assumes that rates of production and consumption of the coupling substances are directly proportional to the growth rates of the populations. Analysis of the above equations indicates that a steady state mutualism can exist only if

$$a_1 a_2 > 1/(Y_{1.p_2} Y_{2.p_1}) \qquad (11.35)$$

This equation, in effect, states that for a mutualism to be established in continuous culture, there must exist a state at which the combined rate of production of the nutrients must exceed the combined rate of consumption. Therefore, two populations having complementary metabolisms may be washed out of a chemostat even though they may form a successful mutualistic association in batch culture.

If the relationship between the growth rates and the concentrations of the limiting substrates is assumed to be of the Michaelis-Menten form

$$M_1 = M_{max_1} [P_2/(k_{1.p2} + P_2)] \qquad (11.36)$$

$$M_2 = M_{max_2} [P_1/(k_{2.p1} + P_1)] \qquad (11.37)$$

then the mutualism as modelled above is unstable. A small perturbation from the steady state will either cause the culture to be completely washed out because consumption of the coupling substances exceeds production or else both populations will grow without bound because production exceeds consumption. This absurd situation arises because there are no external constraints upon the system.

A mechanism that can theoretically produce a stable equilibrium is substrate inhibition; that is, at high concentrations of P_1 and/or P_2 the growth rate of the consumer is inhibited. In order to model this situation (Figure 11.7b), Meyer *et al.* (1975) assumed that Andrew's inhibition expression is valid for both populations, namely:

$$M_1 = M_{e_1} P_2/(k_{1.p2} + P_2 + (P_2)^2/L_1) \qquad (11.38)$$

$$M_2 = M_{e_2} P_1/(k_{2.p1} + P_1 + (P_1)^2/L_2) \qquad (11.39)$$

where M_{e_1} and M_{e_2} are analogous to M_{max}. The model reduces to that of pure mutualism if the constants L_1 and L_2 approach infinity.

However, for finite values of L_1 and L_2 the maximal growth rates are achieved when:

$$P_2 = (k_{1.p2}L_1)^{1/2} \tag{11.40}$$

and:

$$P_1 = (k_{2.p1}L_2)^{1/2} \tag{11.41}$$

Stability analyses by Meyer *et al.* (1975) demonstrate the existence of four possible steady states (other than washout) that permit coexistence. Two of these are stable; either the growth rate of one population is inhibited by an excess of its substrate and that of the other by a scarcity, or *vice versa*. In the event that the growth rate of only one of the partners is subject to substrate inhibition, a stable steady state exists — which is hardly surprising because the above criterion is automatically met. The other two, when either both populations are limited by a lack of substrates or when both are inhibited by excesses, are unstable. The four possible steady states may be nodes or foci depending upon the values that the various parameters assume. In fact, the authors report in their numerical calculations that simply by changing the value of the dilution rate D, stable nodes, foci or an unstable focus may exist. The latter suggests that a stable limit cycle is possible and was confirmed using numerical integration.

Another way to stabilize the system is to introduce two additional substrates (one for each population) with the feed, both of which are potentially growth rate limiting (Figure 11.7c). Equations 11.31 through 11.34 are still applicable, but the incorporation of two new substrates necessitates modifying equations 11.36 and 11.37 and introducing two new mass balance equations, as follows:

$$M_1 = M_{max_1} [S_1/(k_{1.s_1} + S_1)] [P_2/(k_{1.p2} + P_2)] \tag{11.42}$$

$$M_2 = M_{max_2} [S_2/(k_{2.s_2} + S_2)] [P_1/(k_{2.p1} + P_1)] \tag{11.43}$$

and

$$dS_1/dt = D(S_{1f} - S_1) - M_1N_1/Y_{1.s_1} \tag{11.44}$$

$$dS_2/dt = D(S_{2f} - S_2) - M_2N_2/Y_{2.s_2} \tag{11.45}$$

Analyses of the above model by both Meyer *et al.* (1975) and Miura *et al.* (1980) demonstrate the existence of three steady states:

(a) Total washout
(b) An unstable node
(c) A stable node

All are nodes. Total washout needs no comment. In (b) the mutualism is unstable and corresponds to the same as in the model of pure mutualism. In (c) the mutualism is stable because at least one population has its rate of growth limited by a substrate introduced with the feed. This limits the amount of substrate produced for the other partner, controlling the latter's density and thereby stabilizing the system.

Lamb and Garver (1980) investigated a case of mutualism that was apparently stabilized as described in the previous paragraph. An unnamed bacterium (N_1) metabolized methane, so producing higher organic compounds ($C > 1$) which then served as a source of energy and carbon for another unnamed bacterium (a citrate utiliser). The latter produced vitamin B_{12} which was an absolute requirement for the growth of the methane utilizer. Although the rate of utilization of methane appeared to be directly proportional to the rate of growth of N_1, the rate of production of the unknown organic substrates was best described using the model of Luedeking and Piret (1959). Whilst the citrate utilizer, N_2, seemed to be limited only by these compounds, under all steady state conditions N_1 was limited only by the availability of oxygen for oxydizing the methane, which was always in excess. Thus, vitamin B_{12} was in excess (as was the S_2 of the model presented above), whilst the limitation of oxygen stabilized the system. The model predicted the ratio of N_1/N_2 accurately for a number of dilution rates.

Mutualism With Competition

An interesting case is that in which two obligate mutualists compete for a substrate provided in the feed, as in Figure 11.7e. Again, the model of pure mutualism (equations 11.31 through 11.34) holds, but now only one substrate introduced with the feed need be modelled. The necessary modifications of equations 11.39 and 11.44 are:

$$M_2 = M_{max_2} [S_1/(k_{2.s_1} + S_1)] [P_1/(k_{2.p_1} + P_1)] \qquad (11.46)$$

and:

$$dS_1/dt = D(S_{1f} - S_1) - M_1N_1/Y_{1.s_1} - M_2N_2/Y_{2.s_1} \qquad (11.47)$$

Equation 11.42 remains the same, but 11.45 is irrelevant. Analysis of the above model by Meyer *et al.* (1975) and Miura *et al.* (1980) demonstrates the existence of two steady states. The first is unstable and corresponds to that of pure mutualism. The second is stable and is either a node when the growth characteristics of the two populations are similar, or a focus if they are sufficiently different. If both populations are limited by S_1 then one of them must also be limited by the availability of one of the other substrates. Another possibility is that one population is limited by S_1 so that the availability of the product it produces limits the partner. Near equilibrium, the dynamics will be similar to those of commensalism.

In an extension of their work on commensalism with competition, MeGee *et al.* (1972) investigated a mutualism with competition between *Saccharomyces cerevisiae* and *Lactobacillus casei*. As before, the yeast provided the bacterium with riboflavin. However, this time the medium was not buffered to such a great extent, so that the pH dropped due to the synthesis of lactic acid by *L. casei*, thereby stimulating the growth of *S. cerevisae*.

This reciprocal interaction was demonstrated in batch culture, Figure 11.8. MeGee (1971) presented a model of the pH effect that clearly correlates well with the data. However, the demonstration of a mutualistic interaction in a chemostat culture is more complex.

Alone in continuous culture, the density of yeast may exhibit dampened oscillations because of a large maintainance requirement. However, no oscillations of the bacterium occur in pure culture. In a mixed culture in a chemostat the approach to steady state was by dampened oscillations which were particularly noticeable in the bacterial population, as it responded to fluctuations in availability of the riboflavin produced by the yeast. Whether the oscillations were entirely due to the behaviour of the yeast, or whether they were also due to the competitive nature of the system (as described above) is not clear. The catch in the whole argument is that the density of the yeast is lower in the mixed culture than in pure culture. However, a few ingenious calculations by MeGee *et al.* (1972) demonstrated that the density of bacteria is also much lower than would be expected on the basis of a commensalism with competition. Their conclusion was that the yeast was consuming more glucose than in pure culture, and that the mutualistic interaction was caused by a lowering of the pH as in the case of the batch cultures. The reason why both species were of lower density in mixed culture is simply competition.

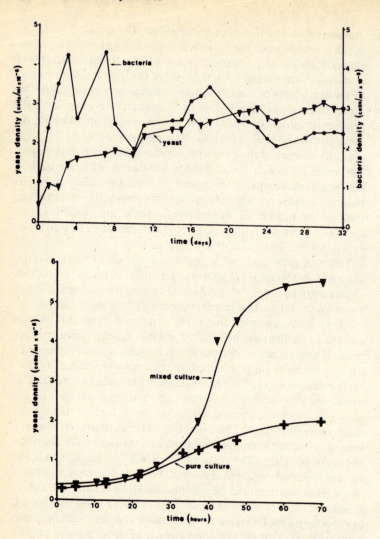

Figure 11.8: Mutualism with Competition. (a) Transient oscillations of *P. vulgaris* found in continuous culture. (b) Yeast density in pure and mixed batch cultures with the solid lines representing predictions of the model.

The Effects of Cross and Product Inhibition

Competition is not the only force potentially capable of generating oscillations in commensalistic systems. Yeoh *et al.* (1968) studied a mutualism involving *Proteus vulgaris* which excreted the biotin that *Bacillus polymyxa* required for growth. The former required the niacin produced by the latter. Surprisingly, in a chemostat the mixed culture displayed what were seemingly sustained oscillations. Because *B. polymyxa* did not grow well in fresh supernatants of *P. vulgaris*, whereas the reverse was true, the authors concluded that the latter produced a substance that was toxic to the former. This inhibitory effect was removed by the addition of chymotrypsin, a proteolytic enzyme, to the chemostat feed, which caused the density of the culture to increase markedly, leading to a steady state with no oscillations.

Yeoh *et al.* (1968) did not provide a mathematical model describing the dynamics of this system. However, their experimental analysis of cross-inhibition contrasts elegantly with the mathematical analysis of substrate inhibition by Meyer *et al.* (1975). Whereas substrate inhibition stabilizes a pure mutualism, cross-inhibition destabilizes any mutualism.

Wilkinson *et al.* (1974) studied some interesting relationships between the members isolated from a methane-utilizing community. A *Pseudomonas* sp., which was limited by oxygen availability, metabolized methane to methanol which inhibited growth at relatively low concentrations. A *Hyphomicrobium* sp. (a denitrifier) metabolized the methanol and in so doing permitted the *Pseudomonas* to grow. However, the growth was not as great as in the original culture. This appeared to be caused by the absence of an *Acinetobacter* sp. and/or a *Flavobacterium* sp. Exactly why these two species enhanced the growth of the mutualists was not established, but the authors suggested that they might remove other toxic side products of metabolism. If so, then these latter two organisms were engaged in a facultative relationship with at least one of the obligate mutualists.

A model, which was essentially a variant of that of pure commensalism described earlier, was presented to account for the behaviour of this system (Figure 11.7d). Equations 11.9, 11.10, 11.11 and 11.13 still apply, 11.12 being irrelevant. The growth rate of the denitrifier was limited only by the availability of methanol (equation 11.20) and that of the methane utilizer by the availability of oxygen, S_1, and by inhibition by P_1:

$$M_1 = M_{max_1} [S_1/(k_{1.s_1} + S_1)] [(1_{1.p_1}/(1_{1.p_1} + P_1)] \qquad (11.48)$$

Although the authors did not analyse this model in any great detail, they demonstrated that it was in qualitative agreement with the data presented.

Chao and Reilly (1972) studied a commensalism between *Acetobacter suboxydans* and *Saccharomyces carlsbergensis* under continuous culture conditions. The bacterium oxidized the mannitol provided in the feed to fructose which was then metabolized by the yeast. Observations indicated that both sugars were limiting the growth rates of their respective consumers so that there was no competition for any other substrate, such as the amino acids provided in the feed. However, when the dilution rate was abruptly changed up or down after a steady state had been reached, the system displayed a series of oscillations of both the population densities and sugar concentrations. No oscillations were found in pure cultures after a change in dilution rate. They supposed that there was cross inhibition from the yeast to the bacteria, as depicted in Figure 11.3d.

Reilly (1974) presented a number of models involving both inhibition and activation of one population by the other to account for the oscillations. The cross inhibition was modelled by:

$$M_i = M_{max_i} [S_i/(k_{i.s_i} + S_i + I_i/L_{i.li})] \qquad (11.49)$$

where I_i is an inhibitor of N_i produced by N_i and $L_{i.li}$ is simply a constant. The effects of metabolism of inhibitors and activators were also modelled. However, only one of the 20 cases investigated yielded dynamics similar to those observed experimentally, and even then there were serious discrepancies, for example, the data indicated that the functions would not regularly pass through the eventual steady state, whereas the model predicted such behaviour.

Equations 11.48 and 11.49 may be compared. Whilst they both describe the inhibitory effects of a substance upon growth rate, they do so in entirely different ways. Reilly's model proposes that the inhibitor increases the value of the Michaelis constant — a situation analogous to the competitive inhibition of the catalytic action of an enzyme. In the model presented by Wilkinson *et al.* (1974), the maximum growth rate of the population is affected by the inhibitor — a situation analogous to mixed competitive inhibition in enzyme kinetics. The two may be distinguished by providing excess substrate to the inhibited population to see if the maximum growth rate can be achieved. Regrettably, there is no indication as to the biological rationale behind each model chosen. Indeed, Reilly (1974) states that the relationship was assumed.

Discussion

Many of the models described contain a rather large number of parameters. This raises the question as to whether such complex models are capable of fitting any and all kinds of data. However, in defence of such an approach, I must point out that constants are frequently estimated in pure cultures before the dynamics of a mixed culture are even monitored. In other words, the behaviour of the mixed culture is predicted from data obtained from pure cultures. This approach has been remarkably successful in predicting both the qualitative and quantitative behaviours of mixed cultures.

In any case, biological systems by their very nature consist of a large number of parameters. Key parameters must be modelled, otherwise any biological realism will be lost. Such models may be quite misleading. By way of example, consider the Lotka-Volterra equations for two competing organism:

$$dN_1/dt = r_1N_1(K_1 - N_1 - \alpha_{12}N_2)/K_1 \tag{11.50}$$

$$dN_2/dt = r_2N_2(K_2 - N_2 - \alpha_{21}N_1)/K_2 \tag{11.51}$$

where r_1 and r_2 are the intrinsic rates of growth, K_1 and K_2 are the equilibrium densities of pure cultures (carrying capacities) and α_{12} and α_{21} reflect the competitive effects of these populations upon one another. The equations predict four outcomes:

(a) if $K_1 > K_2\alpha_{12}$ and $K_2 < K_1\alpha_{21}$ then N_1 wins

(b) if $K_2 > K_1\alpha_{21}$ and $K_1 < K_2\alpha_{12}$ then N_2 wins

(c) if $K_1 < K_2\alpha_{12}$ and $K_2 < K_1\alpha_{21}$ then either N_1 or N_2 may win

(d) if $K_1 > K_2\alpha_{12}$ and $K_2 > K_1\alpha_{21}$ then coexistence results

As is shown in Appendix Two, a pair of Monod equations may be reduced to a form similar to the classical Lotka-Volterra competition equations above namely:

$$dN_1/dt = r_1N_2 [K_1 - N_1 - \alpha_{12}N_2)]/K_1 \cdot$$

$$(k_{1.s_1} + S_{1f})/(k_{1.s_1} + S_1) \tag{11.52}$$

and

$$dN_2/dt = r_2 N_2 [K_2 - N_2 - \alpha_{21} N_1)]/K_2 \cdot$$

$$(k_{2.s_1} + S_{1f})/(k_{2.s_1} + S_1) \qquad (11.53)$$

where:

$$r_1 = M_{max_1} [S_{1f}/(k_{1.s_1} + S_{1f})] - D$$
$$r_2 = M_{max_2} [S_{1f}/(k_{2.s_1} + S_{1f})] - D$$
$$K_1 = Y_{1.s_1} (S_{1f} - S_{1eq})$$
$$K_2 = Y_{2.s_1} (S_{1f} - S'_{1eq})$$

and the prime indicates that the equilibrium value of S_1 may be different for each competitor in pure culture. Although equations 11.52 and 11.53 contain additional terms, note that when equated to zero they generate exactly the same isoclines as the Lotka-Volterra equations. Thus, the final outcome of the competition remains the same.

If one is prepared to accept the above definitions, then the competition coefficients turn out to be nothing more than the ratio of the yield coefficients:

$$\alpha_{12} = 1/\alpha_{21} = Y_{1.s1}/Y_{2.s1}$$

Thus:

$$\alpha_{12}\alpha_{21} = 1$$

This means that events (c) and (d) do not exist; that is, no stable or unstable equilibria exist. Furthermore, a little algebra yields the following inequalities:

(a) if $S_{1eq} < S'_{1eq}$ then N_1 wins
(b) if $S_{1eq} > S'_{1eq}$ then N_2 wins

In other words the carrying capacities of the populations are of no importance. What is important is that one competitor will reduce the concentration of the critical substrate to a level such that the growth rate of the other will be less than the dilution rate. Thus, the second is washed out of the system. In the unlikely event that the equilibrium concentrations of S_1 are the same for both competitors, coexistence should be theoretically possible, if experimentally absurd.

Clearly, the approach of the Lotka-Volterra equations, as expressed in equations 11.50 and 11.51, does not explicitly take the environment into account, with the result that the true nature of the carrying capacities and the competition coefficients is unknown. Thus, when applied to two populations competing for a single growth rate limiting resource, this model makes two false predictions, (c) and (d), whilst describing events (a) and (b) in a misleading manner.

This is not to say that two competitors may never coexist, but rather, if they do, then environmental fluctuations or other interactions must be taking place. For instance, rather than regard commensalisms and mutualisms as being complicated by competitive interactions, one may easily consider the positive interactions as the stabilising forces of an otherwise fundamentally unstable interaction. This is demonstrated in the experimental work of Shindala *et al.* (1965) and MeGee *et al.* (1972), and in the theoretical work of MeGee *et al.* (1972) and Miura *et al.* (1980). When the concentration of the growth factor limiting the density of one partner is sufficiently increased, the result is a concomitant decrease in that of the other population as it is outcompeted. This demonstrates an important point. A positive interaction can only stabilize a competitive one so long as the superior competitor has its activities limited by those of its partner.

If one regards the parameters of the Lotka-Volterra competition equations as in some way 'averaging' over a number of complex biological interactions, then the fact that this model predicts the existence of a stable equilibrium exists might be excused. However, defence by such an argument obscures the true nature of the interactions, leaving us in complete ignorance of the biology of the situation. For excellent reviews of the literature on microbial competition (and also other interactions) which discuss other mechanisms by which competitors may co-exist, see Fredrickson (1977, 1983).

As environmental conditions change, so the manner in which one species affects the other may do so as well. The studies by MeGee *et al.* (1972) clearly demonstrate this. The addition of the growth factor limiting the growth of the commensal changes the system from one of obligate commensalism with competition through one of facultative commensalism with competition to one of pure competition. Furthermore, even if two populations compete in the absence of any other interactions, if the dilution rate is changed, then the equilibrium values of S_i for each species will change, perhaps with the result that the previously inferior competitor becomes superior. This prediction of the Monod model requires that the rates of growth of the two competitors as functions of S_l intersect. This has been experimentally confirmed by Jannasch (1967), Meers (1971) and Harder and Veldcamp (1971). Thus, changes in the abiotic environment may profoundly influence the nature of the interactions between populations.

Gause's hypothesis can now be given a rigorous definition. If two or more populations compete, in the absence of any other interactions or environmental perturbations, for a single, common, unfluctuating growth-rate-limiting resource, then one population will eventually exclude the others. This prediction should be true, not only for microbial populations, but for all populations.

Another more speculative and more interesting point arises. At least to a limited extent, community diversity may be maintained by a plethora of mixed interactions, such as mutualism with competition. Alice Newton has pointed out to me that many marine planktonic species are difficult to grow in pure culture, and yet grow well together in mixed cultures. This suggests, perhaps, that mixed interactions may play some role in maintaining species diversity in this ecosystem. Of course, I do not exclude other explanations such as the keystone predator concept, microniche diversity and environmental perturbations as a means of explaining the paradox of the plankton.

However, the conclusions from studies of micro-organisms inhabiting chemostats are not only useful for speculative explanations of natural ecological phenomena; they can be used to predict events in the field. In an excellent study, Zevenboom (1980) combined detailed laboratory studies of the physiological behaviour of the cyanobacterium *Oscillatoria agardhii* with field studies. The conditions necessary for blooms of this species to occur were predicted and supported by field observations.

A recurring theme throughout this chapter has been the emphasis on the importance of the environment; how it affects the relationships between the inhabitants, and how in turn they modify it. Yet despite the fact that such changes may fundamentally influence both community structure and the nature of interactions between populations, classical theoretical ecology largely ignores the status of the environment, particularly that of the abiotic environment. This is true of even a most basic equation, the logistic, which has been used and modified and served as a foundation, both explicitly and implicitly, for much ecological thought. But as MeGee *et al.* (1972) point out, models that do not account for changes in environmental parameters which significantly influence the dynamics of populations cannot be subjected to meaningful stability analyses.

Mixed microbial populations inhabiting continuous culture devices provide an excellent experimental system for the study of many ecological and evolutionary phenomena. The relatively simple and controllable environments of such apparatus have permitted the dissection of quite complex dynamics of simultaneous interactions in a meaningful manner. Mathematical models can be tested in a rigorous manner unavailable in most ecological experimental systems. With such a system, it may be eventually possible to ask meaningful questions of community diversity, complexity and stability, and to test hypotheses in a manner quite beyond the scope of present day field studies.

Acknowledgements

Work on this chapter was supported by a Public Health Service Grant, GM 30201, from the NIH to Daniel E. Dykhuizen and Daniel L. Hartl. I also wish to express my thanks to them both for their advice and encouragement during the preparation of the manuscript.

References

Chao, C.C. and P.J. Reilly (1972) 'Symbiotic Growth of *Acetobacter suboxydans* and *Saccharomyces carlsbergenesis* in a Chemostat', *Biotechnology and Bioengineering, 14*, 75-92

Dykhuizen, D.E. and D.L. Hartl (1983) 'Selection in Chemostats', *Microbiological Review, 47*, 150-68

Fredrickson, A.G. (1977) 'Behaviour of Mixed Cultures of Microorganisms', *Annual Review of Microbiology, 31*, 63-87

Fredrickson, A.G. (1983) 'Interactions of Microbial Populations in Mixed Culture Systems' in Blanch, H.W., E.T. Papoutsakis and G. Stephanopoulos (eds), *Foundations of Biochemical Engineering: Kinetics and Thermodynamics in Biological Systems*, American Chemical Society Symposium series 207, pp. 210-27

Fredrickson, A.G., J.L. Jost, H.M. Tsuchiya and Ping-Hwa Hsu (1973) 'Predator-Prey Interactions Between Malthusian Populations', *Journal of Theoretical Biology, 38*, 487-526

Harder, W. and H. Veldcamp (1971) 'Competition of Marine Psychrophilic Bacteria at Low Temperatures', *Antonie van Leeuwenhoek, 37*, 51-63

Jannasch, H.W. (1967) 'Enrichments of Aquatic Bacteria in Continuous Culture',

Archives of Microbiology, 59, 165–73

Lamb, S.C. and J.C. Garver (1980) 'Interspecific Interactions in a Methane-Utilizing Mixed Culture', *Biotechnology and Bioengineering, 22*, 2119–35

Lee, I.H., A.G. Fredrickson and H.M. Tsuchiya (1974) 'Diauxic Growth of *Propionibacterium shermanii*', *Applied Microbiology, 28*, 831–5

Lee, I.H., A.G. Fredrickson and H.M. Tsuchiya (1976) 'Dynamics of Mixed Cultures of *Lactobacillus plantarum* and *Propionibacterium shermanii*', *Biotechnology and Bioengineering, 18*, 513–26

Luedeking, R. and E.L. Piret (1959) 'A Kinetic Study of the Lactic Acid Fermentation. Batch Process at Controlled pH', *Journal of Biochemical and Microbiological Technology and Engineering, 1*, 393–412

Marr, A.G., E.H. Nilson and D.J. Clark (1963) 'The Maintenance Requirement of *Escherichia coli*', *Annals of the New York Academy of Science, 102*, 536–48

Meers, J.L. (1971) 'Effects of Dilution Rate on the Outcome of Chemostat Mixed Culture Experiments', *Journal of General Microbiology, 67*, 359–61

Meers, J.L. (1973) 'Growth of Bacteria in Mixed Cultures', *CRC Critical Reviews in Microbiology, 2*, 139–79

MeGee, R.D.III (1971) 'Interactions Between Dissimilar Microbial Populations and Their Environment', PhD thesis, University of Minnesota, Minneapolis

MeGee, R.D.III, J.F. Drake, A.G. Fredrickson and H.M. Tsuchiya (1972) 'Studies in Intermicrobial Symbiosis. *Saccharomyces cerevisiae* and *Lactobacillus casei*', *Canadian Journal of Microbiology, 18*, 1733–42

Meyer, J.S., H.M. Tsuchiya and A.G. Fredrickson (1975) 'Dynamics of Mixed Populations Having Complementary Metabolism', *Biotechnology and Bioengineering, 17*, 1065–81

Miura, Y., K. Sigiura, M. Yoh, H. Tanaka, M. Okazaki and S. Komemushi (1978) 'Mixed Culture of *Mycotorula japonica* and *Pseudomonas oleovorans* on Two Hydrocarbons', *Journal Fermentation Technology, 56*, 339–44

Miura, Y., H. Tanaka and M. Okazaki (1980) 'Stability Analysis of Commensal and Mutual Relations with Competitive Assimilation in Continuous Mixed Culture', *Biotechnology and Bioengineering, 22*, 929–48

Monod, J. (1942) *Recherches sur la Croissance des Cultures Bactériennes*, Paris, Herman et Cie.

Monod, J. (1950) 'La Technique de Culture Continue. Theorie et Applications', *Annales de l'Institut Pasteur de Paris, 79*, 390–410

Nurmikko, V. (1956) 'Biochemical Factors Affecting Symbiosis Among Bacteria', *Experimentia, 12*, 245–84

Reilly, P.J. (1974) 'Stability of Commensalistic Systems', *Biotechnology and Bioengineering, 16*, 1373–92

Shindala, A., H.R. Bungay, N.R. Krieg and K. Culbert (1965) 'Mixed Culture Interactions. I. Commensalism of *Proteus vulgaris* with *Saccharomyces cerevisiae* in Continuous Culture', *Journal of Bacteriology, 89*, 693–6

Wilkinson, T.G., H.H. Topiwala and G. Hamer (1974) 'Interactions in a Mixed Bacterial Population Growing on Methane in Continuous Culture', *Biotechnology and Bioengineering, 16*, 41–59

Yeoh, H.T., H.R. Bungay and N.R. Krieg (1968) 'A Microbial Interaction Involving Combined Mutualism and Inhibition', *Canadian Journal of Microbiology, 14*, 491–2

Zevenboom, W. (1980) 'Growth and Nutrient Uptake Kinetics of *Oscillatoria agardhii*', PhD thesis, Universiteit van Amsterdam, Amsterdam

Appendix One

Under a restricted set of circumstances, the Monod model may be reduced to the logistic (the Verhulst-Pearl) equation. Defining:

$$r_1 = M_{max_1} [S_{1f}/(k_{1.s_1} + S_{1f})] - D$$

then the Monod equation:

$$dN_1/dt = \{M_{max_1} [S_1/(k_{1.s_1} + S_1)] - D\}N_1 \tag{11.54}$$

may be rewritten as:

$$dN_1/dt = r_1 (k_{1.s_1} + S_{1f})/[M_{max_1}S_{1f} - D(k_{1.s_1} + S_{1f})] \bullet$$
$$\{M_{max_1} [S_1/(k_{1.s_1} + S_1)] - D\}N_1 \tag{11.55}$$

Rearranging yields:

$$dN_1/dt = r_1N_1 (k_{1.s_1} + S_{1f})/(k_{1.s_1} + S_1) \bullet$$
$$[(M_{max_1} - D)S_1 - Dk_{1.s_1}]/[(M_{max_1} - D)S_{1f} - Dk_{1.s_1}] \tag{11.56}$$

MeGee (1971) states that for the Monod model:

$$S_1 = (S_1 + N_1/Y_{1.s_1} - S_{1f})e^{-tD} - N_1/Y_{1.s_1} + S_{1f} \tag{11.57}$$

where S_1 and N_1 are the concentrations of critical substrate and the density of micro-organisms immediately after inoculation. Then if:

$$S_1 = S_{1f}$$

and

$$N_1/Y_{1.s_1} \rightarrow 0$$

equation 11.57 will be approximated by:

$$S_1 = S_{1f} - N_1/Y_{1.s_1} \tag{11.58}$$

for all t.
Of course, at equilibrium:

$$S_{1eq} = S_{1f} - N_{1eq}/Y_{1.s_1} \tag{11.59}$$

and

$$M_{max_1} [S_{1eq}/(k_{1.s_1} + S_{1eq})] = D \tag{11.60}$$

Substituting 11.58 and then 11.60 into 11.56 gives:

$$dN_1/dt = rN_1(k_{1.s1} + S_{1f})/(k_{1.s1} + S_1) \bullet$$

$$[Y_{1.s1}(S_{1f} - S_{1eq}) - N_1]/[Y_{1.s1}(S_{1f} - S_{1eq})] \tag{11.61}$$

Rearranging 11.59 to define:

$$K_1 = N_{1eq} = Y_{1.s1}(S_{1f} - S_{1eq})$$

so that 11.61 may be rewritten as:

$$dN_1/dt = r_1N_1(K_1 - N_1)/K_1 \bullet (k_{1.s1} + S_{1f})/(k_{1.s1} + S_1) \tag{11.62}$$

yields an expression similar to the logistic.
If:

$$S_{1f} <<< k_{1.s1}$$

then:

$$(k_{1.s1} + S_{1f})/(k_{1.s1} + S_1) = 1$$

so that equation 11.62 becomes:

$$dN_1/dt = r_1N_1(K_1 - N_1)/K_1 \tag{11.63}$$

the logistic. However, S_{1f} must take a value so ridiculously small that the logistic expression as described in equation 11.63 is absurd for all practical purposes. However, note that even if S_{1f} is large, the initial rate of growth, and the final equilibrium density, K_1, remain unaffected.

Appendix Two

Consider another species, N_2, also growing in a chemostat culture. Then by the same arguments in Appendix One:

$$dN_2/dt = r_2N_2(k_{2.s1} + S_{1f})/(k_{2.s1} + S_1) \bullet$$

$$[(M_{max_1} - D)S_1 - Dk_{2.s2}]/[(M_{max_s} - D)S_{1f} - Dk_{2.s2}] \tag{11.64}$$

where:

$$r_2 = M_{max_2}[S_{1f}/(k_{2.s1} + S_{1f})] - D$$

So far, no assumptions have been made as regards the presence of a competitor because the relationship between the density of the populations and the actual value of S_1 has not as yet been described. Let:

$$S'_{1eq} = S_{1f} - N_{2eq}/Y_{2.s_1} \tag{11.65}$$

where the prime refers to the fact that this is the equilibrium density of S_1 with only N_2 consuming it. Thus:

$$M_{max_2} [S'_{1eq}/(k_{2.s_1} + S'_{1eq})] = D \tag{11.66}$$

Let:

$$S_1 = S_{1f} - N_1/Y_{1.s_1} - N_2/Y_{2.s_2} \tag{11.67}$$

for all t in a competitive system. Then substituting 11.67 and 11.60 into 11.56, and 11.67 and 11.66 into 11.64 yields:

$$dN_1/dt = r_1 N_1 (k_{1.s_1} + S_{1f})/(k_{1.s_1} + S_1) \cdot \tag{11.68}$$
$$[Y_{1.s_1}(S_{1f} - S_{1eq}) - N_1 - (Y_{1.s_1}/Y_{2.s_1})N_2]/[Y_{1.s_1}(S_{1f} - S_{1eq})]$$

and

$$dN_2/dt = r_2 N_2 (k_{2.s_1} + S_{1f})/(k_{2.s_1} + S_1) \cdot \tag{11.69}$$
$$[Y_{2.s_1}(S_{1f} - S'_{1eq}) - N_2 - (Y_{2.s_1}/Y_{1.s_1})N_1]/[Y_{2.s_1}(S_{1f} - S'_{1eq})]$$

Define:

$$K_1 = N_{1eq} = Y_{1.s_1}(S_{1f} - S_{1eq})$$

and:

$$K_2 = N_{2eq} = Y_{2.s_1}(S_{1f} - S'_{1eq})$$

as the equilibrium densities reached by these populations in pure chemostat cultures. Thus 11.68 and 11.69 may be rewritten as:

$$dN_1/dt = r_1 N_2 [K_1 - N_1 - \alpha_{12}N_2)]/K_1 \cdot \tag{11.70}$$
$$(k_{1.s_1} + S_{1f})/(k_{1.s_1} + S_1)$$

and:

$$dN_s/dt = r_2 N_2 [K_2 - N_2 - \alpha_{21}N_1)]/K_2 \cdot \tag{11.71}$$
$$(k_{2.s_1} + S_{1f})/(k_{2.s_1} + S_1)$$

where:

$$\alpha_{12} = 1/\alpha_{21} = Y_{1.s_1}/Y_{2.s_1}$$

Clearly, equations 11.70 and 11.71 are of a form similar to the Lotka-Volterra competition expressions (equations 11.50 and 11.51).
If:

$S_{1f} <<< k_{1.s_1}$ and $S_{1f} <<< k_{2.s_1}$

then equations 11.70 and 11.71 become:

$$dN_1/dt = r_1N_2 [K_1 - N_1 - \alpha_{12}N_2]/K_1 \qquad (11.72)$$

and:

$$dN_2/dt = r_2N_2 [K_2 - N_2 - \alpha_{21}N_1]/K_2 \qquad (11.73)$$

the Lotka-Volterra competition equations. As in Appendix One, S_{1f} must take a value so ridiculously small that equations 11.72 and 11.73 are useless for all practical purposes. However, note that even if S_{1f} is large, the initial rates of growth and the final outcomes are still determined by the carrying capacities and competition coefficients in exactly the same manner as for the Lotka-Volterra competition equations.

12 MUTUALISM, LIMITED COMPETITION AND POSITIVE FEEDBACK

W.M. Post, C.C. Travis and D.L. DeAngelis

Introduction

Direct interspecific mutualism can result from a multitude of interactions involving dispersal, shelter, nutrient cycling, energy provision and reproduction (Faegri and Van der Pijl 1966; Heinrich and Raven 1972; Muscatine and Porter 1977; Whittaker 1975; Howe 1977; Temple 1977). Mutualistic interactions also arise when mutualists mediate competitive or predator-prey interactions (Wright 1973; Janzen 1967; Addicott 1979; Messina 1981; Osman and Haugsness 1981; Heithaus *et al.* 1981). From a broader perspective, complex interactions mathematically analogous to mutualism can arise as a result of temporal and spatial heterogeneity in ecological systems (DeAngelis *et al.* 1979, 1984).

Assignment of the term 'mutualism' to interspecies interaction, even when it involves only a pair of species, can often be ambiguous. The difficulties of defining and analysing mutualistic interactions within a community of several different species are even greater. Given this complexity, an understanding of the role of mutualistic interactions in the structure and evolution of ecological communities requires a strong interplay between field observations, experiments and theory. Our purpose in this chapter is to explore some mathematical foundations that will be essential to any general theory of mutualistic interactions. A particularly useful tool for analysing systems involving mutualism, as well as some communities involving both mutualism and competition, is the mathematical concept of a 'positive feedback system', a system of interacting species in which all feedback loops with two or more links are positive. The theory of such systems will be used here as a framework to analyse several levels of mutualistic relationships found in nature.

Dynamics of Mutualistic Communities

The simplest model of two species mutualism is obtained by modifying the Lotka-Volterra equations for competition so that the interspecies interactions result in mutual benefit (α_{12}, $\alpha_{21} > 0$), rather than harm of other species:

$$\frac{dx_1}{dt} = x_1(r_1 - \alpha_{11}x_1 + \alpha_{12}x_2) \qquad\qquad (12.1)$$

$$\frac{dx_2}{dt} = x_2(r_2 + \alpha_{21}x_1 - \alpha_{22}x_2)$$

This model and a discussion of its salient features have been presented by Pianka (1978), Vandermeer and Boucher (1978), Goh (1979) and Travis and Post (1979). Many of the features of more complicated models of two-species mutualism are exemplified by this simple (though admittedly unrealistic) model. The equilibrium populations $\bar{x}_1$, $\bar{x}_2$ for this two-species mutualistic system are shown in Figure 12.1. Note that $\bar{x}_1 > r_1/\alpha_{11}$, $\bar{x}_2 > r_2/\alpha_{22}$, so that both populations are larger than they would be in the absence of the interaction. This is a characteristic of positive feedback systems, of which systems of mutualists are a special case. Vandermeer and Boucher (1978) examine all possible cases of stability and persistence of two mutualistic populations using the graphical analysis of the Lotka-Volterra equations. However, we present a slightly different analysis which will be useful in examining more general equations and mutualistic systems with more than two species.

In order for equations 12.1 to be stable to small perturbations from equilibrium, it is necessary that slight movements away from the equilibrium position should set up forces tending to restore equilibrium. The mathematical conditions that must be satisfied for a model system to be stable to small perturbations from equilibrium are well known and can be stated in terms of the 'community matrix'. The elements of the community matrix are the coefficients of the linearized system. Equations 12.1 when linearized at the equilibrium, become:

$$\frac{dY_1}{dt} = -\alpha_{11}\bar{x}_1 Y_1 + \alpha_{12}\bar{x}_1 Y_2$$

$$\qquad\qquad (12.2)$$

$$\frac{dY_2}{dt} = \alpha_{21}\bar{x}_2 Y_1 - \alpha_{22}\bar{x}_2 Y_2$$

where $Y_i = x_i - \bar{x}_i$. The equilibrium point is stable if and only if all the eigenvalues of the community matrix:

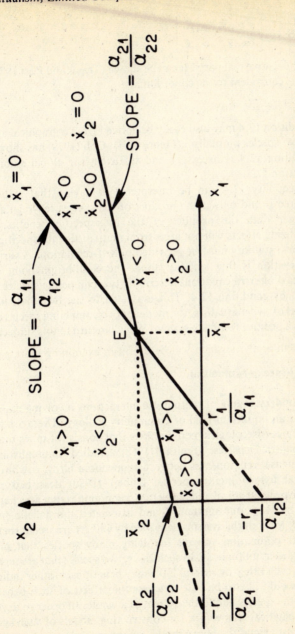

Figure 12.1: Isoclines of a Representative Lotka-Volterra Model of Two Mutualistic Species. The equilibrium E is stable.

$$S = \begin{bmatrix} -\alpha_{11}\bar{x}_1 & \alpha_{12}\bar{x}_1 \\ \alpha_{21}\bar{x}_2 & -\alpha_{22}\bar{x}_2 \end{bmatrix} \qquad (12.3)$$

have negative real parts. It can be shown (Travis and Post 1979) that this is equivalent to the condition:

$$\alpha_{11}\alpha_{22} > \alpha_{12}\alpha_{21} \qquad (12.4)$$

Condition 12.4 may also easily be derived from a graphical analysis of two-species mutualism (Figure 12.1). Goh (1979) has shown that condition 12.4 is necessary and sufficient for global stability of equation 12.1.

Inequality 12.4 can be interpreted as saying that stability is ensured if and only if the product of the species self-regulation is sronger than the product of the interspecific benefits. Since stabilizing effects will not arise from mutualistic interspecific inter-actions, stability must be provided by self-regulation. A surprising observation is that stability about the equilibrium point of the Lotka-Volterrra equations describing two competing species is also given by condition 12.4. This is related to the fact that, like the model of two mutualists, the model of two competing species consti-tutes a positive feedback system. We will return to this subject later.

Multi-species Mutualism

The understanding of mutualistic interactions involving many spe-cies is an important goal of community ecology. There are several examples of ecological communities that consist of more than two mutualistic species. Smith (1968) studied mutualistic and commensal relationships between oropendola birds, cowbirds and several hymenopteran species. Gilbert (1980) described several multispecies tropical and temperate communities that are character-ized by 'keystone mutualist' and associated mutualistic 'mobile links'. Most of the systems described by Gilbert are seed dispersal or flower pollination systems involving many species that may be regarded as mutualistic. In such cases, two species that are separated from each other by several links may benefit each other indirectly. One would like to be able to assess the effects of such benefits to both the species and to the system as a whole. In order to develop a mathematical framework for constructing models of such systems, we must extend the above model (equation 12.1) to more than two populations, and generalize it to more complex types of inter-actions.

We will proceed by considering a general system of differential equations that describes the population growth of a community of n interacting mutualistic species:

$$\frac{dx_i}{dt} = x_i g_i(x_1, x_2, \ldots, x_n) \quad , \quad i = 1, 2, \ldots, n \quad (12.5)$$

where x_i is the population density of the ith species and g_i is the *per capita* growth rate of the ith species. The equilibrium point or points of system 12.5 may be found by solving the algebraic equations:

$$g_i(\bar{x}_1, \bar{x}_2, \ldots, \bar{x}_n) = 0 \quad , \quad i = 1, 2, \ldots, n \quad (12.6)$$

We assume that system 12.6 has at least one positive solution with all $\bar{x}_i > 0$. For equations 12.5 to represent the population dynamics of a mutualistic community, it is necessary that when the community is at equilibrium, the jth species $(j \neq i)$ has a beneficial effect on the *per capita* growth rate of the ith species. A precise mathematical state-ment of this condition is that the elements:

$$a_{ij} = \frac{\partial g_i}{\partial x_j}(\bar{x}_1, \bar{x}_2, \ldots, \bar{x}_n) \quad (12.7)$$

of the interaction matrix A satisfy $a_{ij} > 0$ for $i \neq j$. The model community (12.5) will be stable in a neighbourhood of the equilib-rium $\bar{x}$ if and only if all of the eigenvalues of the community matrix S $= (s_{ij})$ have negative real parts, where the elements s_{ij} are given by:

$$s_{ij} = \frac{\partial}{\partial x_j}(x_i g_i(x_1, x_2, \ldots, x_n)) \quad (12.8)$$

and the partial derivatives in (12.8) are evaluated at the equilibrium point $\bar{x}$. Since:

$$\frac{\partial}{\partial x_j}(x_i g_i(x_1, x_2, \ldots, x_n)) = \begin{cases} x_i \dfrac{\partial g_i}{\partial x_j}(x_1, \ldots, x_n) & \text{if } j \neq i \\[2mm] g_i(x_1, \ldots, x_n) + & \\[2mm] x_i \dfrac{\partial g_i}{\partial x_j}(x_1, \ldots, x_n) & \text{if } j = i \end{cases} \quad (12.9)$$

and $g_i(\bar{x}_1, \ldots, \bar{x}_n)$ 0, the model ecological community 12.5 will be stable in a neighbourhood of the equilibrium x if and only if all of the eigenvalues of the community matrix $S = DA$ have negative real parts, where D is the diagonal matrix defined by $D = \text{diag}(\bar{x}_1, \bar{x}_2, \ldots, \bar{x}_n)$ and A is the interaction matrix composed of the elements defined by 12.7. This stability condition, however, may be simplified for mutualistic systems. A characteristic of mutualistic communities is that stability near a feasible equilibrium can be determined from the interaction matrix A alone (Travis and Post 1979). Moreover, the stability criteria are expressible in relatively simple, biologically interpretable conditions. We will summarize the conditions here. The details are contained in Travis and Post (1979) and Post and Travis (1979).

The stability of a mutualistic community near an equilibrium point is dependent on the strength of the diagonal or intraspecific regulation terms in the population matrix. If these terms are negative and strong compared to the mutualistic benefit a species receives, then the mutualistic community will remain near equilibrium. A precise mathematical statement of this condition is that stability is ensured if and only if there exist positive constants $d_1, d_2, \ldots, d_n$ such that:

$$-d_i a_{ii} > \sum_{\substack{j=1 \\ j \neq i}}^{n} d_j a_{ij}, \qquad i = 1, 2, \ldots, n \qquad (12.10)$$

In the case of two species mutualistic system described by (12.1), these conditions become:

$$-d_1 \alpha_{11} > d_2 \alpha_{12}$$
and:
$$-d_2 \alpha_{22} > d_1 \alpha_{21} \qquad (12.11)$$

The conditions 12.10 or 12.11 are known as quasi-diagonal dominance (Siljak 1975). These conditions roughly state that for a mutualistic community to be stable, the self-regulation of each species must exceed the total *per capita* benefit received from mutualistic interactions.

Although condition 12.10 indicates how stability is maintained in mutualistic communities, it is of little help in actually determining whether a particular mutualistic community will be stable. The problem presented by this condition is that it can be difficult to establish whether or not a vector $d = (d_1, d_2, \ldots, d_n)$ exists such that inequality 12.10 is satisfied. Fortunately there exists an equivalent condition that is verifiable in a finite number of arithmetical steps (see Travis and Post 1979). A necessary and sufficient condition for a mutualistic community to be stable in a neighbourhood of a feasible equilibrium can be stated in terms of the principal minors of the interaction matrix

$$A = \begin{bmatrix} a_{11} & a_{12} & \cdots & a_{1n} \\ a_{21} & a_{22} & \cdots & a_{2n} \\ \vdots & & & \\ a_{n1} & a_{n2} & \cdots & a_{nn} \end{bmatrix}$$

The principal minors of this matrix are

$$D_{11} = \det [a_{11}]$$

$$D_{22} = \det \begin{bmatrix} a_{11} & a_{12} \\ a_{21} & a_{22} \end{bmatrix}$$

$$D_{33} = \det \begin{bmatrix} a_{11} & a_{12} & a_{13} \\ a_{21} & a_{22} & a_{23} \\ a_{31} & a_{32} & a_{33} \end{bmatrix}$$

$$\vdots$$

$$D_{nn} = \det [A]$$

The equilibrium point of the model mutualistic community (12.5) is stable if and only if:

$$(-1)^k D_{kk} > 0 \quad , \quad k = 1, 2, \ldots, n \tag{12.12}$$

For the two species described by equation 12.1, condition 12.12 reduces to inequality 12.4.

Limited Competition

It is clear that powerful mathematical tools can be used in analysing the dynamics of a community consisting solely of populations of mutualistic species. However, few biotic communities can be imagined that consist only of mutualistic populations. Competitive relationships are likely to be at least as common and as important. The question arises as to whether the above techniques can be applied to a mixture of mutualistic and competitive relationships. The answer to this question is that we can apply the results to a larger class of ecological communities than strictly mutualistic ones, but not to all communities.

Consider two competing populations. When the negative effect population 1 has on population 2 increases, for example by heightened aggressive behaviour or increased production of secondary compounds, the equilibrium size of population 2 will decrease. The negative effect population 2 has on population 1 decreases and population 1 increases. Thus, the negative effect population 1 has on population 2 feeds back positively on itself. The converse holds if population 2 is increased. Thus, two competing populations form a simple positive feedback system. Mathematically, the interaction matrix for two competing species is similar to the interaction matrix for two mutualistic species, and similar matrices have identical stability properties. Matrix A* is similar to a matrix A if there exists a nonsingular matrix Q such that

$$A^* = QAQ_{-}1 \tag{12.13}$$

The matrix:

$$Q = \begin{bmatrix} 1 & 0 \\ 0 & -1 \end{bmatrix} \tag{12.14}$$

transforms the interaction matrix for two competing populations:

$$A = \begin{bmatrix} -a_{11} & -a_{12} \\ -a_{21} & -a_{22} \end{bmatrix} \tag{12.15}$$

to the interaction matrix:

$$A^* = \begin{bmatrix} -a_{11} & a_{12} \\ a_{21} & -a_{22} \end{bmatrix} = \begin{bmatrix} 1 & 0 \\ 0 & -1 \end{bmatrix} \begin{bmatrix} -a_{11} & -a_{12} \\ -a_{21} & -a_{22} \end{bmatrix} \begin{bmatrix} 1 & 0 \\ 0 & -1 \end{bmatrix} \tag{12.16}$$

for two mutualistic populations, and has the same stability properties as A. Thus we can study the community matrix of the corresponding mutualistic system to determine whether or not the equilibrium of the two competing populations is stable.

Two competing populations form a positive feedback system because the path from one species to the other (consisting of the links of the effect of one population on the other near equilibrium) and back again consists of an even number of negative effects, resulting in a net positive effect. This concept can be generalized to include any number of interacting populations. In fact, stability criteria 12.12 can be applied to communities which can be divided into two subcommunities satisfying the following conditions.

(1) interactions between populations within each of the two subcommunities are all positive or zero in effect (commensal, mutualistic, or no interaction),

(2) interactions between populations from different subcommunities are all negative or zero (amensal, competitive, or no interaction).

Mathematically, this is equivalent to performing identical row and column permutations on the community matrix to bring it into the form:

$$A = \begin{bmatrix} A_{11} & A_{12} \\ A_{21} & A_{22} \end{bmatrix} \tag{12.17}$$

where the submatricies A_{ij} have the property that A_{11}, A_{22} are square matrice with dimension corresponding to the number of populations in the respective subcommunities. The submatrices also have the sign conditions A_{11}, $A_{22} \geq 0$ (except for the diagonal elements) and A_{12}, $A_{21} \leq 0$. Such communities are called communities with 'limited competition'. Interactions in such communities can be characterized as 'friends of friends are friends, enemies of enemies are friends, and friends of enemies are enemies'. The matrix Q, with which a community of limited competition can be transformed using the similarity transformation 12.13 into one representing a mutualistic community, is given by:

$$Q = \begin{bmatrix} I_1 & 0 \\ 0 & -I_2 \end{bmatrix} \tag{12.18}$$

where I_1, I_2 are identity matrices with dimensions corresponding to A_{11}, A_{22} respectively. We can, therefore extend the methods of analysis of mutualistic communities to these more general systems that

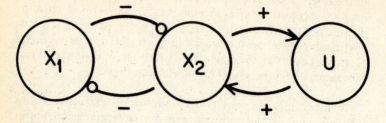

Figure 12.2: Graph of a Model Studied by Rai *et al.* (1983). Populations x_1 and x_2 are competitors and populations x_1 and u are mutualistists.

allow certain patterns of competitive interactions that we call limited competition.

At first glance, this class of communities may seem small, but many ecologically relevant communities may satisfy conditions similar to 12.17. For example, the competitor-competitor-mutualist system considered by Rai *et al.* (1983) is a special case of limited competition. They modelled a system consisting of a pair of competitors x_1 and x_2, and a pair of mutualists, x_1 and u (see Figure 12.2) with the equations:

$$\frac{du}{dt} = uh(u,x_1)$$

$$\frac{dx_1}{dt} = \propto x_1[g_1(u,x_1) - q_1(u,x_1,x_2)] \qquad (12.19)$$

$$\frac{dx_2}{dt} = x_2[g_2(x_2) - q_2(x_1,x_2)]$$

This same basic model can be used to describe interactions between flies, beetles and mites (Springett 1968); beetles, ants and membracids (Messina 1981); hermit crabs and hydroids (Wright 1973) and leaf cutting ants and fungi (Quinlan and Cherret 1978). The interaction matrix A, given by:

$$
A = \begin{bmatrix} \dfrac{\partial h}{\partial u} & \dfrac{\partial h}{\partial x_1} & 0 \\[2ex] \dfrac{\partial}{\partial u}[g_1 - q_1] & \dfrac{\partial}{\partial x_1}[g_1 - q_1] & -\dfrac{\partial q_1}{\partial x_2} \\[2ex] -\dfrac{\partial q_2}{\partial x_1} & \dfrac{\partial}{\partial x_2}[g_2 - q_2] \end{bmatrix} = \begin{bmatrix} - & & 0 \\ - & - & - \\ 0 & - & - \end{bmatrix} \quad (12.20)
$$

has the sign pattern given by 12.17, and thus represents a case of limited competition.

There is a significance in the ability to apply similarity transforms to the interaction matrix that goes beyond pure mathematics. Consider two three-species communities (Figure 12.3). In the first (Figure 12.3a) there are competitive interactions between species 1 and 2 and between species 2 and 3, but a mutualistic relationship between species 1 and 3. The community matrix of this system is transformable into a positive feedback system. The second community (Figure 12.3b) differs in that species 1 and 3 are competitors and not mutualistic. The community of this system is not transformable into a positive feedback system. The difference between these systems embodies an important idea, 'an enemy of an enemy is a friend'. By interacting mutualistically with species 3, species 1 in the system in Figure 12.3a is reinforcing the positive relationship that already exists between these species as mutual competitors of species 2. The net effect of species 1 on species 3 is always positive and vice versa; hence the two species are mutualists, both directly and indirectly. In the system in Figure 12.3b, however, the net effects between species 1 and 3 are ambiguous, because they cannot be represented as pure positive feedback but as a combination of positive and negative feedbacks.

This idea may be more helpful in classifying relationships in larger systems. For example, consider the study of coexisting ant populations by Davidson (1980). Davidson determined that a community of competing ant populations was food limited and measured the magnitudes of both exploitative and interference interactions by observing the ants' diets and behavioural interactions at baits. When estimates of the magnitudes of the competition coefficients are arranged into an interaction matrix, ignoring the elements that are small ($\propto\ \leq\ .2$), the community matrix approximates a community with limited competition. Figure 12.4 depicts the strong inter-

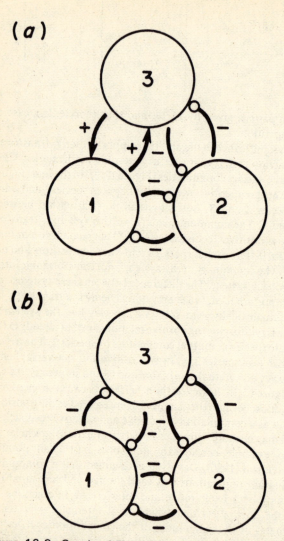

Figure 12.3: Graph of Two Communities Composed of Competitive and Mutualistic Interactions. Community depicted in (a) satisfies the definition of limited competition, while the community depicted in (b) does not.

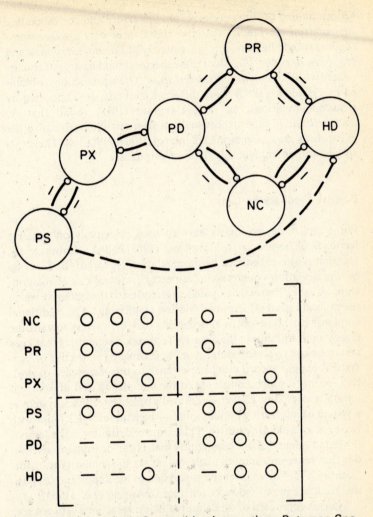

Figure 12.4: The Major Competitive Interactions Between Species in an Ant Community Studied by Davidson (1980) Shows a Community With Limited Competition Except for the Interaction Between PS and HD (dashed). This produces a negative loop of length 5, which is weak compared to the strengths of the feedback loops between PR, PD, NC, and HD. NC = *Novemessor cockerelli*, PR = *Pogonomyrmex rugosus*, PD = *Pheidole desertorum*, PX = *Pheidole xerophila*, and PS = *Pheidole sitarches*.

actions among the ant species. From this figure and corresponding sign matrix, we predict that the abundances of *Pogonomyrmex rugosus* should be positively correlated with *Pheidole xerophila* and *Novemessor cockerelli*. All of these species should be negatively correlated with *Pogonomyrmex desertorum*. These predictions, obtained by Davidson (1980) using different methods, are supported by field manipulations. In fact, Davidson (1980) found that *N. cockerelli* and *P. rugosus* are usually nearest neighbours of each other in the study area, suggesting net interspecific facilitation. These two species are, unambiguously, indirect mutualists.

Positive Feedback Systems

For a long time, ecologists have focused principally on negative feedback or homeostasis. Maruyama (1963) called attention to the fact that positive feedback has been relatively ignored: 'By focusing on the deviation counteracting aspects of mutual causal relationships . . . the cybernetician paid less attention to the system in which the mutual causal effects are deviation amplifying.' The traditional emphasis on homeostasis reflects this bias of interest toward the steady state and thus in the negative feedback mechanisms that keep systems close to steady state. A more complete perspective of ecosystems should accord more importance to the occurrence of positive feedback which contribute to the complexity and diversity of natural ecosystems. Earlier we defined a positive feedback system as a system of interacting populations in which all feedback loops with two or more links are positive. This intuitive definition is the same as the requirement that the community matrix can be transformed into one that represents a mutualistic system via the similarity transformation 12.13. For an extensive discussion of the importance of positive feedback in ecological systems see DeAngelis *et al.* (1984).

Systems composed of direct mutualistic interactions between species are the most striking examples of positive feedback systems in ecology. However, direct positive interactions that are termed mutualism represent only a small fraction of interactions involving positive feedbacks (Gilbert 1980; Howe 1984). One of the most common mechanisms leading to positive feedback, not involving mutualistic interactions in the usual sense of the word, arises from spatial heterogeneity. As an example, we consider the problem of species persistence in a patchy environment.

A Single Species in a Patchy Region

Consider a region consisting of a number of forested patches, or islands, of arbitrary sizes and spacing in a non-forested region. In particular, consider a tree species that, given enough time, would be eliminated by competition (and other factors) from any of the individual patches if there were no transfer of seeds from other patches. Such transfer occurs, however, so that recolonization can take place following local extinction. This being the case, we wish to determine whether or not the species is persistent in a region, where persistence is defined for present purposes to be the tendency of a species to recover from a perturbation that reduces its numbers to nearly zero, that is, the origin is a repellor.

Assume for simplicity that the growth of a single species in a particular forest patch can be described by the logistic equation:

$$\frac{dx_1}{dt} = r_1 x_1 - g_1 x_1^2 \qquad (12.21)$$

where x_1 is the number of trees (seedling size and larger) of the species in the patch, r_1 is the maximum possible rate of increase, and r_1/g_1 is the patch carrying capacity when r_1 is positive, with the subscripts denoting 'patch 1'. The parameter r_1 represents the difference between the rate of recruitment of new trees of the species into the patch resulting from autochthonous seed sources and the mortality of trees in the patch. This parameter is considered to be averaged over time periods long compared to the generation time of the species. In order for the species to disappear from an isolated patch over the long term, it is necessary and sufficient that $r_1 \le 0$. If, for example, the species in question is a pioneer species that invades temporary gaps in the forest canopy, then the local r_1 is negative and its magnitude will reflect the time it takes for the canopy to close over the gap. Parameter g_1 measures the density-dependent effects causing increasing mortality and decreasing successful regeneration within the patch. Admittedly, this is an overly simple model, and we will discuss generalizations later.

A simple but reasonable assumption is that the flux of seeds from one patch to the next is proportional to the population on the source patch. We then have the equations:

$$\frac{dx_1}{dt} = r_1 x_1 - g_1 x_1^2 + k_{12} x_2 \qquad (12.22a)$$

$$\frac{dx_2}{dt} = r_2 x_2 - g_2 x_2^2 + k_{21} x_1 \qquad (12.22b)$$

where x_1 and x_2 are the species population numbers in the two islands and k_{12} and k_{21} are the rates of transfer from one island to the next of seeds that eventually successfully germinate, that is, the colonization rates.

In general, to determine whether the origin (0,0) is a repellor or an attractor, it is necessary to calculate the eigenvalues of the system of equations linearized at (0,0). For the two-island case, this involves finding the eigenvalues of the matrix A:

$$A = \begin{bmatrix} r_1 & k_{12} \\ k_{21} & r_2 \end{bmatrix} \qquad (12.23)$$

The existence of a positive real part of at least one eigenvalue of this matrix implies (0,0) is a repellor and hence the species is persistent. Since A represents a positive feedback system, condition 12.12 can be applied to establish that persistence is equivalent to the conditions r_1 and r_2 negative and:

$$r_1 r_2 < k_{12} k_{21} \qquad (12.24)$$

Many would find it intuitively unreasonable that if a species cannot maintain itself over the long term on any isolated patch, it could do so when the patches are connected by a trickling of inter-patch seed dispersal. One might find it more plausible if looked at in the following way. An isolated population of some plant species in a habitat island that is cut off from outside sources will usually have no trouble providing more than enough seeds to maintain itself. However, suppose that occasionally, because of competition and other factors, there are no available spaces on the island for a seedling of a given species to become established. Then, despite the great fecundity of a species, it may vanish from the island. Later, however, space suitable for the species may open up by chance. At this point, even a very small seed flow from another island may be sufficient to re-establish the species on a given island. The idea of collective maintenance of the species seems more plausible for an environment with many patches, since a fairly steady supply of allochthonous seeds should then be available to each patch. The patches have relationships between each other that are analogous to 'mutualism', at least mathematically, though biologically this single-species inter-patch dispersal is not at all related to mutualism as it is strictly defined.

Multi-species, Multi-island Systems with Competition and Mutualism

In nature, tree species usually occur in close proximity to one or more other tree species, with which they interact to some degree. For the sake of realism, we need to incorporate the effects of inter-specific competition and mutualism among tree species into our model. This could be accomplished in an approximate way by implicitly including the effects of these interactions on the population growth rate, r_i, of the single species, and proceeding as we did in considering a single species. However, since the degree of inter-action with other species depends on the population numbers of the other species in its immediate vicinity, it is best to attempt a more general analysis that considers as variables all important tree species in the system that interact with the species of interest. Note that when one models competition $(-,-)$ and mutualism $(+,+)$, both commensalism $(-,0)$ and amensalism $(-,0)$ are automatically included as special cases.

Let us begin with the simple situation of two competing species on two forest patches. Let x_{1P}, x_{1Q}, x_{2P}, and x_{2Q} be the population numbers of species 1 on patches P and Q, respectively. All the feedback loops are positive feedback loops. The equations for the dynamics of these two competing species can be written as:

$$\frac{dx_{1P}}{dt} = r_{1P}x_{1P} - g_{1P}x_{1P}^2 + k_{1PQ}x_{1Q} - s_{12P}x_{1P}x_{2P} \quad (12.24a)$$

$$\frac{dx_{1Q}}{dt} = r_{1Q}x_{1Q} - g_{1Q}x_{1Q}^2 + k_{1QP}x_{1P} - s_{12Q}x_{1Q}x_{2Q} \quad (12.24b)$$

$$\frac{dx_{2P}}{dt} = r_{2P}x_{2P} - g_{2P}x_{2P}^2 + k_{2PQ}x_{2Q} - s_{21P}x_{2P}x_{2P} \quad (12.24c)$$

$$\frac{dx_{2Q}}{dt} = 2_{2Q}x_{2Q} - g_{2Q}x_{2Q}^2 + k_{2QP}x_{2P} - s_{21Q}x_{2Q}x_{1Q} \quad (12.24d)$$

where k_{1PQ} is the colonization rate of species 1 on patch P from patch Q, and S_{12P} is the competition coefficient of species 1 on species 2 on island P. This set of equations is similar to that considered by Levin (1974). The primary difference is that the transport term is not conservative here. Seeds from a source tree are usually so plentiful that those scattered to other islands do not constitute a significant loss from the source island.

We assume that each of the two species in the two-island system could persist in the absence of the other. It follows from our analysis in the earlier section concerning the persistence of one species on two forest patches that the above assumption is equivalent to the conditions:

$$r_{1P}r_{1Q} < k_{1PQ}k_{2QP}$$

and: (12.25)

$$r_{2P}r_{2Q} < k_{2PQ}k_{2QP}$$

where all r's are negative. Competition decreases the chances of each species persisting. Assume now that the population number of species 1 are close to zero in both islands and that the population numbers of species 2 on the two islands are given by the ordered pair $(x_{2P,E2}, x_{2Q,E2})$, where E2 is the equilibrium point of species 2 in the absence of species 1. To demonstrate the persistence of species 1, it is necessary to show that x_{1P} or x_{1Q} will tend to increase from zero in spite of the competitive presence of species 2. Notice that the community matrix evaluated at the equilibrium point $(0,0,x_{2P,E2},x_{2Q,E2})$, is decomposable; that is, it has the form:

$$A(E2) = \begin{bmatrix} A'_{11} & 0 \\ A'_{21} & A'_{22} \end{bmatrix}$$ (12.26)

where A'_{11} and A'_{22} are square matrices. A consequence of the matrix taking the above form is that the two eigenvalues of A'_{11} and A'_{22} are also the eigenvalues of the matrix $A(E2)$. Since A'_{22} is a diagonal matrix with negative entries the persistence of species 1 is equivalent to the existence, for the submatrix:

$$A'_{11} = \begin{bmatrix} r_{1P} - s_{12P}x_{2P,E2} & k_{1PQ} \\ k_{1QP} & r_{1Q} - s_{12Q}x_{2Q,E2} \end{bmatrix}$$ (12.27)

of at least one eigenvalue with a positive real part. A'_{11} has at least one eigenvalue with a positive real part if and only if $\det(A'_{11}) < 0$. We can repeat the above argument with the population number of species 2 close to zero on both islands and the population number of species 1 being given by $(x_{1P,E1}, x_{1Q,E1})$, where E1 is the equilibrium

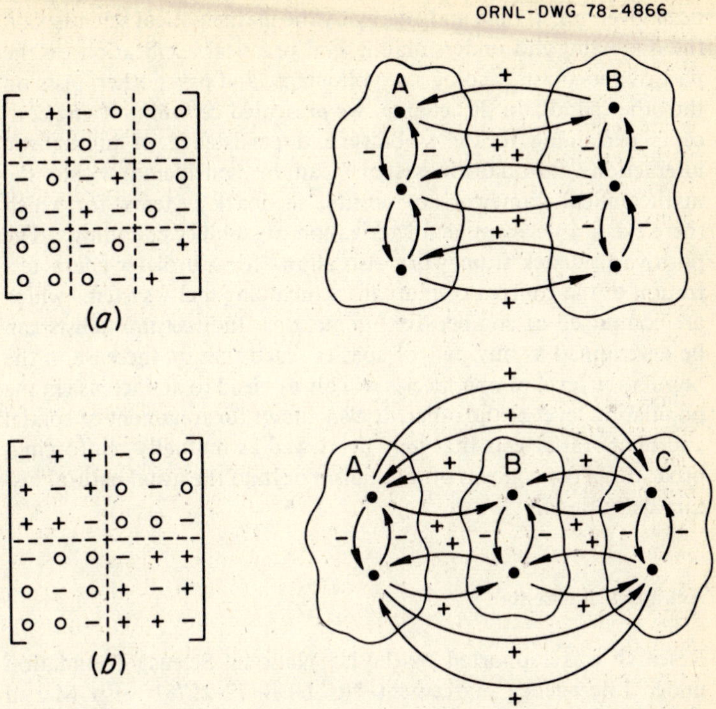

ORNL-DWG 78-4866

Figure 12.5: The Matrix and Graph Representations of (a) Three Competing Species on Two Forest Patches, and (b) Two Competing Species on Three Forest Patches.

point of species 1 in the absence of species 2.

The above concepts apply to N-patch systems as well (see DeAngelis *et al.* 1979). Consider Figures 12.5a and 12.5b. In Figure 12.5a three species interact competitively on two patches, while in Figure 12.5b two species interact competitively in three patches. The matrices in Figure 12.5 show that both models are positive feedback systems. That is, their community matrices can be reduced to positive matrices by means of similarity transformations as discussed earlier.

Conclusion

Maruyama (1963) stated that positive feedback systems 'have not been given much time and energy by the mathematical scientists on the one hand and understanding and practical application on the part of geneticists, ecologists, politicians, and psychotherapists on the other hand'. In this chapter we presented certain unifying concepts pertaining to the stability and persistence of mutualistic interactions. Mutualistic systems can be considered within the mathematical framework of positive feedback systems, for which there exists an elegant characterization of stability conditions. The positive feedback framework also allows for a broader characterization of the concept of mutualism including many systems which are comprised of competitive interactions. Indirect mutualists can be determined as any pair of species, such that an increase in the population level of one species will always lead to an increase in the population level of the other. It also allows for treatment of spatial and temporal effects that may be viewed as mutually reinforcing, broadening the concept of mutualism beyond the usual natural history definitions.

Acknowledgement

Research was supported by the US National Science Foundation under Interagency Agreement No. DEB 77–25781 with Martin Marietta Energy Systems Inc., under Contract No. DE–AC05–840R21400 with the US Department of Energy. Publication No. 2414,
Environmental Sciences Division, Oak Ridge National Laboratory.

References

Addicott, J.F. (1979) 'A Multispecies Aphid-Ant Association: Density Dependence and Species-Specific Effects', *Canadian Journal of Zoology, 57*, 558–69
Davidson, D.W. (1980) 'Some Consequences of Diffuse Competition in a Desert Ant Community', *American Naturalist, 116*, 93–105
DeAngelis, D.L., C.C. Travis and W.M. Post (1979) 'Persistence and Stability of Seed-Dispersal in a Patchy Environment', *Theoretical Population Biology, 16*, 107–25
DeAngelis, D.L., W.M. Post and C.C. Travis (1985) *Positive Feedback Systems in Nature*, Springer-Verlag (in press)
Faegri, K. and L. van der Pijl (1966) *The Principles of Pollination Ecology*, Pergamon Press, Toronto
Gilbert, L.E. (1980) 'Food Web Organization and the Conservation of Neotropical Diversity' in Soule, M.E. and B.A. Wilcox (eds), *Conservation Biology*, Sinauer

Associates, Sunderland, MA

Goh, B.S. (1979) 'Stability in Models of Mutualism', *American Naturalist, 113*, 261–71

Heinrich, B. and P.H. Raven (1972) 'Energetics and Pollination Ecology', *Science, 176*, 597–602

Heithaus, E.R., D.C. Culver and A.J. Beattie (1980) 'Models of some Ant Plant Mutualisms', *American Naturalist, 116*, 347–61

Howe, H.F. (1971) 'Bird Activity and Seed Dispersal of a Tropical Wet Forest Tree', *Ecology, 58*, 539–50

Howe, H.F. (1984) 'Constraints on the Evolution of Mutualisms', *American Naturalist 123*, 764–77

Janzen, D.H. (1969) 'Allelopathy by Myrmecophytes: the Ant *Azteca* as an Allelopathic Agent of *Cecropia*', *Ecology, 50*, 147–53

Levin, S.A. (1974) 'Dispersion and Population', *American Naturalist, 108*, 207–28

Mayurama, M. (1963) 'The Second Cybernetics: Deviation-Amplifying Mutual Causal Processes', *American Scientist, 51*, 164–79

Messina, F.J. (1981) 'Plant Protection as a Consequence of Ant-Membracid Mutualism: Interactions on Goldenrod (*Solidago* sp.)', *Ecology, 62*, 1433–40

Muscatine, L. and J.W. Porter (1977) 'Reef Corals: Mutualistic Symbioses Adapted to Nutrient-Poor Environment', *Bioscience, 26*, 454–60

Osman, R.W. and J.A. Haugsness (1981) 'Mutualism Among Sessile Invertebrates: A Mediator of Competition and Predation', *Science, 211*, 846–8

Pianka, E.R. (1978) *Evolutionary Ecology*, Harper and Row, New York

Post, W.M. and C.C. Travis (1979) 'Quantitative Stability in Ecological Communities', *Journal of Theoretical Biology, 79*, 547–53

Quinlan, R.J. and J.M. Cherret (1978) 'Aspects of the Symbiosis of the Leaf Cutting Ant *Acromyrmex actospinosus* (Reich.) and its Fungus Food', *Ecological Entomology, 3*, 221–30

Rai, B., H.I. Freeman and J.F. Addicott (1983) 'Analysis of Models of Mutualism in Predator-Prey and Competitive Systems', *Mathematical Biosciences, 65*, 13–50

Siljac, D.D. (1975) 'When is a Complex System Stable?', *Mathematical Biosciences, 25*, 25–50

Smith, N. (1968) 'The Advantage of Being Parasitized', *Nature, 219*, 690–4

Springett, B.P. (1968) 'Aspects of the Relationship Between Burying Beetles *Necrophorus* sp. and the mite *Poecilochirus necrophori* Vitz.', *Journal of Animal Ecology, 37*, 417–24

Temple, S.A. (1977) 'Plant-Animal Mutualism: Coevolution with Dodo Leads to Near Extinction of Plant', *Science, 197*, 885–6

Travis, C.C. and W.M. Post (1979) 'Dynamics and Comparative Statics of Mutualistic Communities', *Journal of Theoretical Biology, 78*, 553–71

Vandermeer, J.H. and D.H. Boucher (1978) 'Varieties of Mutualistic Interactions in Population Models', *Journal of Theoretical Biology, 74*, 549–58

Whittaker, R.H. (1975) *Communities and Ecosystems*, Macmillan, New York

Wright, H.O. (1973) 'Effect of Commensal Hydroids on Hermit Crab Competition in the Littoral Zone of Texas', *Nature, 241*, 139–40

13 INDIRECT FACILITATION AND MUTUALISM

John Vandermeer, Brian Hazlett and Beverly Rathcke

Darwin's observations of bumblebees and red clover led him to extrapolate that since field mice, who prey on bumblebee nests, were relatively scarce near villages, they could account for the prevalence of red clover there. The mice are presumably scarce there because of predation by domestic cats. A German scientist then continued to extrapolate that since cats were responsible for the prevalence of red clover, and since red clover was a staple food of cattle and since British sailors thrived on bully beef, one could conclude that Britain's dominant world position as a naval power was ultimately determined by the presence of cats. Thomas Huxley, tongue planted firmly in cheek, went on to note that old maids were the main protectors of cats, thus showing that the British empire owed its existence to the spinsters of England (Farb 1963).

Such stories fire the sense of wonder of students as they open their first ecology text and learn in the preface that ecology is about the complex, diverse forms of interactions among a bewildering array of types, about the surprising and unpredictable consequences of a multitude of indirect linkages. However, their excitement soon fades as they get into the world of the professional ecologist, in which competition is resolved in one of three ways, predator-prey equations oscillate, energy flows, nutrients cycle — and that's pretty much it.

As a step toward recapturing a focus on the complexities of ecology, we offer the following observations on the topic of indirect facilitation and mutualism. In the course of doing so we will of necessity need to mention other types of indirect effects as well, but the intent is to introduce the notion of indirect facilitation as a focus for study. (Facilitation occurs when one population's fitness is increased by the presence of another population. Mutualism is mutual facilitation.)

By way of introduction, a brief historical note is in order. Attempting to develop a recipe for calculating competition coefficients from resource utilization patterns, Robert MacArthur

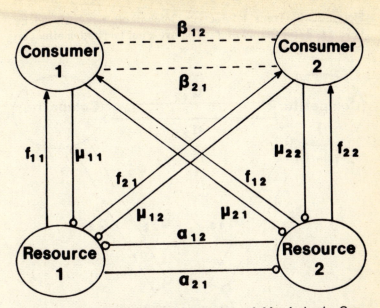

Figure 13.1: Graphical Representation of MacArthur's Consumer-Resource Equations as analysed by Levine (1976) and Vandermeer (1980). μ_{ij} is the utilization of the ith resource by the jth consumer, f_{ij} is the feeding of the ith resource, and x_{ij} is the competitive effect of resource j on resource i. Levine showed that the net effect of consumer 1 on consumer 2 would be facilitative if $\mu_{11}f_{21} + \mu_{21}f_{22} < \mu_{11}\propto_{21}f_{22} + \mu_{21}\propto_{12}f_{21}$. Here as elsewhere, small circles indicate a negative effect and small arrowheads indicate a positive effect.

(1972) formulated the much-used consumer-resource equations, a system of two differential equations for two consumers and two others for their resources. The competitive interaction between the two consumers could easily be computed from the resource utilization functions. Levine (1976) pointed out the non-intuitive fact that under certain circumstances the consumers, assumed to be competitors because they use the same resource, may in fact be mutualistically associated with one another (see also Vandermeer 1980). Levine's analysis is displayed graphically in Figure 13.1.

In the same work Levine showed a perhaps even more nonintuitive result when he demonstrated that under certain conditions,

Figure 13.2: Three Interacting Competitors as Analysed by Levine (1976), Competitor i will have a net facilitative effect on competitor j whenever $\alpha_{ki}\alpha_{jk} > \alpha_{ji}$.

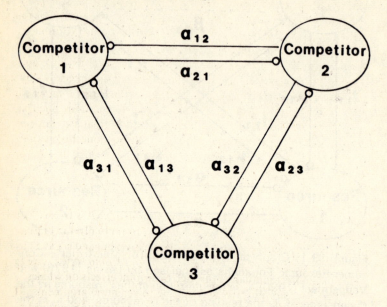

facilitation could result from a system of three competitors. This analysis is displayed graphically in Figure 13.2.

A qualitative examination of Figure 13.1 and 13.2 leads to a very simple biological interpretation of Levine's conditions. From Figure 13.2 we can easily see that there is both a *direct* competitive effect of species 1 on species 2, α_{21}, and an *indirect* effect of species 1 on species 2, namely $\alpha_{31}\alpha_{23}$. That is, species 1 negatively affects species 3 (α_{31}), a species which negatively affects species 2 (α_{23}). Since species 1 negatively affects something else that negatively affects species 2, the consequence is a positive effect of species 1 on species 2. If the negative effect (α_{21}) is smaller than the positive effect ($\alpha_{31}\alpha_{23}$), the overall effect of species 1 on species 2 will be positive, exactly equivalent to Levine's analytical result.

The same result can be seen with reference to Figure 13.1. The effect of consumer 1 on consumer 2 through resource 1 is $\mu_{11}f_{21}$. The same effect through resource 2 is $\mu_{21}f_{22}$. Both effects are negative since consumer 1 negatively affects something which in turn positively affects consumer 2. But a more indirect effect of consumer 1 on consumer 2 occurs through the competitive effect of resource 1 on resource 2. This indirect effect ($\mu_{11}\propto_{21}f_{22}$) is positive. A similar positive effect occurs in the reverse direction, namely $\mu_{21}\propto_{12}f_{21}$. If the total positive effect ($\mu_{11}\propto_{21}f_{22} + \mu_{21}\propto_{12}f_{21}$) is greater than the total negative effect ($\mu_{11}f_{21} + \mu_{21}f_{22}$) the overall effect of consumer 1 on consumer 2 is facilitative, corresponding exactly to Levine's analytical result. The above two examples are summarized in Figure 13.3. It can be seen that a simple graphical analysis can easily demonstrate which interactions must occur and/or be larger than which interactions in order for the result to be in facilitation. Such an analysis is useful for relatively simple systems (that is, those of relatively low dimension), and will be used throughout this chapter.

Some Methodological Questions

Certain cases of indirect facilitation are so obvious it is trivial even to mention them. A plant and predator are mutually facilitative, indirectly through the herbivore which eats the plant and is fed on by the herbivore (Price *et al.* 1980: 58). Krill and *Homo sapiens* are in principle indirect mutualists through their joint effects on whales.

Other cases are unfortunately prone to semantic arguments. If two consumers are *indirectly* facilitative as pictured in Figure 13.1, are they not also *indirectly* competitive? That is, if the effect of consumer 1 on consumer 2 through resources 1 and 2 is indirect, is not the effect of consumer 1 on consumer 2 through resource 1 alone also indirect? For example, in the case of two species of Caribbean anoles, both species eat insects but not exactly the same insects (or at least in exactly the same place). They thus roughly correspond to the diagram in Figure 13.1. Their competitive interaction, well-known to exist for at least some species pairs (Roughgarden *et al.* 1983), is in a sense indirect ($\mu_{11}f_{21} + \mu_{21}f_{22}$ or $\mu_{22}f_{12} + \mu_{12}f_{11}$). If this example is an indirect effect, it might be plausibly argued that *any* case of resource competition is indirect.

Rather than attempt a complete analysis of the proper usage of the notion of indirect, we here take the operational point of view that the direct interaction level is arbitrarily set with respect to which indirect effects are of interest. Thus, in Figure 13.2, if all the competitors are involved in resource competition, one might argue that the competition coefficients as presented in the figure are them-

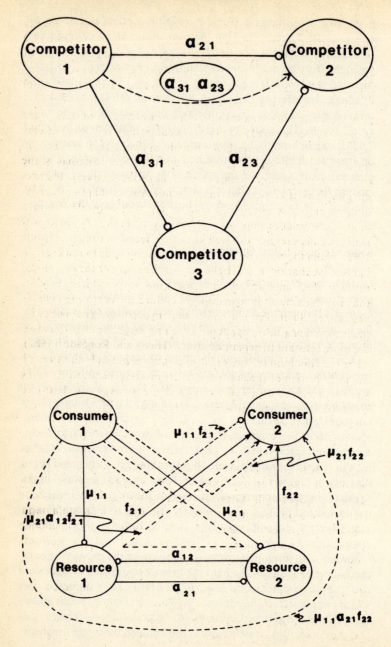

selves indirect. But for the intent of this chapter we can arbitrarily define particular interactions as the direct ones. Thus, in Figure 13.2 the competition coefficients pictured are direct and the effects of one species on another through a third species are indirect, as discussed previously. The notion of an indirect interaction (mutualism or otherwise) is thus a relative one.

A second methodological difficulty is always encountered when dealing with ecological systems at a qualitative level. While it may not be stated, it is nevertheless tacitly assumed that systems are linear, or at least well-behaved in their non-linearities (that is, the sign of the interaction is constant over all variable values). The predictions coming from unwarranted non-linear assumptions can be quite wrong. For example, tomatoes are well known to form a mutualistic relationship with VAM mycorrhizae. A variety of nematode diseases are also well known to decimate tomatoes (Rude 1982). From these two facts alone we could conclude that there is an indirect facilitation in which the mycorrhizae benefit the nematodes through their interaction with tomatoes (mycorrhizae benefits tomato which benefits nematode, therefore, mycorrhizae benefits nematode). But in fact, it is known that mycorrhizae have a negative impact on nematodes, apparently because large, healthy tomatoes are more resistant to nematode attack (Hussy and Roncadori 1982). Thus the condition of the tomato determines not only the strength but also the sign of the indirect interaction. In general, shifts in the sign of an interaction are to be expected in systems of non-destroyed resources (that is, gastropod shells for hermit crabs, pollinators for plants, see below).

Figure 13.3: Summary of Indirect Effects. The indirect effects are the products of associated direct effects, and are represented graphically as dotted lined with either a small arrowhead to indicate a positive effect or a small circle to indicate a negative effect. The dotted lines here, as elsewhere, are positioned so as to follow approximately the pathway of the direct effects whose combination they represent. (a) indirect effect of competitor 1 on competitor 2. The overall effect will be positive if the positive part $(\alpha_{31}\alpha_{23})$ is greater than the negative part (α_{21}), (b) indirect effects of consumer 1 on consumer 2. The overall effect will be positive if the positive part $(\mu_{11}\alpha_{21}f_{22} + \mu_{21}\alpha_{12}f_{21})$ is greater than the negative part $(\mu_{21}f_{22} + \mu_{11}f_{21})$.

A third methodological difficulty has to do with the question of dimensionality, an especially severe problem when dealing with systems qualitatively. It is impossible to consider the system as a whole since ecosystems include large numbers of interacting elements. One must always abstract the system to a lower dimensional form, a fact which can have severe repercussions with regard to its perceived behaviour (Schaffer 1981). For example, if we abstract the well-known ant/homopteran mutualism as consisting of the three dimensions plant-homopteran-ant, we are driven to the conclusion that the ant has a negative effect on the plant (since it protects the homopteran which in turn eats the plant). It is now becoming clear that the assumed negative indirect effect of the ant on the plant frequently does not occur but rather, the ant has a positive effect on the plant because of its protective effect of deterring other potential herbivores (see next section for a more complete discussion). In other words, the incorrect conclusion that the ant must have a negative impact on the plant derives from having conceived of a four-dimensional system (ant-homopteran-plant-herbivores) as a three-dimensional one. Since one is never certain that a particular low dimensional abstraction is appropriate (that is, that the derived indirect effects will in fact have the sign dictated by the abstraction), it is imperative that the indirect effect be experimentally verified.

Demonstrated Examples of Indirect Facilitation

Culver (1982) recognized potential conditions for indirect mutualism in very simple communities of cave isopods. While Culver suggests no specific mechanism to account for the indirect effect, he does provide statistical evidence suggesting that the amphipod *Crangonyx antennatus* and isopod *Caecidotae recurvata* are positively related to one another when the isopod *Lirceus usdagalum* is present, but negatively associated in its absence. This seems to be an exact replica of the three-species competitive case analysed by Levine (1976).

One of the most frequently cited direct mutualisms is that which occurs between homopterans and ants which tend them (Way, 1963), as noted above. Since the ants have a positive effect on the homopterans (by protecting them from potential predators) and the homopterans have a negative effect on the plant, it is natural to expect an indirect negative effect of the ant on the plant, as pictured in Figure 13.4a. That this effect is not universal was first demonstrated in 1972, when it was shown that a mistletoe species experienced more rapid shoot growth in the presence of an ant/homopteran mutualism than in its absence (Room 1972). The mechanism, illustrated in Figure 13.4b, was presumably the protection against other herbivores offered by the ant. Since the function of the ants in their direct mutualism is to drive away other insects, it should come as no surprise that they could easily drive away some of the plllant's enemies. The question becomes one of relative intensity, that is whether or not the indirect negative effect through the homopteran is sufficiently weak in comparison to the indirect positive effect through the other herbivores (see Figure 13.4b).

Subsequent to Room's work several authors have suggested the possibility of ant/plant mutualisms protecting the plant from herbivores in specific cases (Nickerson *et al*. 1977; Laine and Niemala 1980; Jutsum *et al*. 1981; Skinner and Whittaker 1981). Messina (1981) convincingly demonstrated the effect in goldenrod infested with *Formica* ants. But Fritz (1983), failing to find the indirect mutualism in black locust (*Robinia*) emphasized the complex nature of the interaction, and the impossibility of predicting *a priori* whether an indirect mutualism will or will not exist. Fritz noted that for black locust, the ants apparently protected the herbivore from its enemies just as much as they protected the locust from its herbivores, as pictured in Figure 13.4c. In effect, Fritz's system has three indirect effects of the ant on the plant: through the homoptera (negative), through the other herbivores (positive), and third through the natural enemies (negative), as shown in Figure 13.4c. The net effect depends on the relative strengths of these three indirect effects.

Another direct mutualism that has historically received much attention is the relationship between plants and their pollinators. An indirect effect has long been recognized here, in that different plant species may require similar pollinator species, thus setting up a relationship which is formally equivalent to the consumer resource situation. This indirect effect has usually been assumed to be negative, that is, competitive between the two plants (Feinsinger 1983; Waser 1983). But recently it has been suggested that the interaction may be an indirect mutualism (Thompson 1981; Waser and Real 1979; Schemske 1981; Rathcke 1984); the presence of one plant species may increase visits to another.

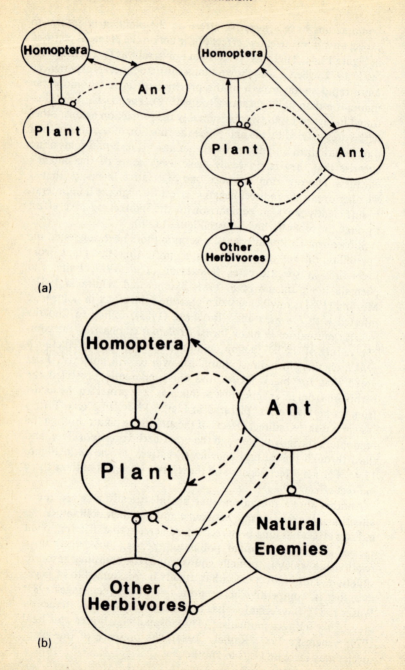

(a)

(b)

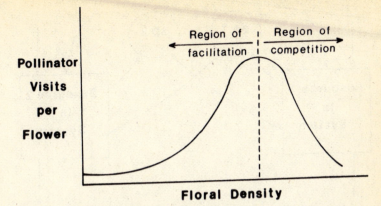

Figure 13.5: Theoretical Relationship Between Pollinator Visitation and Floral Density (after Rathcke 1984)

Abstractly we can visualize the relationship between indirect positive and negative effects as in Figure 13.5, in which the x-axis shows the combined density of several plant species that share pollinators (Rathcke 1984). At low floral densities very few pollinators are attracted to the general region. As floral density increases, more pollinators are attracted and presumably available to all the plant species, thus providing the potentially mutualistic effect. But as floral density increases yet further, the maximum number of pollinators that might be attracted to the region are already there, and any increase in floral density favours individual plants (since the plant with more flowers is likely to attract a greater share of the pollinators) and is thus competitive. Rathcke (1984) has summarized the observational and experimental evidence for this model.

Another example of a system which superficially resembles the

Figure 13.4: Diagrammatic Representation of the Effect of An Ant/Homopteran Mutualism on its Host Plant. (a) The effect as expected from a simple understanding of the natural history of the system, (b) The effect when the ant offers the plant protection against other herbivores, (c) The effect when the ant protects the herbivore from its enemies. As before the dotted lines, which indicate the indirect effects positioned so as to indicate approximately which direct effects are included.

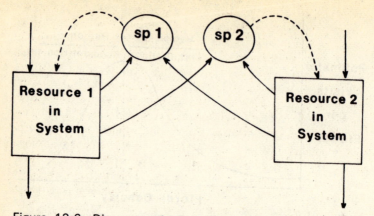

Figure 13.6: Diagrammatic Representation of the Resource Buffer Hypothesis of Indirect Facilitation. Each species potentially acts as a buffer, reducing the rate of critical resource flow out of the system. Dashed and solid lines here represent the flow of nutrients rather than interactive effects as in the other figures. Dashed lines illustrate the recycling of nutrients into the system.

classic case of resource competition is that of hermit crab shell exchange, in which two species of hermit crabs utilize the same resource, gastropod mollusk shells (Hazlett 1981). It has recently been suggested that through the process of shell exchange, some hermit crabs sometimes actually 'negotiate' for this critical resource, exchanging shells most frequently when individuals of both species benefit from the exchange (Hazlett 1978). While this interpretation of shell exchange remains controversial (Dowd and Elwood 1983; Elwood and Glass 1981; Hazlett 1983), its overall theoretical structure suggests a more general mechanism which theoretically might produce indirect mutualisms.

Shell negotiations will produce indirect mutualism if there is a 'flow' of shells into and out of the system. The presence of a second hermit crab species may effectively act as a buffer, holding shells which would otherwise sink to deeper water and thus out of the system, until they can be negotiated to the first species. This is the 'preservation' effect (Hazlett 1981). The two species thus potentially act as resource buffers to one another.

In Figure 13.6 we picture this proposed mechanism in a general way. Any situation in which resources flow into and out of the system (that is, any open ecosystem) can potentially demonstrate this mechanism. For example, terrestrial plants with different detailed nutrient requirements are prime candidates. In any situation in an open ecosystem in which the resource is not destroyed when temporarily utilized by individuals of one species, the buffer effect can lead to facilitation. As depicted in Figure 13.7, whether facilitation of competition between species occurs depends upon the efficiency with which the resource can be acquired by members of one population. Variation in the population size of either consumer relative to the resource flow rate or capture efficiency of the other consumer, can lead to a change in the sign of the interaction between species utilizing a resource in common. We might also note that in such buffered systems, common species can be expected to have facilitatory effects upon rare species while rare species may have slight competitive effects upon the common (since the common species will probably have a higher resource acquisition rate relative

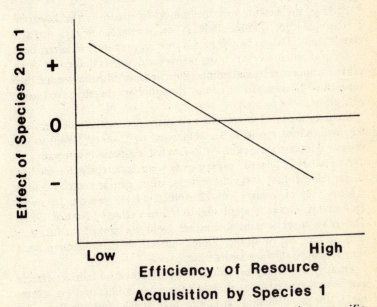

Figure 13.7: Theoretical Relationship Between Interspecific Interaction and the Efficiency of Resource Acquisition, for the Resource Buffering Hypothesis.

to the level of resource flow), that is, there will be 'resource parasitism'.

Circumstantial evidence for the resource buffering mechanism can be found in the vast literature on intercropping with legumes. Frequently it can be shown that the legume has a positive effect on the associated non-legume crop (Trenbath 1976; Snaydon and Harris 1981). It is usually through that this effect is due to the legume's ability to tap a resource not normally available to the non-legume (molecular nitrogen from the atmosphere), later releasing the nitrogen into the system as nitrate, ammonia, or organic nitrogen — that is, in a form which can be utilized eventually by the non-legume. The legume effectively increases the flow of nitrogen into the system, presumably benefiting the non-legume in the process. While not precisely the same as the proposed buffer hypothesis, the legume/non-legume system nevertheless demonstrates its potential operation. The atmospheric nitrogen simply would never enter the system if it were not for the effect of the legume.

Perhaps the best-known facilitation in nature is the keystone predator concept of Paine (1966, 1974). In what is now regarded as a classic study, Paine showed how a starfish predator, *Pisaster*, kept levels of dominant competitors below the densities that could drive other competitors to extinction. The predator thus has an indirect facilitative effect on those competitors that are thereby saved from extinction.

Some of the most interesting cases of experimentally demonstrated facilitations are in fact summarised by Connell (1983) in his review of interspecific competition! What is especially interesting in these studies is that the experiments were apparently designed to detect competition and, surprisingly, often demonstrated positive effects. Out of a survey of 72 studies, 14 (19 per cent) showed facilitation, generally attributed to indirect effects (Connell 1983). This is an impressive number especially in the light of the fact that only 32 per cent of all the studies successfully demonstrated what they set out to show — competition.

Finally, a virtually unlimited array of potential indirect effects can be constructed from already known natural history. For example, in the experimental agriculture area at the Matthaei Botanical Gardens of the University of Michigan, *Myzus persicae* (an aphid) attacks both beans and tomatoes. The very problematic weed *Chenopodium album* attracts large numbers of *Aphis* sp. (another

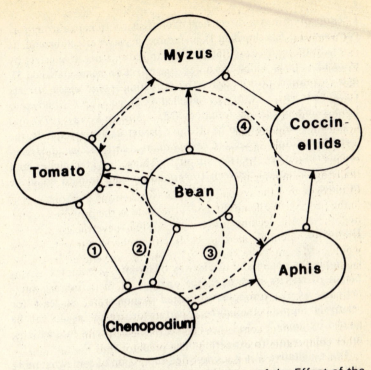

Figure 13.8: Diagrammatic Representation of the Effect of the Weed *Chenopodium* on Tomatoes. Pathway 1 is the direct *negative* effect of the weed on the crop. Pathway 2 is the indirect *negative* effect of the weed on the crop as determined by the interaction of the weed and crop with the beans. Pathway 3 is the indirect *negative* effect of the weed on the crop as determined by the interaction between the weed and the aphid, *Aphis*, the aphid and the bean, and the bean and the crop. Pathway 4 is the indirect *positive* effect of the weed on the crop as determined by the interactions between the weed and *Aphis*, between *Aphis* and the coccinellids, between coccinellids and *Myzus*, and between *Myzus* and the tomato.

aphid), which attacks beans, which in turn have a direct positive effect on tomatoes (unpublished data). The *Aphis* on *Chenopodium* attract large numbers of coccinelids which eat both *Myzus* and *Aphis*. The whole system is diagrammed in Figure 13.8. Thus

Chenopodium has a direct negative effect on tomato through its competitive effect (labelled 1), an indirect negative effect through its competition with beans (labelled 2), an indirect negative effect by attracting *Aphis* which has a negative effect on beans (labelled 3), and an indirect positive effect by attracting *Aphis* which attracts coccinelids which eat *Myzus* (labelled 4). Since *Myzus* is known to have a large effect on tomatoes (Mackauer and Way 1978) and coccinelids are known to be effective predators against aphids, the 5-species indirect chain (*Chenopodium-Aphis*-coccinelid-*Myzus*-tomato) might very well swamp out the shorter negative chains, and the net interaction between *Chenopodium* and tomato could be positive.

Discussion

A simple yet potentially useful way of looking at complex assemblages of species is through the analysis of their indirect interactions. As the examples presented demonstrate, at least for relatively simple communities a certain degree of insight can be gained by going beyond the direct species-to-species interactions, and examining more complex linkages.

This qualitative idea is a simplified expression of somewhat more elaborate quantitative motifs (Levine 1976; Vandermeer 1980) which originally sought to stipulate the mathematical conditions that result in indirect mutualism. In certain respects it also resembles more ambitious mathematical theory which focuses on indirect linkages (Levins 1975; Schaffer 1981), although the latter are more concerned with conditions of neighbourhood stability rather than simply the sign of interactions.

From a strictly empirical point of view, a catalogue of various forms of indirect mutualisms, as presented above, reinforces recent ideas regarding the importance of mutualisms in nature. As has apparently been the case with direct mutualisms, their indirect cousins seem to appear even when they are not being searched for. Might we extrapolate that a more concerned search for mutualisms could reorient our prejudices about the nature of interactions in communities (Risch and Boucher 1976)?

Studying natural communities by focusing on indirect interactions is certainly not new. Early ecologists (recall Darwin's bumblebees) were frequently fascinated with such complexities, and some of the more recent mathematical theory of community structure (for example, May 1976; Lawlor 1979; Levins 1975) inadvertently uses indirect links to analyse stability (for example the representation of the determinant of the community matrix as a sum of 'loops' — indirect interactions — of various lengths). Refocusing the more esoteric mathematical approaches in a way that partially recaptures the earlier qualitative appreciation of complex interactions, as we claim to have done in this chapter, perhaps will provide unique insights into community structure in the future.

References

Connell, J.H. (1983) 'On the Prevalence and Relative Importance of Interspecific Competition: Evidence from Field Experiments', *American Naturalist, 122,* 661–96

Culver, D.C. (1982) *Cave Life: Evolution and Ecology,* Harvard University Press, Cambridge

Dowd, B.A. and R.W. Elwood (1983) 'Shell Wars: Assessment Strategies and the Timing of Decisions in Hermit Crab Shell Fights', *Behaviour, 85,* 1–124

Elwood, R.W. and C.W. Glass (1981) 'Negotiation or Aggression during Shell Fights of the Hermit Crab *Pagurus bernhardus*?', *Animal Behavior, 29,* 1239–44

Farb, P. (1963) *Ecology,* Time-Life Books, New York

Feinsinger, P. (1983) 'Coevolution and Pollination' in Futuyma, D.J. and M. Slatkin (eds), *Coevolution,* Sinauer Associates, Sunderland, Mass., pp. 282–310

Fritz, R.S. (1983) 'Ant Protection of a Host Plant's Defoliator: Consequence of an Ant-Membracid Mutualism', *Ecology, 64,* 789–97

Hazlett, B.A. (1978) 'Shell Exchanges in Hermit Crabs: Aggression, Negotiation, or Both?', *Animal Behavior, 26,* 1278–9

Hazlett, B.A. (1981) 'The Behavioral Ecology of Hermit Crabs', *Annual Review of Ecology and Systematics, 12,* 1–22

Hazlett, B.A. (1983) 'Interspecific Negotiations: Mutual Gain in Exchanges of a Limiting Resource', *Animal Behavior, 31,* 160–3

Hussey, R.S. and R.W. Roncadori (1982) 'Vesicular-Arbuscular Mycorrhizae May Limit Nematode Activity and Improve Plant Growth', *Plant Disease* (Jan. 1982), 9–14

Jutsum, A.R., J.M. Cherritt and M. Fisher (1981) 'Interactions Between the Fauna of Citrus Trees in Trinidad and the Ants *Atta cephalotes* and *Azteca* sp.', *Journal of Applied Ecology, 18,* 187–95

Laine, K.J. and P. Niemala (1980) 'The Influence of Ants on the Survival of Mountain Birches During an *Oporina autumata* (Lep., Geometridae) Outbreak', *Oecologia, 47,* 39–42

Lawlor, L.R. (1979) 'Direct and Indirect Effects of N-Species Competition', *Oecologia, 43,* 355–64

Levine, S.H. (1976) 'Competitive Interactions in Ecosystems', *American Naturalist, 110,* 903–10

Levins, R. (1975) 'Evolution in Communities Near Equilibrium' in Cody, M.L. and J.M. Diamond (eds), *Ecology and Evolution of Communities,* Belknap Press, Cambridge, pp. 16–50

MacArthur, R.H. (1972) *Geographical Ecology,* Harper and Row, New York

Mackauer, M. and M.J. Way (1978) '*Myzus persicae*, an Aphid of World Importance' in Delucchi, V.L. (ed.), *Studies in Biological Control*, IBP9, Cambridge University Press, Cambridge

May, R.M. (1976) 'Patterns in Multi-Species Communities' in May, R.M. (ed.), *Theoretical Ecology*, Blackwell Sci. Publ., Oxford, pp. 142–62

Messina, F.J. (1981) 'Plant Protection as a Consequence of an Ant-Membracid Mutualism: Interactions on Goldenrod (*Solidago* sp.)', *Ecology, 62*, 1433–40

Nickerson, J.C., C.A. Rolph Kay, L.L. Bushman and W.H. Whitcomb (1977) 'The Presence of *Spississtilus festinus* as a Factor Affecting Egg Predation by Ants in Soybeans', *Florida Entomologist, 60*, 193–9

Paine, R.T. (1966) 'Food Web Complexity and Species Diversity', *American Naturalist, 100*, 65–75

Paine, R.T. (1974) 'Intertidal Community Structure', *Oecologia, 15*, 93–120

Price, P.W., C.E. Bouton, P. Gross, B.A. McPheron, J.N. Thompson and A.E. Weis (1980) 'Interactions Among Three Trophic Levels: Influence of Plants on Interactions Between Insect Herbivores and Natural Enemies', *Annual Review of Ecology and Systematics, 11*, 41–65

Rathcke, B. (1984) 'Competition and Facilitation Among Plants for Pollination' in Real, L. (ed.), *Pollination Biology*, Academic Press, New York

Risch, S. and D.H. Boucher (1976) 'What Ecologists Look For', *Bulletin of the Ecological Society of America, 57(3)*, 8–9

Room, P.M. (1972) 'The Fauna of the Mistletoe *Tapinanthus banquensis* Growing on Cocoa in Ghana: Relationships Between Fauna and Mistletoe', *Journal of Animal Ecology, 41*, 611–21

Roughgarden, J., J. Rummel and S. Pacala (1983) 'Experimental Evidence of Strong Present-Day Competition Between the *Anolis* Populations of the Anguilla Bank: a Preliminary Report' in Rhodin, A. and K. Miyata (eds), *Advances in Herpetology and Ecolutionary Biology — Essays in Honor of Ernest Williams*, Harvard University Press, Cambridge, Mass. pp. 499–506

Rude, P.A. (1982) *Integrated Pest Management for Tomatoes*, University of California Statewide Integrated Pest Management Project, Division of Agricultural Sciences, Publication 3274

Schaffer, W.M. (1981) 'Ecological Abstractions: the Consequences of Reduced Dimensionality in Ecological Models', *Ecological Monographs, 51*, 383–401

Schemske, D.W. (1981) 'Floral Convergence and Pollinator Sharing in Two Bee-Pollinated Tropical Herbs', *Ecology, 62*, 946–54

Skinner, G.J. and J.B. Whittaker (1981) 'An Experimental Investigation of Inter-Relationships Between the Wood-Ant (*Formica rufa*) and Some Tree-Canopy Herbivores', *Journal of Animal Ecology, 50*, 313–26

Snaydon, R.W. and P.W. Harris (1981) 'Interactions Below Ground — the Use of Nutrients and Water' in Willey, R.W. (ed.), *International Workshop on Intercropping*, ICRISAT, Hyderabad, India, pp. 188–201

Thompson, J.D. (1981) 'Spatial and Temporal Components of Resource Assessment of Flower-Feeding Insects', *Journal of Animal Ecology, 50*, 49–59

Trenbath, B.R. (1976) 'Plant Interactions in Mixed Crop Communities' in Papendick, R.I., P.A. Sanchez and G.B. Triplett (eds), *Multiple Cropping*, American Society of Agronomy Special Publication 27, pp. 129–70

Vandermeer, J.H. (1980) 'Indirect Mutualism: Variations on a Theme by Stephen Levine', *American Naturalist, 116*, 441–8

Waser, N.M. and L.A. Real (1979) 'Effective Mutualism Between Sequentially Flowering Plant Species', *Nature, 28*, 670–2

Waser, N.M. (1983) 'Competition for Pollination and Floral Character Differences Among Sympatric Plant Species: A Review of Evidence' in Jones, C.E. and R.J.

Little (eds), *Handbook of Experimental Pollination Ecology*, Van Nostrand Reinhold, New York, pp. 277–93

Way, M.J. (1963) 'Mutualism Between Ants and Honeydew-Producing Homoptera', *Annual Review of Entomology*, 8, 307–44

14 A FOOD WEB APPROACH TO MUTUALISM IN LAKE COMMUNITIES

Patricia A. Lane

Introduction

Species survive, reproduce and evolve in ecosystems. These organisms are interconnected with each other and numerous abiotic factors. Many of these interconnections are what are termed biological interactions. For the population level, they have been traditionally defined in two ways: (1) description of the biological mechanism and (2) mathematical analysis of the outcome of the interaction. It has been difficult to characterize the diversity of biological mechanisms rigorously; they defy a pigeon-hole approach. Even the best studied interactions in laboratory cultures have given surprising results. Gause's (1934) yeast cells did not simply exhibit resource competition in food-limited conditions; one species also poisoned the other through alcohol production. Likewise, Park's (1962) *Trilobium* species became cannibalistic and predatory at particular abundance levels. Thus, each species can potentially interact with another in a variety of ways.

Mathematically, the outcome of an interaction for species i and j is usually described at the population level beginning with the equations of change in abundance:

$$dX_i/dt = F_i(X_1, X_2, X_3, \ldots, X_n; C_1, C_2, C_3, \ldots, C_n) \tag{14.1}$$

and:

$$dX_j/dt = F_j(X_1, X_2, X_3, \ldots, X_n; C_1, C_2, C_3, \ldots, C_n) \tag{14.2}$$

The X's are the abundance of variables 1 to n, whereas the C's are parameters 1 to n. Outcomes of interactions between species i and j are found by taking the first partials of F_i and F_j in response to an increase in equilibrium level of the other species ($\partial F_i/\partial X_j; \partial F_j/\partial X_i$). If the first partials are both negative, this is termed competition; if positive, mutualism; and if one is positive and the other negative, predation. The mathematical description reveals little about the mechanism, but knowing the final result is useful for quantitative evaluations and discussions of evolutionary significance. Often initial biological versus later mathematical analysis of a particular species interaction has given contradictory and surprising results (Brinkhurst *et al.* 1972).

Both the biological mechanism and mathematical outcome approaches have centered on one to two species interactions. Although it is recognized that a given species is usually interacting with more than one other species and often many others, ecological theory has been slow to shift from the population to the community level. It has been difficult to understand what a biological interaction means in regard to a species pair positioned in a food web. Competition, predation and mutualism appear to blend into each other at the community level and lose their discreteness.

Weinberg (1975) explained why this is so in a general way when he described small, medium and large number systems. Here I simply translate his ideas into an ecological context. Population ecology essentially represents an analytical approach to small number systems which have few components and few interactions. These systems are amenable to precise mathematical description; for instance, by the differential or difference equations of dynamic systems theory. While traditional systems theory has proven accurate in engineering situations in which exact measurements and complete functional descriptions are possible, its application in ecology has been limited to theoretical and over-simplified systems. This is a result of the near-impossibility of obtaining measurements of all interaction rates and parameters and of constructing a complete functional model of the system. Science, in general, has been successful with small number systems.

Likewise, there has been considerable success with large number systems. These systems have so many components that their average behaviour becomes the key state descriptor. An example would be a flask of ideal gas. The position and velocity of a particular gas molecule is not of interest, but the average properties of volume, temperature and pressure are. Large number systems are usually treated statistically; since the law of large numbers ensures that the error of deviation from average behaviour will become very small as the number of components becomes very large. In addition, when systems possess a large number of components, they tend toward randomness, which is quintessential for statistical treatment. Ecosystems, however, are neither small nor large number systems. They contain too many components to be treated analytically and too few for statistical analysis. A fish and a phytoplankton cell cannot be averaged like ideal gas molecules nor are their behaviours and life history dynamics equivalent to random events.

Ecosystems are medium number systems; they have an intermediate number of components and interactions. As the number of components increases arithmetically, the number of interactions increases geometrically. For $n = 2$ components, there is one interaction term; but for $n = 10$ components, there are $2^n - n - 1$ or 1013 possible interactions of which 90 are pairwise relationships. While in a community of 10 species, many of these interactions could not occur, there would still be a large number of them. Science, and in particular ecology, has few good tools for dealing with this level of complexity. Simplification is needed to solve problems involving medium number systems by preserving the critical interactions that influence system dynamics and ignoring those that do not. This is not easily achieved. Schaffer (1981) gives an excellent discussion of how this simplification can be achieved and what its consequences are. An interconnection can be equated to a direct link or effect between a pair of components in a system. These direct links are equivalent to the mathematical definition of biological interaction given above. In addition, the number of potential paths of effect between any two components in the network also increases geometrically. In the pelagic food webs I have been modelling, there are often several hundred to several thousand paths of effect for networks of 18–22 variables (Lane 1984a,b). Most of these represent what are termed indirect effects.

The distinction between direct and indirect resides solely in how the system is described. A direct effect can be made indirect by inserting a third variable between the two original ones and vice versa. Most forms of resource competition involve an indirect effect if the resource is retained in the model. Indirect effects are usually calculated by multiplying the direct effects along the path linking the two variables of interest. Patten (1983) and Vandermeer (1980) have stressed the importance of indirect paths. Finally the relationship of any two components in a complex network will be some integration of the myriad of all direct and indirect effects. This is termed the community effect.

The purpose of this chapter is to show how direct, indirect and community effects are interrelated in foodwebs using some freshwater examples from the literature. Mutualistic effects are emphasized, but they cannot be considered in isolation at the community level. Loop analysis, a qualitative network technique, is used to represent and analyse the various networks to illustrate what happens to a two-species direct effect as it becomes embedded in increasingly more complex ecological networks. Direct and indirect effects can often be overwhelmed by community effects. For example, a direct and/or indirect path can be negated or reversed by the configuration of the rest of the community not on the path (Lane and Levins 1977), a fact which is often not appreciated. This does not mean that direct effects are less important, but rather that different types of information are provided by different levels of analysis. In addition, some authors have incorrectly equated indirect and community effects. Indirect effects are really only a component of the community effect calculation and as such should not be considered to have the independent existence or importance alluded to by some authors. All three types of effects are totally different entities and cannot be used interchangeably. Loop analysis provides one of the most effective methodologies for computing community effects and for achieving simplification for the medium number systems of community ecology. It is also a good tool for providing community structure-stability descriptors, examining the theoretical outcomes of network alteration, guiding general study and understanding, and designing community level experiments. The subsequent models can be fitted and tested with field or laboratory data.

Mathematical Definitions and Methodology

A form of loop analysis was first used by electrical engineers at MIT in the early 1950s (Mason 1952). Levins (1973, 1975) developed the theory of loop analysis for biological and ecological systems. Berryman (1981), Hutchinson (1975), Roughgarden (1979) and Vandemeer (1981) have given introductions to the technique. The methodology has been applied to several systems including: human physiology and disease (Levins 1973); community evolution (Levins 1975); population genetics (Desharnais and Costantino 1980); agriculture (Boucher 1985); species interactions (Boucher et al. 1982, Henry 1980); freshwater plankton communities (Briand and McCauley 1978; Lane and Levins 1977; Lane and Blouin 1984, 1985); and marine mesocosms and environments (Lane 1982; Lane 1984a,b,c,d; Lane and Collins 1984). Lane (1984e) and Wright and

Lane (1984) have described some recent theoretical developments.

Loop analysis follows the methods of dynamic systems theory but uses only the signs of the first partials. Thus, the advantage of this technique is that it relies only on knowledge of the system structure rather than on detailed knowledge of the functional relationships between system components and measurements of all coefficients involved. By relinquishing the need for precise measurement and mathematical expression, loop analysis provides a methodology for medium number systems. Another feature of loop analysis is that it allows the model to be represented as a directed graph or diagraph. This permits the concepts and terminology of graph theory to be used. The analogy of loop analysis to graph theory provides a way to calculate feedbacks for these systems in a way which makes it easy to visualize system interaction paths.

Linear system dynamics and the assumption of steady state is crucial for the application of loop analysis to a particular system. Because loop analysis is related to dynamic systems theory, interactions are based on linear functions which are represented by the signs of the slopes of such functions. Thus, it is assumed that variable change is proportional to the levels of other variables in the system. Any interactions which have non-linear aspects will be represented by a linear approximation. Such non-linear aspects include curved-line relationships (quadratic, exponential, logarithmic, reciprocal) with non-continuous slope or levelling-off functions, and relationships where the combined effect of two variables is different from their sum. In many cases, the linear approximations reflect the dynamics of the real system accurately enough that there is no difference in the results. If there are significant differences, they will be ignored by the analysis. Problems may arise, however, when interactions are involved whose functions change sign over some range of interest, usually in the neighbourhood of steady state. If the researcher is careful to make sure that his model is based on functions which reflect system behaviour around the current steady state, there will rarely be a problem with the analysis. The system, however, may not follow the same behaviour away from the steady state or a different steady state.

The Community Matrix and Direct Effects

Loop analysis begins with a community matrix which corresponds to the Jacobian matrix used in dynamic systems analysis. Where a system has n variables, the matrix will be of order n and its elements,

Differential Equations

$$\frac{dx_1}{dt} = b_1 - \rho_{12}x_2 - k_1 x_1 - \theta$$

$$\frac{dx_2}{dt} = b_2 x_2 + r_{12}\,\rho_{12}\,x_1 x_2$$

Community Matrix   Loop Diagram

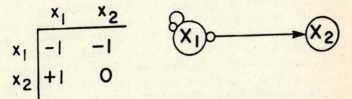

$$\begin{array}{c|cc}
 & x_1 & x_2 \\
\hline
x_1 & -1 & -1 \\
x_2 & +1 & 0
\end{array}$$

Figure 14.1: Three Mathematical Representations of a Predator (X_2)–prey (X_1) Interaction. The prey is self-damped and enhances the predator. The predator decreases the prey. In the differential equations, b_1, b_2, p_{12}, k_1 and θ are parameters.

a_{ij}, will be defined as $A = \{a_{ij}\}$ for i and j from 1 to n where:

$a_{ij} = +1$ if $\partial F_i/\partial X_j > 0$; -1 if $\partial F_i/\partial X_j < 0$; and 0 if $\partial F_i/\partial X_j = 0$.

In particular, for self-damping variables, the diagonal element $a_{ii} = -1$ while for self-reproducing variables, $a_{ii} = 0$. Wright and Lane (1984) elaborate on the distinctions between self-reproducing and self-damping variables. The a_{ij} estimates the signs of the direct, linear effects of variable j on the instantaneous rate of change of variable i at steady state. While changes in variable abundances are related through interactions of the whole system, the a_{ij} or the links, refer to direct effects on rates of change. Links can often be inferred from the specific biological mechanisms by which one variable affects the rate of change of another.

The community matrix may be thought of as the adjacency matrix of an n-node digraph where each link from node j to node i has the sign of a_{ij}. Self-damping variables are negatively linked to themselves. Typically, a negative link is drawn with a circlehead (--o) while a positive link is drawn with an arrowhead (→). The direct effect is on the variable the arrow- or circle-head touches. Figure 14.1 gives three different representations of the same two variable system of a predator feeding on its prey.

Definitions Used in Loop Analysis and Indirect Effects

In loop analysis, a path from variable i to variable j is the algebraic product of elements which traverse a series of links from variable i to variable j, in which no variable is ever passed through twice. This path product is also termed an indirect effect ($P_{ij}{}^k$). A loop is a path from a given variable back to itself. A path of length k which is not a loop, traverses k variables and has k-1 links. A loop of length k traverses k variables and include k links. A spanning path or loop passes through all variables in a given system. Thus, a two-variable link is a path of length 1 where k = 2 while a self-damping link is a loop of length 1. A spanning path and a spanning loop each must have a length of n for an n-node system. Finally, a set of disjunct loops is a set of one or more loops in which no two loops have variables in common and a spanning set of disjunct loops is a set of disjunct loops in which each variable in the system is passed through once and only once. The feedback of a system of n variables in a community matrix can be expressed as,

$$\Sigma F_n = (-1)^{m+1} L(m,n) \tag{14.3}$$

The feedback at level k with k less than n, is defined as the sum of feedbacks for all submatrices of the given order k. If there are no disjunct loop sets in a submatrix, then the feedback for that submatrix is zero. As an algebraic convenience, feedback at the zero level, $F_o = -1$. F_o occurs when the path is a spanning one. If all sets of disjunct loops which span the system have an odd number of negative loops when calculating feedback, then the overall feedback will be entirely negative. That the feedback at all levels is negative is a necessary condition for system stability. This requires that at least a majority of disjunct loop sets at any level are negative. That feedback at all levels is negative, however, is not sufficient condition for stability since there is an additional Routh-Hurwitz criterion that feedback at higher levels cannot be too negative compared to lower level feedbacks (Levins 1973, 1975; Wright and Lane 1984). Because mutualism involves positive feedback, it has sometimes been characterized as destabilizing in community structure. This is not necessarily so. Positive feedback at higher levels can help dilute excess negative feedback in relation to the Routh-Hurwitz criterion and thus promote overall system stability.

Parameter Input to Steady State Communities and Community Effects

Given the structure of the community and the assumption of steady state, it can be computed how an alteration in the rate of change of one component caused by some outside factor or parameter input will affect the levels of all other variables. This is termed the community effect. Given a system of n variables $X_1, \ldots, X_n$ each with rates of change $dX_i/dt = F_i(X_1, \ldots, X_n)$, and that the matrix $(\partial F_i/\partial X_j)$ is known for each pair of variables i and j, then assume there are n system parameters $C_1, \ldots, C_n$ for which $\partial F_i/\partial C_j = 1$ if i = j and 0 if i $\neq$ j. Thus, all direct effects must be known and it is assumed that each parameter has a proportional effect on the rate of change function of its corresponding variable. Then,

$$dX_i/dt = F_i(X_1, \ldots, X_n; C_1, \ldots, C_n) = 0 \qquad (12.4)$$

Notice that the partial of a variable's rate of change function with respect to a system parameter, $\partial F_i/\partial C_h$, is quite distinct from the partial of the variable's actual level with respect to the same parameter, $\partial X_i/\partial C_h$. A change in a given parameter may affect the rate of change functions for one or more variables and the corresponding partials $\partial F_i/\partial C_h$ will be non-zero. This condition is called a 'parameter input' to the system at the given variable. Even if the parameter input is only at one variable, however, the actual levels of all variables may be affected because of the interactions in the system structure.

Loop analysis theory (Levins 1973, 1975) shows that for any given elements of this matrix, community effects can be calculated as follows,

$$\frac{\partial X_i}{\partial C_j} = \Sigma \; \frac{\partial F_j}{\partial C_j} \; \cdot \; \text{path}_{ij}^{(k)} \; \cdot \; F_{n-k}(\text{compl } P_{ij}^k)/F_n \qquad (14.5)$$

where $\text{path}_{ij}^{(k)}$ is the product of community matrix elements corresponding to the links in a path of length k from variable j to variable i in the loop model of the community. The mathematical basis of this formula is given in Levins (1973, 1975) and Wright and Lane (1984). The sum is over all such paths from variable j to i of any length. Note that in the sum over paths for community effects, only valid paths with a non-zero feedback in the complement are counted. These are defined as paths which either span the system so that the complement feedback is $F_o = -1$ or paths whose complement contains at least one disjunct loop set which includes all variables not on the path. The term $F_{n-k}(\text{compl } P_{ij}^k)$ refers to the feedback of the path complement; that is, the feedback of the subsystem remaining after all variables involved in the associated path are removed. This subsystem is of order n-k and as such the feedback is said to be of order n-k. The term F_n refers to the feedback of the whole system.

In an ecosystem with four trophic levels, the increase of nutrient concentrations might serve to increase the rates of change of algal species. This would stimulate the abundance of herbivores and the increased consumption would restore the standing levels of nutrients. Similar effects would propagate up the food-chain so that the only species whose population level would increase might be the top-level carnivore. Therefore, it is of interest to know how a set of parameter inputs for a given parameter C_h, of one or more of the functions describing system variables, F_i, given by the vector of $(\partial F / \partial C_h)$, will be related to changes in the actual levels of system variables X_i, given by the vector $(\partial X / \partial C_h)$, under steady state conditions.

The ratio of complement feedback to overall feedback is equivalent to an amplification factor. This ratio measures the capacity of the system as a whole to absorb the effects of changes as they move along a path. If the path is not valid as defined above, then the beginning variable will not be able to affect the terminal variable. If the complement feedback has the same sign as the overall feedback, then effects will not be absorbed over the path and the end effect will be as indicated by the path product. If the complement feedback is of opposite sign to the overall feedback, the effects of the path may be reversed. This is because the path variables are linked to other variables in the network. If there were no such links then the path and its complement would be independent systems.

The indirect effects of paths indicate how the variables involved transmit changes from variable to variable, without considering how links with other variables in the system will affect the end result. If the path were removed from the system by breaking all links with the complements, then the indirect effect would indicate the start-to-end effect. If the whole system is to be considered, however, then any measure of community effect from one variable to another must include the feedbacks. Indirect and community effects will be equivalent only when (1) the indirect effect is represented by a spanning path, or (2) if not a spanning path, then the feedback ratio is positive. In addition, if there is more than one path, then all must have the same sign. The probability of these conditions being met decreases as the number of components and average connectance of the network increase. Even when the above equivalence conditions are met, indirect effects should not be relied upon in routine analysis. If one ignores the amplification or feedback ratio of the community effect equation, there is always the danger that F_n is positive, which indicates system instability and often reversal of the sign of the indirect effect. For example, Levine's (1976) models have $F_n = 0$. Direct effects are in units of reciprocal time and indirect effects are in units of reciprocal time to the k power, where k is the length of the path. Community effects are dimensionless.

Some Examples

In all of the following examples (Figures 14.2, 14.9) the community effects are presented as +, –, 0 and ? values in matrices termed *effects matrices*. The community effects are read across each row of the matrix associated with a particular parameter input in the left hand column. Only positive parameter inputs are calculated. Negative community effects can be found by simply reversing the signs of the predictions. Zero values remain the same, as do ambiguities. Ambiguous effects are listed as ?'s in the effects matrix. They arise when two (path times complement) products have opposite signs. Ambiguity can be partially resolved using the semi-quantitative method of Lane and Levins (1977).

At the right of the community effects matrix is the list of model characteristics for each figure. Variables, links and loops are enumerated. Valid paths are those that have at least one non-zero complement. Valid complements are those used to identify valid paths and they are also listed in the table. Finally, overall feedback (F_n) is given with its sign and numerical value which is equal to the number of excess negative terms in relation to positive ones. The computer

Loop Diagram

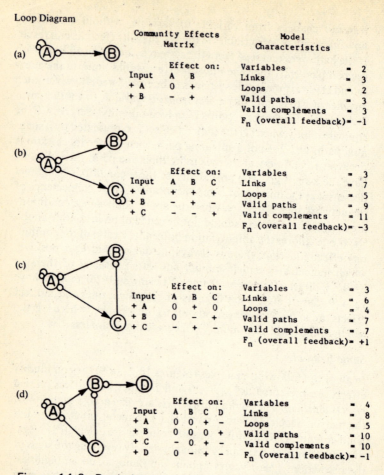

	Community Effects Matrix			Model Characteristics	

(a)

Input	Effect on:			Variables	= 2
	A	B		Links	= 3
+ A	0	+		Loops	= 2
+ B	−	+		Valid paths	= 3
				Valid complements	= 3
				F_n (overall feedback)	= −1

(b)

Input	Effect on:			Variables	= 3
	A	B	C	Links	= 7
+ A	+	+	+	Loops	= 5
+ B	−	+	−	Valid paths	= 9
+ C	−	−	+	Valid complements	= 11
				F_n (overall feedback)	= −3

(c)

Input	Effect on:			Variables	= 3
	A	B	C	Links	= 6
+ A	0	+	0	Loops	= 4
+ B	0	−	+	Valid paths	= 7
+ C	−	+	−	Valid complements	= 7
				F_n (overall feedback)	= +1

(d)

Input	Effect on:				Variables	= 4
	A	B	C	D	Links	= 8
+ A	0	0	+	−	Loops	= 5
+ B	0	0	0	+	Valid paths	= 10
+ C	−	0	+	−	Valid complements	= 10
+ D	0	−	+	−	F_n (overall feedback)	= −1

Figure 14.2: Predation and Exploitation Competition. A is a resource variable, B and C are competing species, and D is a predator.

printout gives actual numbers of positive and negative terms which are not shown in the figures. Each term contains a product of the direct interactions and because their magnitudes were not originally specified, the magnitude of these terms cannot be determined. Thus, the numerical value for F_n used here is simply a count of signed terms and not a quantitative measure of their magnitude.

Predation and Exploitation Competition. In a simple representation of predation (Figure 14.2a) A is a self-damped prey species and B is the predator. This system behaves according to our intuition in that a positive input to the prey results in an increase in the predator while a positive input to the predator reduces the prey species. When two predators (B and C) feed on the same prey (A) (Figure 14.2b), exploitation competition occurs which involves an indirect path. A positive input to either predator reduces the other, by way of the intervening prey variable.

To illustrate how small changes in a network may alter the community effects, Figure 14.2c shows two species (B and C) feeding on the same resource (A). Neither predator is self-damped, but C interferes directly with B. The difference from the case in Figure 14.2b is that the overall feedback is negative in Figure 14.2b but positive in Figure 14.2c making this system unstable. In this case, positive inputs to either predator will increase the other, which is a counter-intuitive result. Systems with positive overall feedback (F_n), however, are unstable and should not be used. Such systems are given in a few of the examples used here to illustrate that even in the simplest networks, community effects can be counter-intuitive if stability is not checked first. Loop analysis provides a convenient way to calculate stability and to alter unstable model networks and make them stable. For example, stability may be restored by adding predator D (Figure 14.2d) resulting in negative overall feedback. In this case, however, inputs to B or C have no effect on each other as a result of the satellite variable D, which creates an invalid complement for the paths between B and C. A satellite has only a single input. If it is undamped, it buffers its nearest neighbour against variation.

Interference Competition. Consider the simple example of two species exhibiting interference competition (Figure 14.3). Species A and B have negative direct effects on each other, represented by negative links (Figure 14.3a). The effects of inputs to this simple system are, however, counter-intuitive as a result of the community effect. First, a positive input to species A results in no change since the path does not have a valid complement because B is not in a loop. Second, a positive input to A results in an increase in B, not a decrease as might be expected, since the parameter input (+) times the path (–) must be multiplied by the feedback of the complement (–) (negative by definition since the complement contains no variables) and this product is divided by the overall feedback (+).

Loop Diagram

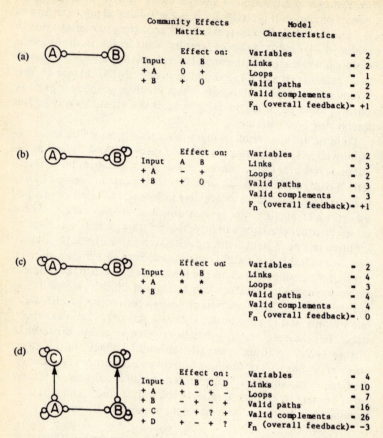

	Community Effects Matrix			Model Characteristics	
(a)	Input	A	B	Variables	= 2
	+ A	0	+	Links	= 2
	+ B	+	0	Loops	= 1
				Valid paths	= 2
				Valid complements	= 2
				F_n (overall feedback)	= +1

(b)	Input	A	B	Variables	= 2
	+ A	−	+	Links	= 3
	+ B	+	0	Loops	= 2
				Valid paths	= 3
				Valid complements	= 3
				F_n (overall feedback)	= +1

(c)	Input	A	B	Variables	= 2
	+ A	*	*	Links	= 4
	+ B	*	*	Loops	= 3
				Valid paths	= 4
				Valid complements	= 4
				F_n (overall feedback)	= 0

(d)	Input	A	B	C	D	Variables	= 4
	+ A	+	−	+	−	Links	= 10
	+ B	−	+	−	+	Loops	= 7
	+ C	−	+	?	+	Valid paths	= 16
	+ D	+	−	+	?	Valid complements	= 26
						F_n (overall feedback)	= −3

Figure 14.3: Interference Competition. A and B are competing species; C and D are predators.

In Figure 14.3b, a positive input to A does result in a change in A since the path has a valid complement because B is self-damped and therefore in a loop. In this case, the effect on A is negative, which is again counter-intuitive as a result of the community effect. This system is also unstable. When both species are self-damped (Figure 14.3c) no predictions can be made since the overall feedback (F_n) is zero. In this example, the positive loop of length two is exactly balanced by two negative loops of length one.

These simple models are inadequate representations of interference competition since an increase in either species results in an increase in the other as a result of overall positive feedback (F_n). Thus, as a result of community effects, the relationship is mutualistic (+ +) rather than competitive (− −). When embedded in a larger network, however, the effects may change to conform to intuitively-expected results (Figure 14.3d). In this example, a parameter input to A results in a decrease in B and vice-versa, as expected for competing species. The difference from this result and those for models 14.3a–14.3c is that the overall feedback in the network is now negative. Thus, even the simplest models of biological interactions can be erroneous and/or confusing if they are not first checked for total network effects. In particular, models of interference competition need to be carefully interpreted in a network context. In model 14.3d variables C and D interact as community mutualists. Vandermeer (1980), building on earlier results by Levine (1976), demonstrated community mutualism in a similar network using an algebraic analysis.

Mutualism. A simple representation of mutualism (Figure 14.4a) produces counter-intuitive predictions, as did the simple interference competition model in Figure 14.2a. Self-damping of A (Figure 14.4b) allows a prediction for the effect of a positive input to B, but the result is again counter-intuitive. The reason, as before, is that positive overall feedback (F_n), makes these systems unstable. If both variables are self-damped (Figure 14.4c) no predictions can be made since F_n = 0. As in the case of interference competition, this system can operate when part of a more complex loop network. Adding predators (Figure 14.4d) produces a stable network with a mutualistic community effect between A and B; that is, a positive parameter input to either A or B results in an increase in the other. Negative parameter inputs would simply reverse the prediction signs to negative, as would be expected for mutualists. These are still very simple examples because there are only single paths between any two variables.

Multi-Trophic Level System. In a more complicated model of a food web (Figure 14.5) mutualistic relationships arise via intervening variables on different trophic levels. In a general network of predator-prey pairs, paths which travel upwards by an even number of trophic levels and then back down by an even number of trophic

Loop Diagram

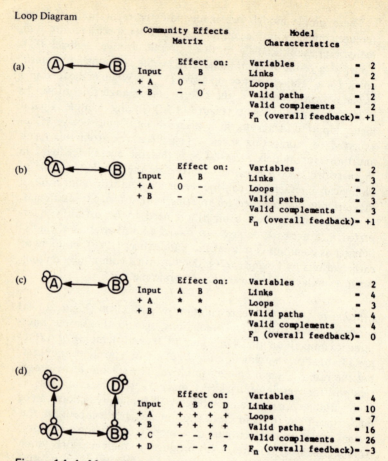

<table>
<tr><td></td><td></td><td colspan="3">Community Effects
Matrix</td><td>Model
Characteristics</td></tr>
</table>

(a)

		Effect on:		Variables	= 2
Input	A	B		Links	= 2
+ A	0	-		Loops	= 1
+ B	-	0		Valid paths	= 2
				Valid complements	= 2
				F_n (overall feedback)= +1

(b)

		Effect on:		Variables	= 2
Input	A	B		Links	= 3
+ A	0	-		Loops	= 2
+ B	-	-		Valid paths	= 3
				Valid complements	= 3
				F_n (overall feedback)= +1

(c)

		Effect on:		Variables	= 2
Input	A	B		Links	= 4
+ A	*	*		Loops	= 3
+ B	*	*		Valid paths	= 4
				Valid complements	= 4
				F_n (overall feedback)= 0

(d)

	Effect on:				Variables	= 4
Input	A	B	C	D	Links	= 10
+ A	+	+	+	+	Loops	= 7
+ B	+	+	+	+	Valid paths	= 16
+ C	-	-	?	-	Valid complements	= 26
+ D	-	-	-	?	F_n (overall feedback)= -3

Figure 14.4: Mutualism. A and B are mutualistic species; C and D are predators.

levels, or the reverse, result in mutualistic community effects between components on the same trophic level. In this system, H_1 and H_2 are mutualistic (positive input to either one increases the other) as a result of the paths of effect through the invertebrate predators (P_1, P_2) and vertebrate carnivore (C). The paths through R have no effect since they do not have valid complements. Similarly, P_1 and P_2 exhibit mutualistic community effects by the paths through H_1, H_2 and R. The negative paths through C have no valid

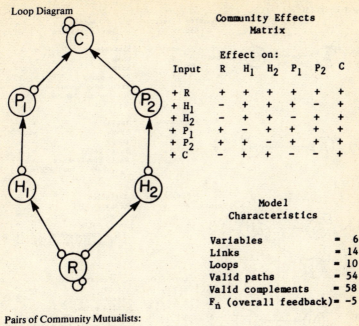

Loop Diagram

Community Effects
Matrix

Effect on:

Input	R	H₁	H₂	P₁	P₂	C
+ R	+	+	+	+	+	+
+ H₁	-	+	+	+	-	+
+ H₂	-	+	+	-	+	+
+ P₁	+	-	+	+	+	+
+ P₂	+	+	-	+	+	+
+ C	-	+	+	-	-	+

Model
Characteristics

Variables	= 6
Links	= 14
Loops	= 10
Valid paths	= 54
Valid complements	= 58
F_n (overall feedback)	= -5

Pairs of Community Mutualists:

R, P₁	C, H₁
R, P₂	C, H₂
P₁, P₂	H₁, H₂

Figure 14.5: Multi-Trophic Level System. R is a resource variable, H_1 and H_2 are herbivores, P_1 and P_2 are invertebrate predators, and C is a vertebrate carnivore.

complements and do not produce a community ⸝⸝⸝ ⸝t. For both pairs of community-level mutualists, (P_1, P_2 and H_1, H_2), the paths of effect producing the relationship traverse variables on two intervening trophic levels before returning to the original level. This is a common way (Type A) in which mutualism via community effects arises in complex food webs. Note also that variables two trophic levels apart from each other (C and H_1, H_2; R and P_1, P_2) exhibit + + community effects but these are termed type B community mutualism. If interference competition links occur between components on the same trophic level, then community mutualism can result from paths traversing an odd number trophic levels. This is termed type C community mutualism.

Loop Diagram

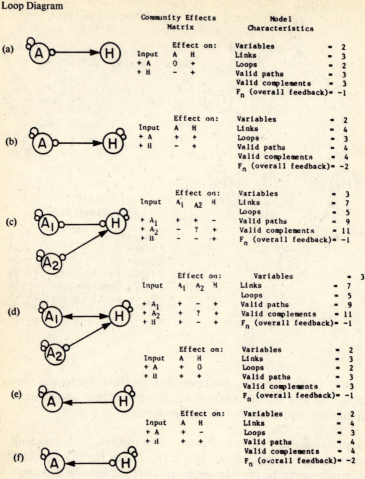

Figure 14.6: Algal-Herbivore Interactions. A, A_1, A_2 are algal species; H is a herbivore (after Porter 1976, 1977).

Direct, Indirect and Community Effects in Freshwater Communities

Algal-Herbivore Interactions

In a simple representation of an algal-herbivore interaction (Figure 14.6a), the herbivore benefits by consuming the algae which are decreased by the herbivore. An input to algae does not affect their level since changes are absorbed by the undamped herbivore. If the herbivore is self-damped, representing a cause of mortality not in the model, such as an omitted carnivore, then algae increase as a result of a positive input to themselves (Figure 14.6b). In some cases, algae are consumed which are toxic to the herbivore, with the result that both suffer from an increase in (positive input to) the other (Porter 1977) (Figure 14.6c). Moreover, some algae are able to pass relatively intact through the digestive tracts of herbivores. The algae benefits by absorbing some nutrients released by the herbivore and the herbivore benefits by digesting some of the algae (Porter 1976). This is modelled in Figure 14.6d, in which A_1 is able to withstand ingestion while A_2 is digested. The result is that A_1 and the herbivore are directly mutualistic, with each benefiting by a positive input to the other. A_2, however, suffers when the input is to either A_1 or H. In this example the direct effects are the same as the community effects. In some cases, the algae benefit from ingestion while no effect occurs for the herbivore (Porter 1976) (Figure 14.6e). Changes in the algae do not affect H, while A benefits from an increase in H. Finally, certain gelatinous green algae have a type of parasitic relationship with herbivores since they benefit from ingestion while the herbivores decline when the algae are dominant (Porter 1977) because of loss of nutrition (Figure 14.6f).

Oligochaete Interactions

Brinkhurst *et al.* (1972) reported some interesting results from a study involving three tubificid oligochaete species that coexist in Toronto harbour. Laboratory studies of growth of the species *Limnodrilus hoffmeisteri* (L), *Tubifex tubifex* (T) and *Peloscolex multisetosus* (p) in various combinations showed that T benefited by the presence of L (Figure 14.7a), and that both L and T benefited from the presence of P (Figures 14.7b and c). L produced a resource (R_1) in its faecal material which was utilized by T (Figure 14.7d). T benefited from this interaction while L was unaffected. P produced a second resource variable (R_2) which was used by both L and T (Figure 14.7e). Both L and T benefited from an input to P but were decreased by positive inputs to each other. Thus, they appeared to be in competition for R_2. When the three species and two resources are combined in a single model (Figure 14.7f), no pair of species can be shown to be community mutualists since no pair of species both benefit from a positive input to the other. The effect of an input to L

Loop Diagram

	Community Effects Matrix	Model Characteristics

For (a):

Effect on:

Input	L	T
+ L	+	+
+ T	0	+

Variables	= 2
Links	= 3
Loops	= 2
Valid paths	= 3
Valid complements	= 3
F_n (overall feedback)	= -1

For (b):

Effect on:

Input	P	L
+ P	+	+
+ L	0	+

Variables	= 2
Links	= 3
Loops	= 2
Valid paths	= 3
Valid complements	= 3
F_n (overall feedback)	= -1

For (c):

Effect on:

Input	P	T
+ P	+	+
+ T	0	+

Variables	= 2
Links	= 3
Loops	= 2
Valid paths	= 3
Valid complements	= 3
F_n (overall feedback)	= -1

For (d):

Effect on:

Input	L	R_1	T
+ L	+	+	+
+ R_1	0	+	+
+ T	0	-	+

Variables	= 3
Links	= 6
Loops	= 4
Valid paths	= 7
Valid complements	= 8
F_n (overall feedback)	= -2

Figure 14.7: Oligochaete Interactions. L. is *Limnodrilus hoffmeisteri*, T is *Tubifex tubifex*, P is *Peloscolex multisetosus*, and R_1 and R_2 are detrital resource variable (after Brinkhurst *et al.* 1972).

on T is ambiguous as a result of multiple paths of effect in the network (through R_1 as opposed to R_2). Thus, the direct one-link effect of Figure 14.7a does not operate as expected in the context of community effects in the food web. This is another example of how important it is to take the total network into account.

Herbivore Interactions in a Food Web

In a study of interactions between *Daphnia pulex* (DO and *Ceriodaphnia reticulata* (C), Lynch (1978) noted that traditional

Loop Diagram

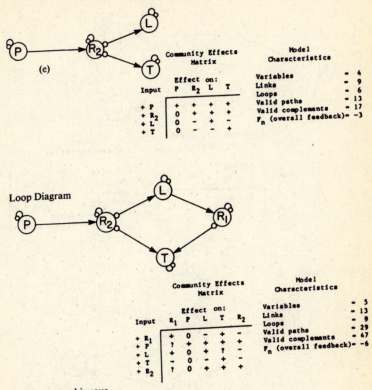

Community Effects Matrix

Input	Effect on:			
	P	R_2	L	T
+ P	+	+	+	+
+ R_2	0	+	+	+
+ L	0	−	+	−
+ T	0	−	−	+

Model Characteristics

Variables	= 4
Links	= 9
Loops	= 6
Valid paths	= 13
Valid complements	= 17
F_n (overall feedback)	= −3

Loop Diagram

Community Effects Matrix

Input	Effect on:				
	R_1	P	L	T	R_2
+ R_1	+	0	−	+	−
+ P	?	+	+	+	+
+ L	+	0	−	+	−
+ T	−	0	−	+	−
+ R_2	?	0	+	+	+

Model Characteristics

Variables	= 5
Links	= 13
Loops	= 9
Valid paths	= 29
Valid complements	= 47
F_n (overall feedback)	= −6

?- outcome ambiguous

competition theory may be inappropriate for species whose resources interact; and indeed that, 'coexploitation of interacting resources may actually lead to improved conditions for some exploiters'. This may be demonstrated in a simple loop model. Herbivores, D and C, each depend on separate algal resources (A_1 and A_2) which are both dependent on the same nutrient (N) (Figure 14.8a). A_1 and A_2 are competitors and both suffer as a result of positive inputs to the other. D and C, however, are mutualists by virtue of the interaction between their resources. This is similar to Levine's (1976) and Vandermeer's (1980) results. Positive inputs to either

Loop Diagram

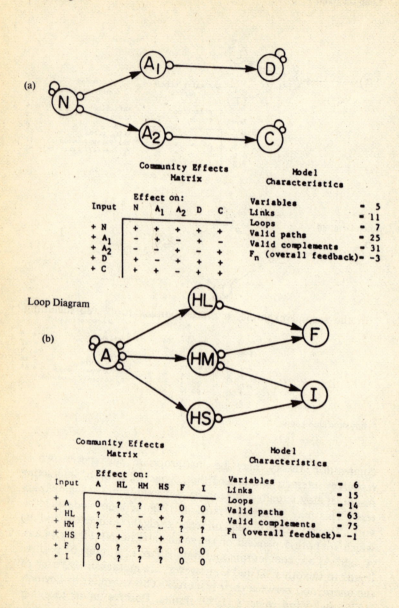

Community Effects
Matrix

Input	Effect on:				
	N	A₁	A₂	D	C
+ N	+	+	+	+	+
+ A₁	−	+	−	+	+
+ A₂	−	−	+	−	+
+ D	+	−	+	+	+
+ C	+	+	−	+	+

Model
Characteristics

Variables	= 5
Links	= 11
Loops	= 7
Valid paths	= 25
Valid complements	= 31
F_n (overall feedback)	= −3

Loop Diagram

(b)

Community Effects
Matrix

Input	Effect on:					
	A	HL	HM	HS	F	I
+ A	0	?	?	?	0	0
+ HL	?	+	−	+	?	?
+ HM	?	−	+	−	?	?
+ HS	?	+	−	+	?	?
+ F	0	?	?	?	0	0
+ I	0	?	?	?	0	0

Model
Characteristics

Variables	= 6
Links	= 15
Loops	= 14
Valid paths	= 63
Valid complements	= 75
F_n (overall feedback)	= −1

species increase the abundance of the other. In this case, the mutualism arises from paths passing through variables on two intervening trophic levels, and these indirect effects are the same as the community effects. Lynch's (1978) conclusion based on his data analysis is supported by the theoretical results from the loop model.

Dodson (1970) reported that size-selective predation may serve to provide feeding niches for other predators. The vertebrate predator (salamander) preyed selectively on larger herbivorous zooplankton, which were in competition for algal resources with the small herbivores. The small zooplankton were the preferred food of the invertebrate predator (*Chaoborus* sp.), which benefited from an increased food supply in the presence of the salamanders. This system is equivalent to the model in Figure 14.8a. The large and small herbivores (A_1, A_2) show competitive community effects, while the salamanders and phantom midge *Chaoborus* sp. (D, C) show mutualistic community effects. A positive input to the salamanders increases the amount of small herbivores available, which benefits *Chaoborus* sp. as expected. An input to *Chaoborus* sp. can also benefit the salamanders through their food resource, the large herbivores. This is a case in which direct effects and community effects are equivalent.

In the second example, a more complex food web including predators was studied (Lynch 1979) (Figure 14.8b). A large herbivore H_L (*Daphnia pulex*), an intermediate size herbivore H_M (*Ceriodaphnia reticulata*) and a small herbivore H_S (*Bosmina longirostris* and rotifers) each utilized the same algal resource. H_L and H_M were subject to predation by planktivorous fish, while H_M and H_S were preyed upon by an invertebrate predator (*Chaoborus* sp.). Indirect effects obtained by multiplying links along paths would indicate that H_L, H_M and H_S should be competitors, each being negatively affected by a positive input to the other. This is an example, however, of how indirect link-by-link effects may be reversed by community effects. Instead of being competitors, H_L and H_S exhibit community mutualism in the effects matrix since

Figure 14.8: Herbivore Interactions in a Food Web. A, A_1 and A_2 are algal variables; N is a nutrient resource; D is *Daphnia* sp.; C is *Ceriodaphnia* sp.; H_s, H_m and H_l are small, medium and large herbivores respectively; I is an invertebrate predator (*Chaoborus* sp.); and F is a planktivorous fish (after Lynch 1978, 1979).

each benefits from a positive input to the other. H_L and H_M, and H_S and H_M are still competitive pairs in the community effects table.

F and I have no effect on each other in this food web since the paths between them do not have valid complements. There are four paths through the system from F to I. Note that the network is symmetrical around the A to H_M linkage and thus the F on I effects correspond to those for I on F. The shortest path from F to I is through H_M to I and it is of length 3 and negative. Its complement consists of the variables A, H_L and H_S and there is no set of disjunct loops which does not leave out either H_L or H_S; thus, there is no valid complement. The other three paths are all of length 5, all positive and all pass through A. One is from F through H_M to A to H_S to I while the other two go from F through H_L to A, then one through H_M to I and one through H_S. These paths have complements consisting of H_L, H_S and H_M respectively and, as none of these variables is self-damped, none of the complements is valid. As we have seen in previous, simple examples, variables in complements like these which are not self-damped and do not take part in stabilizing loop sets tend to change themselves from parameter inputs and thus absorb the effects of such inputs and prevent variables further in the path from being affected. In this diagram, variable A is highly buffered by its own self-damping and by its loops with variables H_L, H_M and H_S. These three variables tend to absorb any inputs from F or I from being transmitted through to the other side of the network, to I or F respectively. The actual fate of these variables in such circumstances is, however, ambiguous, because there are conflicting forces increasing and decreasing all three variables. This is an example of a model in which community effects dominate some relationships, and must be considered for correct prediction of outcomes. Note with this level of network complexity, there are 63 valid paths.

Gull Lake Pelagic Food Web

The pelagic food web from Gull Lake, Michigan is given in Figure 14.9. The variables and links in this model represent the community in May, 1973 and were determined by fitting the loop diagram to field data. The quantitative field data are first characterized as loop variables using standard correlation techniques and biological intuition. Next, the quantitative abundances over time of each loop variable are calculated and then converted into qualitative signs. These signs ($+$, $-$, 0) are entered on a loop model template and the models are fitted to the data signs by hand. Additional verification

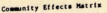

Community Effects Matrix

Effect on:

Input	SI	F	NN	ED	LP	RG	ID	ME	LE	BG	DC	GR	XX	BS	CY	X
+ SI	+	-	+	-	+	+	+	+	+	-	-	+	-	+	+	-	0
+ P	-	+	-	0	-	-	-	-	-	-	-	-	-	-	-	0	0
+ NN	0	0	+	0	0	0	0	0	0	0	0	0	0	0	0	0	0
+ ED	-	+	-	+	+	+	+	+	+	+	-	+	+	+	+	-	0
+ LP	+	-	+	-	+	+	+	+	+	+	+	-	+	+	+	-	0
+ RG	+	+	-	+	-	-	+	+	+	+	+	-	+	+	+	-	0
+ ID	+	+	-	+	-	-	+	-	-	+	+	-	-	-	-	-	0
+ ME	-	-	+	-	-	-	-	+	-	-	-	+	+	+	+	-	0
+ LE	+	+	?	-	-	-	-	-	-	+	+	-	+	-	+	-	0
+ BG	+	-	?	-	-	-	-	-	-	+	+	-	+	+	+	-	0
+ DC	-	-	+	-	+	+	+	+	+	+	-	+	-	+	+	-	0
+ GR	+	-	?	-	-	-	-	-	-	-	-	-	-	+	+	+	0
+ XX	+	-	+	-	+	+	+	+	+	+	+	+	+	+	+	-	0
+ BS	-	+	?	+	+	+	+	+	+	+	+	+	+	+	+	+	0
+ CY	0	0	+	0	0	0	0	0	0	0	0	0	0	-	+	-	+
+ X	-	+	?	+	+	+	+	+	+	+	+	+	+	+	+	-	+

Model Characteristics

Variables	=	16
Links	=	40
Loops	=	24
Valid paths	=	227
Valid complements	=	1392
Fn (overall feedback)	=	-30

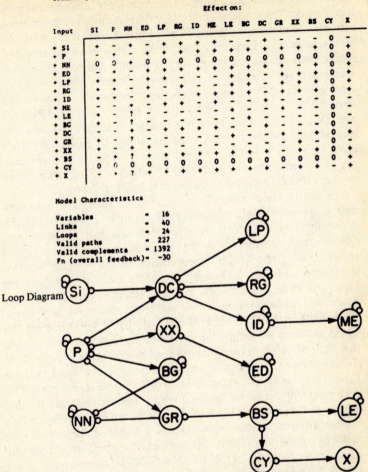

Loop Diagram

Figure 14.9: Gull Lake Pelagic Food Web. Key: Si-Silica concentration; P-Phosphorus concentration; NN-Nitrogen concentration; DC-Diatoms and non-diatom chrysophytes; XX-Group of miscellaneous large algal species; BG-Bluegreen algae; GR-Green algae; LP-*Daphnia longeremis* and *D. pulex*; R.G.-*Daphnia retrocurva* and *D. galeata*; ID-Immature *Diaptomus oregonensis* and *D. minutus*; ED-Adult *Epischura lacustris, Diaptomus oregonensis* and *D. minutus*; BS-*Bosmina longirostris*; ME-*Mesocyclops edax*; LE-*Leptodora kindtii*; CY-*Cyclops bicuspidatus*; X-Unknown predator.

came from experimental work on predator-prey interactions in the community. The model includes 16 variables and thus the community effects matrix contains 256 elements. Three of the variables represent nutrients, so only 169 of the matrix elements represent interactions among species or species groups. Thirteen matrix elements are self-interactions, leaving 156 matrix elements representing 78 direct interactions of species of types (+ , −) and (− , 0). One of the variables in the network (CY) has a satellite variable (X), which buffers it against changes. Eleven of its pair-wise interactions are zero-zero. Of the remaining 67 pairs, 16 (24 per cent) are (− , −), 14 (21 per cent) are (+ , +) and 37 (55 per cent) are (+ , −) in terms of community effects. Thus, mutualistic community effects are almost as prevalent as competitive ones.

Mutualistic community effects occur when an input to each variable in the pair benefits the other. These interactions arise in this network when the paths of effect between two variables pass upwards and downwards in the network by an even number of trophic levels (type A). This phenomenon results from the fact that the majority of linkages between variables are of the predator-prey type, and in such a network, effects alternate sign when passing downward through successive trophic levels. Examples of this type are LP-ED, RG-ED, ID-ED, BS-ED, BS-LP, BS-RG and BS-ID. These variables all occupy the same trophic level (herbivores), and the paths of effect between them travel down two trophic levels in the network to P and then up two trophic levels. Note that the herbivores in these pairs specialize on different algal resources. This may represent a strong selective drive for resource allocation, since species at the same trophic level usually do not compete when they choose different resources, but actually enhance each other as a result of community effects. In the case of the predator X, its mutualistic interactions (X-ED, X-LP, X-RG, X-ID) involve paths which travel down four trophic levels to nutrient P and then back up two trophic levels. Mutualistic relationships also arise between variables occupying different trophic levels with one trophic level intervening. These pairs are DC-ME, GR-LE and X-BS (type B). The species at the highest trophic level enhances the other by controlling the intervening predator, while the one at the lower trophic level serves as food for the top carnivores' prey. Note that in this network, there are 227 paths of effect.

The notion of a single indirect path is clearly obsolete in this context. Indirect effects actually have very little ecological significance except as a component of the community effect. Because of the potential for inaccuracy and confusion, the term should probably not be used. Even the simplest networks have community effects which can yield surprising results. The notion that variables not on a given path can nevertheless influence the effect of the path requires more attention in ecology. The more variables are removed from each other, the greater is their sensitivity to the rest of the network.

General Comments on Pelagic Food Webs

More than 100 loop models have been constructed for four lake communities using field data sets (Land and Blouin 1985, Lane and Levins unpublished). These models exhibit approximately the same level of complexity illustrated in the last model (Figure 14.9) for the Gull Lake plankton community. Lane (1984c) described how to fit loop models to large data sets. The following comments represent some general observations on how direct, indirect and cummunity effects are interrelated in these networks.

Variables

To date, about 200–300 species of plankton have been included in the lake loop models. The methodology has been developed so that each loop variable can be defined in terms of the species comprising it. The level of variable aggregation used has yielded about 90–95 per cent accuracy in fitting the loop models to data, however, variable aggregation is always controversial. Each loop variable represents a functional group of species that have the same links to the rest of the network. Generally, the more species per loop variable, the more difficult the problem of persistence becomes for the variable as a functional component in a set of networks. Using this level of aggregation, loop analysis throws away the information on how the species within variables may be coexisting. They can be considered to be competitors and co-prey. They may be allocating resources at a finer level of sophistication than is portrayed in the models. Satellite variables are particularly important to identify since they nullify any effects of paths to adjacent variables they are attached to. Some of the best candidates are the blue-green algae which are usually inedible. When the cladoceran *Diaphanosoma leuchtenbergianum* is present, however, these algae lose their satellite function since this herbivore appears to consume them.

Links and Direct Effects

In the introduction, the results of Gause and Park were mentioned as demonstrating that more than one types or biological mechanism was operating in their laboratory competition experiments. A direct effect will represent the mathematical outcome of the dominant mechanism. In loop analysis, the direct effects between a pair of variables can be represented by nine types of link including the absence of a relationship. Over an annual cycle, many links have only one to two states; however, several exhibit 4-5 states. Thus, biological intuition of species interactions even on a pairwise basis can be misleading. Too often, biological interactions and even whole food webs are depicted by static representations. For the pelagic zone of lakes, about 80–85 per cent of the direct effects are standard predator-prey links, 10–15 per cent are self-damping terms and 1–3 per cent are one-way relationships such as luxury consumption. In all of the present models, there has never been a classical mutualism link ($\leftrightarrow$) between two loop variables. This is not to say that, for example, the algal-herbivore mutualism documented by Porter (1976, 1977) does not occur, but it does not appear to be a critical part of the system dynamics at the level of resolution used here with loop analysis. It may be that the models need further refinement in that the biology she described is a special case phenomenon. Connectance for these networks is approximately 15–25 per cent; this represents the per centage of non-zero alphas in the community matrix including diagonal terms.

Food Webs and Community Effects

Because the link types can change over an annual cycle, a set of loop diagrams are needed to characterize the seasonal dynamics of a lake community. These food web models essentially resonate about the dominant network of the most prevalent links and variables, which is termed the core model (Lane 1984a,b,d). Individual loop diagrams give weak information about system structure; however, many of them put together yield fairly robust results. Loop analysis has been criticized for only providing weak, qualitative information but this is proving not to be a valid criticism if enough characterizations are completed.

Not only do link types change, but also the signs and entry points of the parameter inputs shift over an annual cycle. In both oligotrophic freshwater and marine communities, approximately 85 per cent of all parameter inputs enter at the nutrient or algal level. The fluctuation of parameter inputs and their level in the trophic hierarchy have some immediate consequences for the interpretation of community effects. First, the fluctuating nature of the inputs coupled with the changes in links, means that there will be a diffuse and fluctuating control in aquatic food webs. Thus, a good case for long-term coevolution and adaptation mechanisms cannot be made for pairs of variables except in restricted cases. Howe (1984) gives an overview of how two species' mutualistic interactions are affected by community evolution. He suggests that most mutualistic adaptations are generalized rather than species-specific. He states, that

in a world of ecological flux, this hinders the evolution of finely-tuned mutualism. What specificity exists is likely to be at taxonomic levels higher than the species, and will reflect mutually compatible biological groundplans rather than pairwise reciprocal evolution of species pairs. Pairwise coevolution should be largely restricted to depauperate communities, and will almost certainly be rare in highly diverse biotas.

Second, if most of the inputs are low in the trophic hierarchy, for instance a positive input at the nutrient level, then these inputs will cause different algal groups and their associated herbivores to increase if their respective complements are negative. Thus, these variables will be positively correlated with each other but will not exhibit mutualistic community effects because the inputs are not entering at the herbivore levels like the C and H_1, H_2; and R and P_1, P_2 pairs of Figure 14.5. Thus going up (or down) an even number of trophic levels is fundamentally different from the zigzag pattern of going up and down or down and up an even number of trophic levels.

The prevalence of nutrient-algal inputs indicates that there may be fewer mutualistic community effects produced as zigzag patterns. Community effects produced by these zigzag paths should most likely yield mutualistic relationships between variables on the same trophic level that are either specializing on different resources and/or subject to different major invertebrate predators that have the same vertebrate predator. To date, most of the inputs to herbivores and their predators have been sporadic and without discernible pattern. Thus, neither direct nor community effect mutualism appears to be prevalent in freshwater plankton communities in the northern temperate zone. In the example for Gull Lake (Figure 14.9) there appear to be more mutualistic community effects than indicated here, because interactions were calculated as percentages of all possible parameter inputs. If these interactions were weighted by the actual parameter inputs used to fit the models then community effect mutualism would be a much smaller per centage of all potential community effects, perhaps as little as 3 to 5 per cent for oligotrophic systems. Thus, for many lakes community mutualism may not be of great importance. As environments become nutrient enriched, parameter inputs enter higher in the trophic hierarchy and thus, ensure that community mutuaiism will be more prevalent than in oligotrophic networks.

Acknowledgments

Richard Levins offered many valuable suggestions in the planning stages; he has been a long-term collaborator in applying loop analysis to aquatic communities. John Wright greatly facilitated the mathematical analysis and theoretical development. He wrote the improved loop calculation programme and critiqued the manuscript. Roger Day wrote the first version of the loop calculation programme. Anthony Blouin assisted in the literature review, in construction of the sample loop models, data analysis and manuscript preparation. Lyn Wilson typed the manuscript and Terrance Collins drafted the figures.

The Department of Mathematics, Statistics and Computing Science at Dalhousie University generously provided computer time on their VAX/UNIX system. The study was supported by grants from the Natural Sciences and Engineering Research Council of Canada (G0233, G0749) and the US Environmental Protection Agency (R-810520-01-0) to P. Lane.

References

Berryman, A.A. (1981) *Population Systems: A General Introduction*, Plenum Press, New York
Boucher D.H. (1985) 'Competition and the Mutualism Between Legumes and *Rhizobium* (submitted)

Boucher, D.H., S. James and K.H. Keeler (1982) 'The Ecology of Mutualism', *Annual Review of Ecology and Systematics*, 13, 315–47

Briand, F. and E. McCauley (1978) 'Cybernetic Mechanisms in Lake Plankton Systems: How to Control Undesirable Algae', *Nature, 273*, 228–30

Brinkhurst, R.O., K.E. Chua and N.K. Kaushil (1972) 'Interspecific Interactions and Selective Feeding by Tubificid Oligochaetes', *Limnology and Oceanography, 17*, 122–33

Desharnais, R.A. and R.F. Costantino (1980) 'Genetic Analysis of a Population of *Tribolium*. VII. Stability: Response to Genetic and Demographic Perturbations', *Canadian Journal of Genetics and Cytology, 22*, 577–89

Dodson, S.I. (1970) 'Complementary Feeding Niches Sustained by Size-Selective Predation', *Limnology and Oceanography, 15*, 131–7

Gause, G.F. (1934) *The Struggle for Existence*, William and Wilkins, Baltimore, Maryland

Henry, C. (1980) 'Competitive Interactions Studied by the Mathematical Method of Loop Analysis', *Comptes Rendus des Séances de l'Académie des Sciences, Serie D: Sciences Naturelles, 290*, 787–90

Howe, H.F. (1984) 'Constraints on the Evolution of Mutualisms', *American Naturalist, 123*, 764–77

Hutchinson, G.E. (1975) 'Variations on a Theme by Robert MacArthur' in Cody, M.L. and J.M. Diamond (eds), *Ecology and Evolution of Communities*, The Belknap Press of Harvard University Press, Cambridge, Mass., pp. 492–521

Lane, P.A. (1982) 'Using Qualitative Analysis to Understand Perturbations in Marine Ecosystems in the Field and Laboratory' in Archibald, P. (ed.), *Environmental Biology State of the Art Seminar*, Office of Exploratory Research, US Environmental Protection Agency, Washington, DC, EPA-600/9-82-007, pp. 94–122

Lane, P.A. (1984a) 'Peering Through Nature's Kaleidoscope at a Marine Plankton Community' in *Environmental Biology State of the Art Seminar*, Office of Exploratory Research, US Environmental Protection Agency, Washington, DC (in press)

Lane, P.A. (1984b) 'A Comparison of Marine and Freshwater Plankton Food Webs' MS.

Lane, P.A. (1984c) 'Preparation of Data for Loop Analysis' *Ecology*, in Press

Lane, P.A. (1984d) 'Symmetry, Change, Perturbation and Observing Mode in Natural Communities' *Ecology*, in press

Lane, P.A. (1984e) 'Once upon a Googol: The Quest for Biologically Reasonable Networks' MS.

Lane, P.A. and A.C. Blouin (1985a) 'Qualitative Analysis of the Pelagic Food Webs of Three Acid Impacted Lakes', *Internationale Revue der Gesamten Hydrobiologie' 69*, in press

Lane, P.A. and A.C. Blouin (1984) 'Plankton of an Acid-Stressed Lake (Kejimkujik National Park, Nova Scotia, Canada). Part 3, Community Network Analysis', *Verh. Internat. Verein. Limnol. 22*, 406–11

Lane, P.A. and T.M. Collins (1984) 'Foodweb Models of a Marine Plankton Community: An Experimental Approach', *Journal of Experimental Marine Biology and Ecology* (in press)

Lane, P.A. and R. Levins (1977) 'Dynamics of Aquatic Systems. II. The Effects of Nutrient Enrichment on Model Plankton Communities', *Limnology and Oceanography, 21*, 454–71

Levine, S.H. (1976) 'Competitive Interactions in Ecosystems', *American Naturalist, 110*, 903–10

Levins, R. (1973) 'The Qualitative Analysis of Partially Specified Systems', *Annals of the New York Academy of Science, 231*, 123–38

Levins, R. (1975) 'Evolution in Communities Near Equilibrium' in Cody, M.L. and J.M. Diamond (eds), *Ecology and Evolution of Communities*, The Belknap Press of Harvard University Press, Cambridge, Mass., pp. 16–50

Lynch, M. (1978) 'Complex Interactions Between Natural Coexploiters — *Daphnia* and *Ceriodaphnia*', *Ecology, 59*, 552–64

Lynch, M. (1979) 'Predation, Competition and Zooplankton Community Structure: An Experimental Study', *Limnology and Oceanography, 24*, 253–72

Mason, S.J. (1952) 'Some Properties of Signal Flow Graphs', *Proceedings of the I.R.E., 41*, 1144–56

Park, T. (1962) 'Beetles, Competition and Populations', *Science, 138*, 1369–75

Patten, B.C. (1983) 'On the Quantitative Dominance of Indirect Effects in Ecosystems' in Lauenroth, W.K., G.V. Skogerboe and M. Elug (eds), *Analysis of Ecological Systems: State-of-the-Art in Ecological Modelling*, Elsevier Scientific Publishing Company, pp. 27–37

Porter, K.G. (1976) 'Enhancement of Algal Growth and Productivity by Grazing Zooplankton', *Science, 192*, 1332–4

Porter, K.G. (1977) 'The Plant-Animal Interface in Freshwater Ecosystems', *American Scientist, 65*, 159–70

Roughgarden, J. (1979) *Theory of Population Genetics and Evolutionary Ecology: an Introduction*, MacMillan Publishing Co. Inc., New York

Schaffer, W.M. (1981) 'Ecological Abstraction: The Consequences of Reduced Dimensionality in Ecological Models', *Ecological Monographs, 51*, 383–401

Vandermeer, J. (1980) 'Indirect Mutualism: Variations on a Theme by Stephen Levine', *American Naturalist, 116*, 441–8

Vandermeer, J. (1981) *Elementary Mathematical Ecology*, John Wiley and Sons Inc., New York

Weinberg, G.M. (1975) *An Introduction to General Systems Thinking*, John Wiley and Sons, New York

Wright, J. and P.A. Lane (1984) 'Theory of Loop Analysis', *Ecology* (in press)

15 MUTUALISM IN AGRICULTURE

Douglas H. Boucher

Introduction — The Rat's Five-year Plan

I'd like to tell Delio a story of my home town that will interest him. I'll tell it to you briefly and you can recount it to him and Giuliano.

A baby is sleeping. Nearby there's a big cup of milk ready for when he wakes up. A rat drinks the milk. The baby, waking up and not finding the milk, screams; his mother screams. The rat bats his head desperately against the wall but realizes that doesn't accomplish anything, so he rushes to a goat to get some milk. The goat says she'll give him milk if she can have some grass to eat. The rat goes to the countryside for grass, but the dry countryside begs water. The rat goes to the fountain. The water in the fountain, which was ruined during the war, leaks away — it needs a master mason.

The rat goes to the master mason; he needs stones. The rat runs to the mountain, and at this point there's a sublime dialogue between the rat and the mountain, which has been so devastated by deforestation on the part of speculators that all its ridges and humps show. The rat tells his whole story and promises that the baby, when he grows up, will replant the pines, oaks, chestnuts, and other trees. Convinced, the mountain donates the stones, and so on, and the baby has so much milk he can even wash in it. He grows up, plants the trees; then everything changes. The skeleton of the mountain disappears beneath new soil, the atmospheric precipitation becomes regular once again and the trees hold on to their moisture and prevent torrents from devastating the plains. In other words, the rat conceives nothing less than a genuine five-year-plan! This is truly a story of a country ruined by deforestation. My darling Giulia, please tell the story to the children and then let me know what impression it makes on them. A tender hug.

Antonio (Gramsci 1979: 196–7)

This story from a peasant village in Sardinia, recounted by the Italian Communist leader Antonio Gramsci in a letter from a fascist prison in 1931, can serve as a metaphor for the many mutualisms involved in human productive activities. Both the value of different mutualisms in increasing the productivity of our domesticated plants and animals, and the mutualistic nature of agriculture and forestry in general, have long been recognized in many cultures, and economically valuable organisms were important in much of the early scientific study of mutualisms.

The last few decades of the nineteenth century, in which the concept of mutualism was crystallized (Chapter 1, this volume), were also a period of major advance in the study of economically valuable mutualisms. Thus the 1880s and 1890s saw the elucidation of mutualistic nature of legume nodules (Schneider 1903, Fred *et al.* 1932) and mycorrhizae (Schneider 1897) and the first major successes in using beneficial insects to protect crops (Egerton 1976). The discovery of new mutualisms was often the work of economic biologists; thus Stephen Forbes (of 'The Lake as a Microcosm' fame) established the dependence of the corn aphid (*Aphis maidis*) on ants (*Lasius alienus*) in the course of searching for a way to control it (Comstock 1887). The mutualism between humans and their crop plants and animals was often mentioned in contemporary reviews, for example by Roscoe Pound:

> Wheat is cultivated by man and enabled to grow in quantities, and in localities which, under ordinary conditions, would be impossible. It gains this partial exemption from the struggle for existence only at the expense of an immense number of individuals sacrificed, but it is nevertheless, a great advantage which it gains. This may be called mutualism. (Pound 1893: 509–10)

Humans can make practical use of mutualisms for agriculture and forestry in several ways. We can encourage mutualisms which already exist by manipulation of the environment or by introducing more efficient partners. Conversely we can control pests by controlling their mutualists, We can also create new indirect mutualisms by encouraging or introducing beneficial organisms as biological controls. Finally, we can use the concept of agriculture as a mutualism, as a conceptual tool to guide ecosystem management. There is, however, an alternative approach, which seems to have become more and more powerful as agriculture has become industrialized.

This is the elimination of dependence on mutualisms (and indeed on biological interactions in general) in food and fibre production, and the substitution of genetic, physical and chemical manipulation in their place. It remains to be seen whether the importance of mutualism in agriculture and forestry will grow or diminish.

Nutrition/Digestion

Uptake and use of nutrients is a fundamental requirement for growth, and the productivity of plants and animals is often nutrient-limited. Mutualistic solutions to this problem have often been used by humans, starting with the simple incorporation of mutualists into agroecosystems (for example, legumes as green manure, rotation crops and intercrops), but often going further by isolating, selecting and inoculating one of the partners. The most important example of this is the use of mutualistic nitrogen fixation.

Nitrogen Fixation

Nodulation. In the century since it was discovered that a bacteria fixes nitrogen within legume nodules, the manipulation of this mutualism has become an important and sophisticated part of agriculture. Legumes have long contributed nitrogen to agroecosystems as elements of rotation, as green manures, in mixed cropping and in mixed pastures, and about half of the nitrogen fixed on earth comes from agricultural soils (Hardy and Havelka 1975). Inoculation of seed with *Rhizobium* is a standard practice for soybeans, which account for about half of world grain legume production, and is also common with temperate forage legumes such as clovers. On the other hand, inoculation is rare with other important grain legumes such as beans, peas, peanuts and cowpeas. This is partly due to the fact that these crops are characteristic of Third World nations (inoculant production is heavily concentrated in the developed world; the Nitragin Company of Milwaukee, Wisconsin is the leading producer). However responses to inoculation in beans and other grain legumes are often disappointing; this may be a simple reflection of the fact that sufficient *Rhizobium* are generally already present in the soil.

The bulk of the nitrogen fixed by grain legumes is removed when their seeds are harvested; thus transfer to associated crops is unlikely to be important, though it may be substantial in pastures. While cereal-legume combinations such as maize and beans are very common in tropical countries, and generally outyield monocultures (Trenbath 1974), these do not usually represent plant-plant mutualisms, since both crops produce less seed in association; however the value of their combined yield is greater. Leguminous trees such as black locust (*Robinia*) may also be useful in reforestation of eroded soils (Moiroud and Capellano 1982), as are actinorhizal nitrogen fixers such as *Alnus* and *Casuarina* which form nodules with the actinomycete *Frankia* (Stewart 1982). The actinorhizal mutualism is unlikely to be important in agriculture, as only one actinorhizal crop plant (the raspberry *Rubus ellipticus*; Becking 1979) is known. However there exists considerable potential in forestry, using mixed stands or short-rotation coppicing (Gordon *et al.* 1979).

Azolla. The water fern *Azolla* and its blue-green algal symbiont *Anabaena* have long been used as a green manure for rice in China and Vietnam (Lumpkin and Plucknett 1980; Watanabe 1982), and has attracted interest in the US as well (Talley and Rains 1980). It regularly gives increases of several hundred kilos of rice per hectare (Watanabe 1982). Given the closeness of the association, isolation, selection and reintroduction of the algae have not yet been achieved; thus management involves use of the complete association rather than inoculation with one partner as in legumes. In effect, it is the *Azolla*-rice mutualism, not the *Azolla-Anabaena* mutualism, which is managed.

Associative Fixation. Other crop-microbe mutualisms tend to be associative rather than symbiotic; in the terms of Lewis (Chapter 2, this volume), they fall toward the less integrated end of continuum 8. As a consequence, the amount of fixed nitrogen transferred to the crop is considerably less than with legumes (Stewart 1982; Dobereiner 1983). However associations have the advantage of involving many important crop plants, including maize, sugar cane, rice, sorghum, millet and various pasture grasses (Dobereiner 1983; van Berkum and Bohlool 1980). The bacteria involved (*Beijerinkia, Azotobacter, Azospirillum*) are often found in a free-living state, and the associations vary considerably in degree and specificity. Inoculation with the bacteria can give substantial yield increases (Okon 1982), although this may be due to other effects and interactions with other rhizosphere organisms as well as to nitrogen fixation (Stewart 1982). The microbial communities of plant roots may be exceedingly complex, and processes as detoxification of hydrogen sulfide (*Beggiatoa* on rice; Joshi and Hollis 1977) may be the bases of additional mutualisms.

Mycorrhizae

The plant root-fungus mutualisms known as mycorrhizae are found in most higher plants, including almost all crops (crucifers and sugar beets being the major exception) and timber trees. Thus, except in disturbed situations (strip mining, greenhouses) it is a question of improving the mutualism rather than creating it. While considerable work has been done with inoculation of vesicular-arbuscular (V-A) mycorrhizae (Menge 1983; Hayman 1980, 1981), results have often been disappointing. As with grain legumes, the presence of other mycorrhizal strains in the soil is a major obstacle, and researchers have gone so far as to destroy the natural mycorrhizal fungi by fumigation to be able to introduce improved strains.

The major advantage of mycorrhizae is more complete uptake of soil nutrients, particularly phosphorus; thus, as expected, responses are less or even negative in phosphorus-rich soils (Menge 1983). Other less consistent or more speculative advantages include protection against pathogens (Zambolin and Schenck 1983), ability to use organic nitrogen and phosphorus sources (Stribley and Read 1980; Herrera *et al.* 1978), and transfer of nutrients between plants (Whittingham and Read 1982). V-A mycorrhizae also consistently increase nitrogen fixation in nodulated legumes (Cluett and Boucher 1983).

Ectomycorrhizae are less widely distributed than V-A mycorrhizae, but their plant associates include many important temperate timber species (Fogel 1980). While here again, most naturally growing trees will have mycorrhizae already (Malloch and Malloch 1981), disturbed sites and greenhouses may often require inoculation.

Since many mycorrhizal fungi are hypogeous (producing spores below ground) they may depend on small mammals for dispersal (Maser *et al.* 1978). Thus reforestation of cut-over sites may be impeded by small-mammal elimination, despite the mammals' consumption of tree seeds. Such complications demonstrate the importance of considering indirect interactions as well as direct ones in deciding on management strategies (Chapters 13 and 14, this volume).

Ruminants

The guts of all animals, including domesticated ones, are incredibly complex ecosystems, with a large number of microbial species interacting in a variety of ways. Ruminants are worthy of special mention because they have evolved a gut morphology and flora which give them access to a major new source of food: grass (Hungate 1966). This makes possible many major features of modern agriculture: pastures, haying, year-round milk production, and indirectly, destruction of tropical rainforest for conversion to grassland (Parsons 1983). Livestock production using non-ruminants has become very dependent on feed grain production and extreme concentration of animals for feeding. Ruminants potentially can have a fundamentally different relation to plant production, since they can utilize vegetative as well as reproductive biomass, as well as habitats which are marginal for cereal production. However post-war developments in the livestock industry (feedlots, grain and soybean export) have tended to lessen these differences, so that for economic reasons the advantages offered by the rumen mutualism are rather incompletely utilized. The tendency seems to be toward *less* efficient utilization of ruminants (in ecological terms) rather than more.

Protection

Defence of crop plants and animals against their predators, parasites, and diseases can make use of mutualism in two different ways. On the one hand, associated plants and animals which offer protection can be encouraged or introduced; on the other hand, the pest organism can be attacked indirectly by attacking its mutualists. The second method is mostly speculative, but the first has a long history of success.

A variety of mechanisms are known by which crop damage can be reduced through combining plants so as to reduce pest incidence (Atsatt and O'Dowd 1976); they range from feeding deterrents (Freeman and Andow 1983) to predator and parasite attractants (for example, nectar) to the simple confusion causing by a mixture of species (Risch *et al.* 1983). Depending on the economic value of the associated plants, this may be regarded as mixed cropping, companion planting, or integrated pest and weed management (Zandstra and Motooka 1978). Soils may be made more 'disease-suppressive' by favouring root-colonizing species of *Pseudomonas* which produce siderophores (Schroth and Hancock 1982); mycorrhizae, as mentioned above, may also be helpful.

Relatively few crop plants attract ants to defend them by producing extrafloral nectar, although vetches (Koptur 1979) and the tropical plant *Bixa orellana* (Bentley 1977) seem to use this mechanism rather efficiently. Ants can also be manipulated in less species-specific ways to control arthropods and weeds (Risch and Carroll 1982), the 'green islands' of undamaged trees around ant nests during defoliating insect outbreaks suggest similar possibilities in forests (Laine and Niemela 1980). Here we are edging into classic biological control (use of natural enemies to protect plants), which although reasonably considered an indirect mutualism (Price *et al.* 1980), is an enormous subject in and of itself.

Many pest organisms seem heavily dependent on mutualists for their survival, but I know of few attempts to control them by attacking their mutualists. Nevertheless several possibilities can be suggested. Homoptera can be controlled by eliminating or luring away their ants (Leston 1970; Adenuga 1975). Leaf-cutter ants, in turn, could be attacked with fungicide rather than insecticide (Weber 1972). The dispersal agents of spores of fungi and bacteria may be a weak link in disease transmission (Jennersten 1983; Gilbert 1980). When parasites of domestic animals interact mutualistically, control of one may help control the other (Ewing *et al.* 1982). Detailed natural history studies may reveal many cases in which a mutualist is more easily controlled than the pest itself.

Pollination

Cultivated plants are not generally dependent on mutualists for pollination, being either reproduced vegetatively, self-pollinated or wind-pollinated. Most animal-pollinated crop species are trees, although squash bees are important for fruit production in various cucurbitaceous vines (Hurd *et al.* 1971). The pollinators of tree crops range from tiny ceratopogonid midges in cacao (Purseglove 1968) to bats (Baker 1963), but bees (both wild and honeybees) are probably the most important. The economic value of honeybees as pollinators can far outweight the value of the honey they produce (de Oliveira 1983). Nevertheless selection *against* mutualistic pollination must have often been strong in the process of domestication; thus domesticated *Phaseolus* have their keel petal twisted through 360° or more, making pollen release impossible, and cultivated figs escape dependence on highly host-specific fig wasps (Agaonidae) through parthenocarpy (Janzen 1979).

Seed Dispersal

Even more so than with pollination, domestication has generally involved selection against relationships with mutualistic (or any other) seed dispersal agents. Farmers want to control any and all seed dispersal themselves, and fruit-consuming animals will almost always be seen as pests. Even in forestry, reforestation by aerial seeding and from nursery stock is more and more the rule, despite its cost; the heavy consumption of seeds by dispersing animals, and their lack of spatial precision (affecting later competition and thus wood growth rates) is too high a price to pay.

Humanity and Its Mutualists

While the view of agriculture as a mutualism between humans and their domesticated plants and animals is a long-established one (Espinas 1878), one can reasonably ask whether its value is anything more than metaphorical. Indeed, the analogy should not blind us to the fact that agriculture is fundamentally different from other mutualisms in at least two ways. First, the definition of 'benefit' for the human partners is neither fitness nor population growth rate but rather economic criteria such as profits. These criteria will vary from society to society according to their social, political and economic organization. Second, agriculture involves planning and foresight, in a way that natural selection does not. The human actors in the ecosystem can decide consciously what the consequences of different courses of action will be, and make their choices accordingly — at least potentially.

This being said, however it must be recognized that human agricultural activities often have unforeseen effects, both on our domesticated organisms and on their mutualists. The selection pressures involved in domestication are strong and have led to rapid evolutionary changes (Doyle 1983; Price 1984). While some of these are deliberate (artificially selected), many are the unplanned consequences of natural selection operating in an artificial environment.

Effects on mutualists, such as the elimination of mycorrhizal fungi when fumigating for nematodes or the poisoning of bees by pesticides, are generally unplanned and often unforeseen.

The development of agriculture can be viewed as a change in humanity's relation with its food organisms from predation to mutualism. Hunter-gatherers found it to their advantage to aid their food organisms in reproduction, protect them from other predators, increase the resources available to them, and in general increase their fitnesses and population sizes. Starting perhaps as simply prudent predation, this has developed into facultative mutualism and for many domesticated organisms, obligate mutualism. It has become obligate for the majority of humans as well.

One way to aid domesticated organisms has been to aid their mutualists, as reviewed here. However an alternative path has also been followed, and indeed seems to be the predominant trend in agriculture in recent years. This is simply to eliminate the mutualists and replace them with direct human intervention.

The earlier phases of this development generally involved genetic manipulation, through selection of organisms which would not require (and indeed, would exclude) mutualistic pollinators and seed dispersers. However in the twentieth century, and especially since the Second World War, such manipulation has been supplemented by direct physical and chemical intervention. Thus, insects are controlled by pesticides rather than natural enemies, soil tilth is maintained by ploughing rather than by earthworms, nitrogen is provided by fertilizer rather than by biological fixation, and ruminants are fed more feed grains and less cellulose.

The long-term implications of these choices are disturbing, and it is worth asking whether the trend toward elimination of mutualists in agriculture is inevitable or even desirable. To take a currently popular example, consider the idea of genetically engineering nitrogen fixation in cereals by introducing *nif* genes from bacteria (Marx 1977). This is touted as an alternative to nitrogen fertilizer, but in fact there is also the third, mutualistic alternative, and comparison with it is instructive. In brief, the question is whether a nitrogen-fixing maize plant will be a better kind of soybean. In other words, will a plant which contains *nif* genes within its own genome, be as efficient a fixer as one which closely associates with *nif* bacteria? Quite apart from the technical difficulties of transferring the necessary genes, there is 'the prospect that increasing biological nitrogen fixation (by gene transfer), for which the plant supplies the energy, may actually decrease crop yields by taking energy that would otherwise go into producing useful parts of the plant' (Marx 1977: 638). If the limitations to nitrogen fixation are energy and environmental conditions rather than lack of proper genes, then the solution lies in ameliorating the mutualism rather than in gene transfer.

This is but one example of the fundamental choices that we must make in agriculture. On the one hand we can continue to emphasize genetic manipulation and chemical products, with their attendent economic and ecological costs. Or we can choose the direction of the rat's five-year-plan — to design a web of mutualistic interactions that is beneficial both to nature and to humanity.

References

Adenuga, A.O. (1975) 'Mutualistic Association Between Ants and some Homoptera — Its Significance in Cocoa Production', *Psyche*, *82*, 24–8

Atsatt, P.R. and D.J. O'Dowd (1976) 'Plant Defense Guilds', *Science*, *193*, 24–9

Baker, H.G. (1963) 'Evolutionary Mechanisms in Pollination Biology', *Science*, *139*, 877–83

Becking, J.H. (1979) 'Nitrogen Fixation by *Rubus ellipticus* J.E. Smith', *Plant and Soil*, *53*, 541–5

Bentley, B.L. (1977) 'The Protective Function of Ants Visiting the Extrafloral Nectaries of *Bixa orellana* L. (Bixaceae)', *Journal of Ecology*, *65*, 27–38

Cluett, H.C. and D.H. Boucher (1983) 'Indirect Mutualism in the Legume-*Rhizobium*-Mycorrhizal Fungus Interaction', *Oecologia*, *59*, 405–8

Comstock, J.H. (1887) 'Relations of Ants and Aphids', *American Naturalist*, *21*, 382 and 579–80

DeOliveira, D.D. (1983) 'Importance Economique de la Pollinisation par les Insectes', *Annales de la Société Entomologique du Québec*, *28*, 40–50

Dobereiner, J. (1983) 'Dinitrogen Fixation in Rhizosphere and Phyllosphere Associations' in Lauchli, A. and R.L. Bieleski (eds), *Encyclopedia of Plant Physiology*, vol. 15, Springer-Verlag, Berlin, pp. 330–50

Doyle, R.W. (1983) 'An Approach to the Quantitative Selection in Aquaculture', *Aquaculture*, *33*, 167–85

Egerton, F.N. (1976) 'Ecological Studies and Observations in America before 1900' in Taylor, B.J. and T.J. White (eds), *Issues and Ideas in America*, University of Oklahoma Press, Norman, OK, pp. 311–51

Espinas, A.V. (1878) *Des Sociétés Animales*, Ballière, Paris

Ewing, M.S., S.A. Ewing, M.S. Keener and R.J. Mulholland (1982) 'Mutualism among Parasitic Nematodes: a Population Model', *Ecological Modelling*, *15*, 353–66

Fred, E.B., I.L. Baldwin and E. McCoy (1932) *Root Nodule Bacteria and Leguminous Plants*, University of Wisconsin Press, Madison

Freeman, A.B. and D.A. Andow (1983) 'Plants Protecting Plants: the Use of Insect Feeding Deterrents', *Scientific Horticulture*, *34*, 48–53

Fogel, R. (1980) 'Mycorrhizae and Nutrient Cycling in Natural Forest Ecosystems', *New Phytologist*, *86*, 199–212

Gilbert, D.G. (1980) 'Dispersal of Yeasts and Bacteria by *Drosophila* in a Temperate Forest', *Oecologia*, *46*, 135–7

Gordon, J.C., C.T. Wheeler and D.A. Perry (eds.) (1979) *Symbiotic Nitrogen Fixation in the Management of Temperate Forests*, Oregon State University, Corvallis, OR

Gramsci, A. (1979) *Letters from Prison*, Quartet Books, London

Hardy, R.W.F. and V.D. Havelka (1975) 'Nitrogen Fixation Research: A Key to World Food?', *Science, 188*, 633–43

Hayman, D.S. (1980) 'Mycorrhiza and Crop Production', *Nature, 287*, 487–8

Hayman, D.S. (1981) 'Mycorrhiza and its Significance in Horticulture', *The Plantsman, 2*, 214–24

Herrera, R., T. Merida, N. Stark and C.F. Jordan (1978) 'Direct Phosphorus Transfer from Leaf Litter to Roots', *Naturwissenschaften, 65*, 208–9

Hungate, R.E. (1966) *The Rumen and its Microbes*, Academic Press, New York

Hurd, R.D.Jr, E.G. Linsley and T.W. Whitaker (1971) 'Squash and Gourd Bees (*Pepenapis, Xenoglossa*) and the Origin of the Cultivated *Cucurbita*', *Evolution, 25*, 218–34

Janzen, D.H. (1979) 'How to Be a Fig', *Annual Review of Ecology and Systematics, 10*, 13–51

Jennersten, O. (1983) 'Butterfly Visitors as Vectors of *Ustilago violacea* Spores between Caryophyllaceous Plants', *Oikos, 40*, 125–30

Joshi, M.M. and J.P. Hollis (1977) 'Interaction of *Beggiatoa* and Rice Plant: Detoxification of Hydrogen Sulfide in the Rice Rhizosphere', *Science, 195*, 179–80

Koptur, S. (1979) 'Facultative Mutualism between Weedy Vetches Bearing Extrafloral Nectaries and Weedy Ants in California', *American Journal of Botany, 66*, 1016–20

Laine, K.J. and P. Niemala (1980) 'The Influence of Ants on the Survival of Mountain Birches During an *Oporinia autumnata* (Lep., Geometridae) Outbreak', *Oecologia, 47*, 39–42

Lane, P. (1985) 'A Food Web Approach to Mutualism in Lake Communities', this volume, Chapter 14

Leston, D. (1970) 'Entomology of the Cocoa Farm', *Annual Review of Entomology, 15*, 273–94

Lewis, D.H. (1985) 'Symbiosis and Mutualism: Crisp Concepts and Soggy Semantics', this volume, Chapter 2

Lumpkin, T.A. and D.L. Plucknett (1980) '*Azolla*: Botany, Physiology and Use as a Green Manure', *Economic Botany, 34*, 111–53

Malloch, D. and B. Malloch (1981) 'The Mycorrhizal Status of Boreal Plants: Species from Northeastern Ontario', *Canadian Journal of Botany, 59*, 2167–72

Marx, J.L. (1977) 'Nitrogen Fixation: Prospects for Genetic Manipulation', *Science, 196*, 638–41

Maser, C., J.M. Trappe and R.A. Nussbaum (1978) 'Fungal-Small Mammal Interrelationships with Emphasis on Oregon Coniferous Forests', *Ecology, 59*, 799–809

Menge, J.A. (1983) 'Utilization of Vesicular-Arbuscular Mycorrhizal Fungi in Agriculture', *Canadian Journal of Botany, 61*, 1015–24

Moirud, A. and A. Capellano (1982) 'Le Robinier, *Robinia pseudoacacia* L. une Espèce Fixatrice d'Azote Intèressante?', *Annales de Science Forestière, 39*, 407–18

Okon, Y. (1982) '*Azospirillum*: Physiological Properties, Mode of Association with Roots and Its Application for the Benefit of Cereal and Forage Grass Crops', *Israel Journal of Botany, 31*, 214–20

Parsons, J.J. (1983) 'Beef Cattle (Ganado)' in Janzen, D.H. (ed.), *Costa Rican Natural History*, University of Chicago Press, Chicago, pp. 77–9

Pound, R. (1893) 'Symbiosis and Mutualism', *American Naturalist, 27*, 509–20

Price, E.O. (1984) 'Behavioral Aspects of Animal Domestication', *Quarterly Review of Biology, 59*, 1–32

Price, P.W., C.E. Bouton, P. Gross, B.A. McPheron, J.N. Thompson and A.E.

Weis (1980) 'Interactions among Three Trophic Levels: Influence of Plants on Interactions Between Insect Herbivores and Natural Enemies', *Annual Review of Ecology and Systematics, 11*, 41–65

Purseglove, J.W. (1968) *Tropical Crops: Dicotyledons*, vol. 1, John Wiley, New York

Risch, S.J. and C.R. Carroll (1982) 'The Ecological Role of Ants in Two Mexican Agroecosystems', *Oecologia, 55*, 114–9

Risch, S.J., D.A. Andow and M. Altieri (1983) 'Agroecosystem Diversity and Pest Control: Data, Tentative Conclusions and New Research Directions,' *Environmental Entomology, 12*, 625–9

Schneider, A. (1897) 'The Phenomena of Symbiosis', *Minnesota Botanical Studies, 1*, 923–48

Schneider, A. (1903) 'Outline of the History of Leguminous Root Nodules and Rhizobia with Titles of Literature Concerning the Fixation of Free Nitrogen by Plants — III', *Minnesota Botanical Studies, 3*, 133–9

Schroth, M.N. and J.G. Hancock (1982) 'Disease-Suppressive Soil and Root-Colonizing Bacteria', *Science, 216*, 1376–81

Stewart, W.D.P. (1982) 'Nitrogen Fixation — its Current Relevance and Future Potential', *Israel Journal of Botany, 31*, 5–44

Stribley, D.P. and D.J. Read (1980) 'The Biology of Mycorrhiza in the Ericaceae VII. The Relationship between Mycorrhizal Infection and the Capacity to Utilize Simple and Complex Organic Nitrogen Sources, *New Phytologist, 86*, 365–71

Talley, S.N. and D.W. Rains (1980) '*Azolla filiculoides* Lam. as a Fallow-Season Green Manure for Rice in a Temperate Climate', *Agronomy Journal, 72*, 11–8

Trenbath, B.R. (1974) 'Biomass Productivity of Mixtures', *Advances in Agronomy, 26*, 177–210

Van Berkum, P. and B.B. Bohlool (1980) 'Evaluation of Nitrogen Fixation by Bacteria in Association with Roots of Tropical Grasses', *Microbiological Reviews, 44*, 491–517

Vandermeer, J., B. Hazlett and B. Rathcke (1985) 'Indirect Facilitation and Mutualism', this volume, Chapter 13

Watanabe, I. (1982) '*Azolla-Anabaena* Symbiosis — its Physiology and Use in Tropical Agriculture' in Dommergues, Y.R. and H.G. Diem (eds), *Microbiology of Tropical Soils and Plant Productivity*, Martinus Nijhoff/Dr W. Junk, The Hague, pp. 169–85

Weber, N.A. (1972) *Gardening Ants, the Attines*, American Philosophical Society, Philadelphia

Whittingham, J. and D.J. Read (1982) 'Vesicular-Arbuscular Mycorrhiza in Natural Vegetation Systems III. Nutrient Transfer Between Plants with Mycorrhizal Interconnections', *New Phytologist, 90*, 277–84

Zambolin, L. and N.C. Schenck (1983) 'Reduction of the Effects of Pathogenic Root-Infecting Fungi on Soybean by Mycorrhizal Fungus '*Glomus mosseae*', *Phytopathology, 73*, 1402–5

Zandstra, B.H. and P.S. Motooka (1978) 'Beneficial Effects of Weeds in Pest Management — A Review', *PANS, 24*, 333–8

INDEX